W9-BMQ-856

K	The number of levels of the independent variable
LL	Lower limit of a confidence interval
MS	Mean square; ANOVA term for the variance
N	The number of scores or observations
O	In chi square, the observed frequency
r	Pearson product-moment correlation coefficient
r_s	A correlation coefficient for ranked data; named for Spearman
r^2	The coefficient of determination
S	The standard deviation of a sample; describes the sample
$\hat{s}$	The standard deviation of a sample; estimates σ
s^2	Variance of a sample; estimates σ^2
s_D	Standard deviation of a distribution of differences between correlated scores
$s_{\bar{D}}$	Standard error of the difference between correlated means
$s_{\bar{X}_1 - \bar{X}_2}$	Standard error of a difference between means
$s_{\bar{X}}$	Standard error of the mean (population σ estimated from sample)
SS	Sum of squares; the sum of the squared deviations from the mean
T	Wilcoxon matched-pairs signed-ranks T statistic for correlated samples
t	t test statistic
t_α	Critical value of t; level of significance $= \alpha$
U	Mann-Whitney U statistic for independent samples
UL	Upper limit of a confidence interval
X	A score
x	A deviation score
$\bar{X}$	The mean of a sample
$\bar{\bar{X}}$	The mean of a set of means
X_H	The upper limit of the highest score in a distribution
X_L	The lower limit of the lowest score in a distribution
Y'	The Y value predicted for some X value
$\bar{Y}$	The mean of the Y variable
z	A score expressed in standard deviation units; a standard score
z	Test statistic when sampling distribution is normal
z_X	A z value for a score on variable X
z_Y	A z value for a score on variable Y

6

Basic Statistics

TALES OF DISTRIBUTIONS

For Thea

Even if our statistical appetite is far from keen, we all of us should like to know enough to understand, or to withstand, the statistics that are constantly being thrown at us in print or conversation—much of it pretty bad statistics. The only cure for bad statistics is apparently more and better statistics. All in all, it certainly appears that the rudiments of sound statistical sense are coming to be an essential of a liberal education.

Robert Sessions Woodworth

Basic Statistics

TALES OF DISTRIBUTIONS

Chris Spatz

Hendrix College

Brooks/Cole Publishing Company

I(T)P® An International Thomson Publishing Company

Pacific Grove • Albany • Belmont • Bonn • Boston • Cincinnati • Detroit
Johannesburg • London • Madrid • Melbourne • Mexico City • New York • Paris
Singapore • Tokyo • Toronto • Washington

Sponsoring Editor: *Jim Brace-Thompson*
Consulting Editor: *Roger E. Kirk*
Marketing Team: *Gay Meixel, Romy Taormina*
Editorial Assistant: *Terry Thomas*
Production Editor: *Nancy L. Shammas*
Manuscript Editor: *Carol Dondrea*
Permissions Editor: *Cathleen S. Collins*
Cover and Interior Design: *E. Kelly Shoemaker*

Interior Illustration: *Alexander Teshin Associates, Accurate Art, Inc.*
Cover Photo: *Ed Young*
Art Editor: *Lisa Torri*
Typesetting: *Techset Composition, Ltd.*
Cover Printing: Phoenix Color Corporation, Inc.
Printing and Binding: Courier Westford, Inc.

COPYRIGHT © 1997 by Brooks/Cole Publishing Company
A division of International Thomson Publishing Inc.
I(T)P The ITP logo is a registered trademark under license.

For more information, contact:

BROOKS/COLE PUBLISHING COMPANY
511 Forest Lodge Road
Pacific Grove, CA 93950
USA

International Thomson Publishing Europe
Berkshire House 168–173
High Holborn
London WC1V 7AA
England

Thomas Nelson Australia
102 Dodds Street
South Melbourne, 3205
Victoria, Australia

Nelson Canada
1120 Birchmount Road
Scarborough, Ontario
Canada M1K 5G4

International Thomson Editores
Seneca 53
Col. Polanco
México, D. F., México
C.P. 11560

International Thomson Publishing GmbH
Königswinterer Strasse 418
53227 Bonn
Germany

International Thomson Publishing Asia
221 Henderson Road
#05–10 Henderson Building
Singapore 0315

International Thomson Publishing Japan
Hirakawacho Kyowa Building, 3F
2–2–1 Hirakawacho
Chiyoda-ku, Tokyo 102
Japan

All rights reserved. No part of this work may be reproduced, stored in a retrieval system, or transcribed, in any form or by any means—electronic, mechanical, photocopying, recording, or otherwise—without the prior written permission of the publisher, Brooks/Cole Publishing Company, Pacific Grove, California 93950.

Printed in the United States of America

10   9   8   7   6   5   4   3   2

Library of Congress Cataloging-in-Publication Data
Spatz, Chris, [date]
    Basic statistics: tales of distributions/Chris Spatz. – 6th
  ed.
      p. cm.
    Includes bibliographical references (p.   –   ) and index.
    ISBN 0-534-26424-7 (hardcover)
    1. Statistics.   I. Title.
QA276.12.S66   1996
519.5–dc20   96-20147
                                              CIP

All product names are trademarks or registered trademarks of their respective companies.

BRIEF CONTENTS

CONTENTS

3   Variability   55

TRANSITION PAGE   75

4   Correlation and Regression   77

9 Analysis of Variance: One-Way Classification 213

10 Analysis of Variance: Factorial Design 239

11 Analysis of Variance: One-Factor Correlated Measures 271

Appendixes 345

Basic Statistics: Tales of Distributions, Sixth Edition, is designed for a one-term, introductory course in statistics. In addition to traditional statistical topics, some experimental design terms and issues are covered. My goal has been to produce a book that is comprehensible and complete for students who take only one statistics course and comprehensible and preparatory for students who will take additional courses.

Although detailed directions and examples are given for each statistical procedure, this book concentrates heavily on conceptualization and interpretation of statistical results. In many places the reader is invited to stop and think, or stop and do an exercise. Some problems simply ask the student to decide which statistical technique is appropriate. Because of this book's emphasis and style, I am confident that it will reinforce an instructor's efforts to promote critical thinking.

My expectation for students who work through this book is that they will be able to:

- solve statistical problems
- understand statistical reasoning
- write explanations that are congruent with statistical analyses
- choose proper statistical techniques to analyze data from simple experimental designs

Students who meet these expectations will be able to understand the statistical concepts in many journal articles as well as be able to analyze data from their own research. I also expect they will use this statistical knowledge in the years that follow.

I think you will like this book. Most students find it relatively engaging because it is written in an informal, personal style. There are examples and problems from many fields, including some that have been worked on by the pioneers in statistics.

In addition to the writing style and the varied problems, this book has a number of features that are designed to make statistics easier to learn. For example, the problems are an integral part of the text. The answers, complete with all necessary steps or explanations, are in Appendix E. Concepts that will be important in later chapters are identified as "Clues to the Future" and set off in boxes. The "Error Detection" boxes note ways to detect or prevent mistakes. At the beginning of each chapter, a list of objectives provides orientation. This same list also serves as a review exercise for the chapter. Three glossaries are provided: the Glossary of Words and the Glossary of Formulas are Appendixes C and D, respectively; the Glossary of Symbols is printed on the inside covers of the book.

Two ancillary publications supplement this textbook. For students, there is a softcover Study Guide that provides additional explanations, problems, and answers. For professors, there is an Instructor's Manual that contains teaching suggestions and test items. Test items are also available in electronic format.

This sixth edition differs in a number of ways from the last edition. I added an *effect size index* to the analyses of the one-sample t test, both two-sample t tests, and the one-way ANOVA. An emphasis on effect size was also incorporated into other statistical descriptions. The section on confidence intervals about a mean difference was eliminated from the chapter on two-sample t tests. All three basic ANOVA designs are covered now that a new chapter has been added, Analysis of Variance: One-Factor Correlated Measures (Chapter 11). I rewrote all of Chapter 6, Samples, Sampling Distributions, and Confidence Intervals; most of Chapter 7, Hypothesis Testing and Effect Size: One-Sample Tests; and many other paragraphs and sections. My aim was clarity, especially on the topics of sampling distributions and hypothesis testing. Responding to the recommendations of teachers and reviewers, the sample estimator of σ was changed from s to $\hat{s}$, which will distinguish it easily from S (sample standard deviation). Chapter 14, Choosing Tests and Writing Interpretations, was modified to encourage students to create overall summaries of elementary statistics. My answers are given in table form (Table 14.1) and as decision trees (Figure 14.1 and Figure 14.2). I brought all contemporary data sets up to date.

I am pleased to acknowledge all the help I have received from students, colleagues, and Hendrix College. Students identified some 42 errors in the previous two editions. (I pay $2.00 to the first to report an error.) Colleagues at Hendrix and elsewhere have made many useful suggestions. Bob Eslinger produced accurate graphs of the F, t, χ^2, and normal distributions. Rob Nichols wrote a sampling program for me. Hendrix librarians JoAnn McMillen and Delores Thompson provided enthusiastic and competent help for all six editions of this book. Roger E. Kirk of Baylor University, my consulting editor for all six editions, deserves a special thanks. Over the years he has saved me from several errors, taught me some statistics, and always had a note of encouragement for me. Jim Brace-Thompson, Nancy Shammas, and the staff at Brooks/Cole contributed in many ways.

I also want to acknowledge the help of reviewers for this edition: Eugene Chao, Berea College; David Chattin, St. Joseph's College; Dennis Cogan, Texas Tech University; Anupa K. Doraiswami, Morris Brown College; James Overton, Coker College; Joel Royalty, Murray State University; Anthony Santucci, Manhattanville College; Randolph Smith, Ouachita Baptist University; Boyd Spencer, Eastern Illinois University; Philip Tolin, Central Washington University; Mahlon Wagner, State University of New York—Oswego; and Donald Walter, University of Wisconsin—Parkside.

I am grateful to the Longman Group UK Ltd., on behalf of the Literary Executor of the late Sir Ronald A. Fisher, F.R.S. and Dr. Frank Yates, F.R.S., for permission to reproduce Tables III, IV, and VII from their book *Statistical Tables for Biological, Agricultural, and Medical Research*, Sixth Edition, (1974).

I especially want to acknowledge James O. Johnston, my good friend and former co-author, who contributed to the first three editions of this book. Without his efforts, there never would have been a first edition.

My most important acknowledgment goes to my wife and family, who have helped and supported me in many ways during the almost 25 years of this project.

I've been a teacher for much of my life—first as an older sibling, then as a parent, and now as a professor. Education is a task of the first order, in my opinion. I hope that my writing conveys both my enthusiasm for this task and my philosophy of teaching. (By the way, if you are a student who is thoroughly reading the *whole* preface, you should know that in a number of places in this book I included phrases or examples with *your kind* in mind.)

Chris Spatz

Basic Statistics

TALES OF DISTRIBUTIONS

Introduction

O B J E C T I V E S   F O R   C H A P T E R   1

After studying the text and working the problems in this chapter, you should be able to:

1. Distinguish between descriptive and inferential statistics
2. Define the words *population, sample, parameter, statistic,* and *variable* as these terms are used in statistics
3. Distinguish between quantitative and qualitative variables
4. Identify the lower and upper limits of a quantitative measurement
5. Identify four scales of measurement and give examples of each
6. Distinguish between statistics and experimental design
7. Define some experimental-design terms—*independent variable, dependent variable,* and *extraneous variable*—and identify these variables when you are given a description of an experiment
8. Describe the relationship of statistics to epistemology
9. Identify a few events in the history of statistics

This is a book about statistics that is written for people who do not plan to become statisticians. In fact, I was not trained as a statistician myself; my graduate school education was in psychology. I studied statistics, however, so I could analyze, understand, and explain the data from experiments (and also because the courses were required).

Besides psychology, many other disciplines use quantitative data. In each of these disciplines, understanding statistics is at least helpful and may even be necessary. Lawyers, for example, sometimes have cases in which statistics can help establish a more convincing case for their clients. I have assisted lawyers at times, occasionally testifying in court about the proper interpretation of data. (You will analyze and interpret some of these data in later chapters.) Statistics is a powerful method for getting answers from data, and it is sometimes the best way to persuade others that your conclusions are correct.

1

Whatever your current feelings about your future as a statistician, I believe you will benefit from this course. When a statistics course is successful, students learn to identify the questions that a set of data can answer, determine the statistical procedures that will provide the answers, carry out the procedures, and then, using plain English and graphs, tell the story the data reveal.

The best way for you to acquire all these skills (especially the part about telling the story) is to use a *generative* approach while you are studying. What should you generate? Generate questions for topics you don't yet understand. When you can, generate answers. When you aren't able to generate an answer, plan how you can find an answer. Decide what sections are important. Generate a summary for important sections. And, of course, you should generate answers to the problems scattered throughout the chapter. This generative approach to learning is guaranteed to pay handsome dividends (see Wittrock, 1986). For me, writing my thoughts is a better way to generate understanding than just thinking. Is writing also better for you?

WHAT DO YOU MEAN, "STATISTICS"?

The *Oxford English Dictionary* says that the word *statistics* came into use more than 200 years ago. At that time, statistics referred to a country's quantifiable political characteristics—characteristics such as population, taxes, and area. Statistics meant "state numbers." Tables and charts of those numbers turned out to be a most satisfactory method for understanding a country better, for comparing different countries, and for making projections about the future. Later, the word *statistics* was used to refer to tables and charts compiled by people studying trade (economics) and natural phenomena (science).

Today the word *statistics* has two meanings. A **descriptive statistic**[1] is an index number that summarizes or describes a set of data. You are already familiar with at least one of these index numbers, the arithmetic average, called the **mean**. You have probably known how to compute a mean since elementary school—just add up the numbers and divide by the number of numbers. As you already know, the mean is a number that indicates the central tendency of a collection of numbers. You will learn about other descriptive statistics starting with Chapter 2. The basic idea is simple: A descriptive statistic expresses with one number a characteristic of a set of data.

Starting about the middle of the 19th century, scientists began to use reasoning and concepts that are today recognized as the beginnings of **inferential statistics**. Like earlier statisticians, these scientists were trying to understand some phenomenon that had caught their interest. Their situation was different, however. Descriptive statisticians could gather *all* the data on a particular topic, but these scientists were forced to work with *samples*. Samples show variation that is due to the effects of chance, and this variation left the scientists with some uncertainty. Inferential statistics was then developed as a way to measure and account for the effects of chance that go with any sample.

Here's the textbook definition of inferential statistics: a method that takes chance factors into account when samples are used to reach conclusions about populations.

---

[1] Words and phrases printed in boldface type are defined in Appendix C: Glossary of Words.

Like most textbook definitions, this one will be much more meaningful after you understand the concept. And keep in mind that this concept is one of *the big ones*, so pay careful attention whenever it comes up.

Inferential statistics has proved to be a very useful method in scientific disciplines. Many other fields use inferential statistics, too, so I have selected examples and problems from a variety of disciplines for this text and its study guide.

Here is an example of the use of inferential statistics in psychology. Today there is a lot of evidence that uncompleted tasks are remembered better than completed tasks. This is known as the *Zeigarnik effect*. Let's go back to a time in Berlin when this phenomenon was about to be discovered by Bluma Zeigarnik.[2] Zeigarnik asked the participants in her experiment to do about 20 tasks, such as to work a puzzle, make a clay figure, and construct a box from cardboard. For each participant, half the tasks were interrupted before completion. Later, when participants were asked to recall the tasks they had worked on, they listed more of the interrupted tasks (about seven) than the completed tasks (about four).

So, should you conclude that interruption improves memory? Not yet. It might be that interruption actually has no effect, but that several *chance factors* happened to favor the interrupted tasks in this particular experiment. One way to meet this objection is to run the experiment again to see if you still get a difference in favor of interruption. Another, better, solution is to use inferential statistics.

Inferential statistics works by giving you the probability of obtaining the difference you actually observed if, in fact, interruption has no effect on memory. If the probability you get is very low, then the "interruption has no effect on memory" hypothesis can be rejected. This rejection leaves you with an "interruption improves memory" explanation.

A conclusion based on inferential statistics might be a statement such as, "The difference in the memory of the two groups is most likely due to interruption and not to chance because chance *by itself* would rarely produce two samples this different." Probability, as you can see, is an important element of inferential statistics.

Here is a general overview of how statistics is used today. Researchers start with some phenomenon, event, or process that they want to understand better. They translate it into numbers, manipulate the numbers according to the rules of statistics, translate the numbers back into the phenomenon, and, finally, tell the story of their new understanding of the phenomenon, event, or process. You can understand why many textbooks refer to statistics as a tool.

CLUE TO THE FUTURE

The first part of this book is devoted to descriptive statistics (Chapters 2–4) and the second part to inferential statistics (Chapters 5–13). The two parts are not independent, though, because to understand and explain inferential statistics you must first understand descriptive statistics.

[2] A summary of this study can be found in Ellis (1938). The complete reference, and all others in the text, can be found in the References section at the back of the book.

WHAT'S IN IT FOR ME?

What's in it for me? is a reasonable question. Decide if any of the following answers apply to you and then add any others that are true in your own case. One thing that is in it for you is that, when you have successfully completed this course, you will understand how statistical analyses are used to *make decisions*. You will understand both the elegant beauty and the ugly warts of the process. H. G. Wells, a novelist who wrote about the future, said, "Statistical thinking will one day be as necessary for efficient citizenship as the ability to read and write." So chalk this course up under the heading "general education about decision making, to be used throughout life."

Another result of the successful completion of this course is that you will know how to use a tool that is basic in many disciplines, including psychology, education, medicine, sociology, biology, forestry, economics, and others. Although people involved in these disciplines are interested in different phenomena, they all use statistics. As examples, statistics has been used to determine the effect of reducing cigarette taxes on smoking among teenagers, the safety of a new surgical anesthetic, and the memory of young school-age children for pictures (which is as good as that of college students). Statistics show what diseases have an inheritance factor, how to improve short-term weather forecasts, and why giving intentional walks in baseball is a poor strategy. All these examples come from *Statistics: A Guide to the Unknown* (3d ed., 1989), a book edited by Judith M. Tanur and others. This book, written for those "without special knowledge of statistics," has 29 essays on many specific uses of statistics.

Statistics is used in places that might surprise you. In American history, for example, the authorship of 12 of the Federalist Papers was disputed for a number of years. (*The Federalist Papers*, a group of 85 short essays, were written under the pseudonym "Publius" and published in New York City newspapers in 1787 and 1788. Written by James Madison, Alexander Hamilton, and John Jay, they were designed to persuade the people of the state of New York to ratify the Constitution of the United States.) To determine authorship, each of the 12 papers was graded with a quantitative *value analysis* in which the importance of such values as national security, a comfortable life, justice, and equality was assessed. The value analysis scores were compared with value analyses of papers known to have been written by Madison and Hamilton (Rokeach et al., 1970). Another study, by Mosteller and Wallace, analyzed the Papers using the frequency of words such as *by* and *to* (reported in Tanur et al., 1989). Both studies concluded that Madison wrote all 12 papers.

Here is an example from law. Rodrigo Partida was convicted of burglary in Hidalgo County, a border county in southern Texas. A grand jury rejected his motion for a new trial. Partida's attorney filed suit, claiming that the grand jury selection process discriminated against Mexican-Americans. In the end (*Castaneda* v. *Partida*, 430 U.S. 482 [1976]), Justice Harry Blackmun of the U.S. Supreme Court, wrote, regarding the number of Mexican-Americans on the grand jury, "If the difference between the expected and the observed number is greater than two or three standard deviations, then the hypothesis that the jury drawing was random (is) suspect." In Partida's case the difference was approximately 12 standard deviations, and the Supreme Court ruled that Partida's attorney had presented *prima facie* evidence. (*Prima facie* evidence is so good that you win the case unless the other side rebuts the evidence, which in this case did not happen.) *Statistics: A Guide to the Unknown* includes two essays on the use of statistics by lawyers.

Not only will your study of statistics help you understand presentations by others in a variety of disciplines and professions, it will also help you understand your own observations and experiments. Knowing statistics will help make your own discoveries more fun.

I hope that this course will encourage you or even teach you to ask questions about the statistics you hear and read. Consider the questions, one good and one better, asked of a restaurateur in Paris who served delicious rabbit pie. He served rabbit pie even when other restaurateurs could not obtain rabbits. A suspicious customer asked this restaurateur if he was stretching the rabbit supply by adding horse meat.

"Yes, a little," he replied.

"How much horse meat?"

"Oh, it's about 50–50," the restaurateur replied.

The suspicious customer, satisfied, began eating his pie. Another customer, who had learned to be more thorough in asking questions about statistics, said, "50–50? What does *that* mean?"

"One rabbit and one horse" was the reply.

To end this section, I have put into one sentence my answer to your question "What's in it for me?" *When you learn the basics of statistics you will understand data better, you will be able to communicate with others who use statistics, and you will be better able to persuade others*. (You will be able to present *prima facie* evidence.) In short, I am sure there will be lots of lasting value for you in this course.

SOME TERMINOLOGY

Like most courses, statistics introduces you to many new words. In statistics, most of the terms are used over and over again. Your best move, when introduced to a new term, is to *stop*, *read* the definition carefully, and *memorize* it. As the term continues to be used, you will become more and more comfortable with it.

■ Populations and Samples

A **population** consists of all members of some specified group. A **sample** is a subset of a population. The population is the thing of interest, is defined by the investigator, and includes all cases. The following are some populations:

Family income of college students in the fall of 1995
The proportion of salmon that actually return to their hatching stream
Gestation time for human beings
Depressed Alaskans
Persistence of pigeons
The memory of human beings[3]

---

[3] I didn't pull these populations out of thin air; they are all populations that researchers have been interested in. Studies of these populations are described in this book.

Investigators are always interested in a population. However, you can see from these examples that populations are often so large that not all the members can be measured. The investigator must often resort to measuring a sample that is small enough to be manageable but large enough to be representative of the population. A sample taken from the college student family income population might include only 40 students. From the last population of the list, Zeigarnik used a sample of 164. Throughout the book I discuss ways to increase your confidence that samples are indeed representative of their populations.

Resorting to the use of samples introduces uncertainty into the conclusions because different samples from the same population nearly always differ from one another in some respects. Inferential statistics is used to determine whether or not such differences should be attributed to chance. You will meet this problem head-on in Chapter 6.

Most authors of research articles carefully explain the characteristics of the samples they use. Often, however, they do not identify the population, leaving that task to the reader. For example, consider a study on a drug therapy program involving a sample of 14 teenage male acute schizophrenics at Hospital A. Suppose the report of the study doesn't identify the population. Left to you, to what population are you willing to generalize? Among the possibilities are: all teenage male acute schizophrenics at Hospital A; all teenage male acute schizophrenics; all teenage acute schizophrenics; or all schizophrenics. The answer to this question is not addressed in statistics books because answers depend on the specifics of a research area, but many researchers generalize generously. For example, some are willing to generalize from the results of a study on rats to "all mammals." Remember, the reason for gathering data on a sample is to generalize the results to some larger population.

■ Parameters and Statistics

A **parameter** is some numerical (number) or nominal (name) characteristic of a population. An example is the mean IQ score of *all* first grade pupils in the United States. A **statistic** is some numerical or nominal characteristic of a sample. The mean IQ score of 50 first graders is a statistic and so is the observation that all are females. A parameter is constant; it does not change unless the population itself changes. Only one number can be the mean of a population. Unfortunately, it often cannot be computed because the population is unmeasurable. So, we use a statistic as an estimate of a parameter although, as suggested before, a statistic tends to differ from one sample to another. If you have five samples from the same population, you will probably have five different sample means. In sum, parameters are constant; statistics are variable.

■ Variables

A **variable** is something that exists in more than one amount or in more than one form. The essence of *measurement* is to assign numbers on the basis of observed variation.

Height and gender are both variables. The variable height is measured using a standard scale of feet and inches. For example, the notation 5′7″ is a numerical way to identify a group of persons similar in height. Of course, there are many other height groups, each with an identifying number. Gender is also a variable. With gender, there are (usually) only two groups of people. Again, we may identify each group by assigning a number to it, usually 0 for females and 1 for males. All subjects represented by 0 have the same gender. I will often refer to numbers like 5′7″ and 0 as *scores* or *test scores*. A score is simply the result of a measurement.

■ Quantitative Variables

Many of the variables you will work with in this text will be **quantitative variables**. When a quantitative variable is measured, the scores tell you something about the amount or degree of the variable. At the very least, a larger score indicates more of the variable than a smaller score does.

Take a closer look at an example of a quantitative variable, IQ. IQ scores come only in whole numbers like 94 and 109. However, it is reasonable to think that of two persons who scored 103, one just barely made it and the other scored almost 104. Picture the variable IQ as the straight line in **Figure 1.1**.

Figure 1.1 shows that a score of 103 is used for a range of possible scores: the range from 102.5 to 103.5. The number 102.5 is the **lower limit** and 103.5 is the **upper limit** of the score 103. The idea is that IQ can be any value between 102.5 and 103.5, but that all scores in that range are rounded off to 103.

In a similar way, a score of 12 seconds is used to express all scores between 11.5 and 12.5 seconds; 11.5 seconds is the lower limit and 12.5 is the upper limit of a score of 12 seconds.

Sometimes scores are expressed in tenths, hundredths, or thousandths. With a good stopwatch, for example, time can be measured to the nearest hundredth of a second. Scores such as 6.00 and 12.05 result. These scores also have lower and upper limits, and the same reasoning holds: Each score represents a range of possible scores. The score 6.00 is the middle of a range of scores running from 5.995 to 6.005. Lower and upper limits of the score 12.05 are 12.045 and 12.055.

In summary, to find the lower and upper limits of any score, subtract half the unit of measurement from and add half the unit of measurement to the last digit on the right of the score. Thus, for the score 6.1 seconds, the lower limit is 6.1 − 0.05 = 6.05 seconds and the upper limit is 6.1 + 0.05 = 6.15 seconds.

The basic idea is that the limits of a score have one more decimal place than the score itself, and the number in this last decimal place is 5. A good way to be sure you understand this is to draw a few number lines like the one in Figure 1.1, put on them scores such as 3.0, 3.8, and 5.95, and find the lower and upper limits.

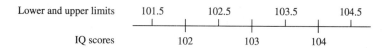

Lower and upper limits   101.5   102.5   103.5   104.5

IQ scores   102   103   104

FIGURE 1.1 The lower and upper limits of IQ scores of 102, 103, and 104

CLUE TO THE FUTURE

The lower and upper limits of a score will be important in Chapter 2, "Organization of Data, Graphs, and Central Tendency."

■ Qualitative Variables

Qualitative variables do not have the continuous nature that quantitative variables have. For example, gender is a qualitative variable. The two measurements on this variable, female and male, are different in a qualitative way. Using a 0 and a 1 instead of names doesn't change this fact. Another example of a qualitative variable is political affiliation, which has measurements such as Democrat, Republican, Independent, and Other.

Some qualitative variables have the characteristic of *order*. The qualitative variable college classification has the ordered measurements of senior, junior, sophomore, and freshman. Another example is the qualitative variable military rank, which has the measurements sergeant, corporal, and private.

PROBLEMS AND ANSWERS

At the beginning of this chapter, I urged you to take up the generative approach to learning statistics. Did you? For instance, did you draw those lines and label them, as suggested several paragraphs back? If not, try it now for the score 0.05. What is its lower limit? If you didn't get 0.045, you need to reread that material. If you still have trouble, get some help from your instructor or from another student who understands it. Be active. Do you understand the word *nominal* that was used in the last section? How near is a dictionary? Make it your policy throughout this course never to let anything slip by you.

The generative approach means that you read actively. You will know you are doing this if you reread a paragraph that you did not understand the first time, or wrinkle your brow in concentration, or say to yourself, "Hmm, I really don't understand that; I'll ask my instructor about it tomorrow" or, best of all, nod to yourself, "Oh yes."

From time to time I will interrupt with a set of problems so that you can practice what you have just been reading about. Working these problems correctly is also evidence of active reading. You will find the answers at the end of the book in Appendix E. Here are some suggestions for *efficient* learning.

1. Buy yourself a spiral notebook for statistics. Work all the problems for this course in it because, for several problems, I refer you back to a problem you worked previously. When you make an error, don't erase it—circle it and work the problem correctly below. Seeing your error later serves as a reminder of what not to do on a test. If you find that *I* have made an error, write to me with a reminder of what not to do in the next edition.

2. Never, *never* look at an answer before you have worked the problem (or at least tried twice to work the problem).
3. When you come to a set of problems, work the first one and then immediately check your answer against the answer in the book. If you make an error, find out why you made it—faulty understanding, arithmetic error, or whatever.
4. Don't be satisfied with just doing the math. If a problem asks for an interpretation, write out your interpretation.
5. When you finish a chapter, go back over the problems immediately, reminding yourself of the various techniques you have learned.
6. Use any blank spaces near the end of the book for your special notes and insights.

Now, here is an opportunity to see how actively you have been reading.

PROBLEMS

1. Give the lower and upper limits of the following measurements that are of quantitative variables. For the qualitative variables, write "qualitative."
   a. 8, gallons of gasoline
   b. 5, a category of daffodil in a flower show
   c. 10.00, a time required to run 100 meters
   d. 811.1, the identification of American poetry authors in the Dewey Decimal System
   e. 4.5, millions of dollars
   f. 2.95, tons of grain
2. Write a paragraph that gives the definitions of and the relationships among the following terms: *population, sample, parameter,* and *statistic.*
3. Classify each statement as being related to either descriptive or inferential statistics.
   a. Based on a sample, the pollster predicted Demosthenes would be elected.
   b. The population of the American colonies in 1776 was 2 million.
   c. Demosthenes received 54.3 percent of the votes cast.
   d. The world population is projected to be 6.1 billion in the year 2000.
   e. A generative approach to learning works; the 29 students who generated summaries and inferences performed 20 percent better than those who just memorized (Wittrock, 1991).   ■

SCALES OF MEASUREMENT

Numbers mean different things in different situations. Consider three answers that appear to be identical but are not:

"What number were you wearing in the race?"	"5"
"What place did you finish in?"	"5"
"How many minutes did it take you to finish?"	"5"

The 5's all look the same, but the situations they refer to are different (identification number, finish place, and time). As a result, each 5 carries a *different set* of information.

To illustrate this, consider another person whose answers to the same three questions were 10, 10, and 10. Taking the first question by itself, and knowing that the two people had scores of 5 and 10, what can you say? You can say that the first runner was different from the second, but *that is all*. (Think about this until you agree.) On the second question, with scores of 5 and 10, what can you say? You can say that the first runner was faster than the second and, of course, that they were different. Comparing the 5 and 10 on the third question, you can say that the first runner was twice as fast as the second runner (and, of course, was faster and different).

The point of this is to draw the distinction between the *thing* you are interested in and the *number* that stands for the thing. Much of your experience with numbers has been with pure numbers or with the measurement of things such as time, length, and amount. "Four is twice as much as two" is true for the pure numbers themselves and for time, length, and amount, but it is not true for finish places in a race. Fourth place is not twice anything in relation to second place—not twice as *slow* or twice as *far behind* another runner.

S. S. Stevens's 1946 article is probably the best-known effort to focus attention on these distinctions. He identified four different kinds of *scales of measurement*, each of which carries a different set of information. Each scale uses numbers, but the information that can be inferred from the numbers differs. The four scales are nominal, ordinal, interval, and ratio. The information carried in the first of these, the nominal scale, is quite simple. In the **nominal scale**, numbers are used simply as names and have no real quantitative value. Numerals on sports uniforms are an example; here, 45 is *different* from 32, but that is all you can say. The person represented by 45 is not "more than" the person represented by 32, and certainly it would be meaningless to calculate a mean. Psychological diagnoses, personality types, and political parties are three examples of nominal variables. The variable called psychological diagnoses consists of a set of categories. A person is simply classified into one of the categories, which may then be given a number. With a nominal scale, you could even reassign all the numbers and still maintain the original meaning, which is only that things with different numbers are different. Of course, all things that are alike have the same number.

A second kind of scale, the **ordinal scale**, has the characteristic of the nominal scale (different numbers mean different things) plus the characteristic of indicating "greater than" or "less than." In the ordinal scale, the object with the number 3 has less or more of something than the object with the number 5. Finish places in a race are an example of an ordinal scale. The runners finish in rank order, with 1 assigned to the winner, 2 to the runner-up, and so on. Here, 1 means less time than 2. Judgments about anxiety, quality, and recovery often correspond to an ordinal scale. "Much improved," "improved," "no change," "worse" are levels of an ordinal recovery variable. Ordinal scales are characterized by rank order.

The third kind of scale is the **interval scale**, which has the properties of both the nominal and ordinal scales, plus the additional property that intervals between the numbers are equal. "Equal interval" means that the distance between the things represented by 2 and 3 is the same as the distance between the things represented by 3 and 4. The Celsius thermometer is based on an interval scale. The difference in temperature between 10° and 20° is the same as the difference between 40° and 50°.

The Celsius thermometer, like all interval scales, has an arbitrary zero point. On the Celsius thermometer, this zero point is the freezing point of water at sea level. Zero degrees on this scale does not mean the complete absence of heat; it is simply a convenient starting point. With interval data, there is one restriction: You may not make simple ratio statements. You may not say that 100° is twice as hot as 50° or that a person with an IQ of 60 is half as intelligent as a person with an IQ of 120.[4]

The fourth kind of scale, the **ratio scale**, has all the characteristics of the nominal, ordinal, and interval scales, plus one other: It has a true zero point, which indicates a complete absence of the thing measured. On a ratio scale, zero means "none." Height, weight, and time are measured with ratio scales. Zero height, zero weight, and zero time mean that no amount of these variables is present. With a true zero point, you can make ratio statements such as "16 kilograms is four times heavier than 4 kilograms."[5]

Knowing the distinctions among the various scales of measurement will help you in two tasks in this course. The kind of *descriptive statistics* you can compute from your numbers depends, in part, on the scale of measurement the numbers represent. For example, it is senseless to compute a mean on numbers from a nominal scale. Calculating a mean social security number, a mean telephone number, or a mean psychological diagnosis is either a joke or evidence of not understanding numbers.

Scales of measurement are also important in choosing the kind of *inferential statistic* that is appropriate for a set of data. If the dependent variable (see next section) is a nominal variable, then a chi square analysis is appropriate (Chapter 12). If the dependent variable is a set of ranks (ordinal data), a nonparametric statistic is required (Chapter 13). Most of the inferential statistics you will study (Chapters 6–11) will analyze interval or ratio scale data.

To end this section, I must tell you that it is sometimes difficult to classify some of the variables used in the social and behavioral sciences. Often they appear to fall between the ordinal scale and the interval scale. For example, a score may provide more information than simply rank, but equal intervals cannot be proved. Intelligence test scores and **Likert scale** scores are examples. In such cases, researchers generally treat the scores as if they were from an interval scale.

STATISTICS AND EXPERIMENTAL DESIGN

Here is a story that will help distinguish between statistics (applying straight logic) and experimental design (observing what actually happens). This is an excerpt from a delightful book by E. B. White, *The Trumpet of the Swan* (1970).[6]

> The fifth-graders were having a lesson in arithmetic, and their teacher, Miss Annie Snug, greeted Sam with a question.
> "Sam, if a man can walk three miles in one hour, how many miles can he walk in four hours?"

[4] Convert 100°C and 50°C to Fahrenheit ($F = 1.8 C + 32$), and suddenly the "twice as much" relationship disappears.
[5] Convert 16 and 4 kilograms to pounds (1 kg = 2.2 pounds), and the "four times heavier" relationship is maintained.
[6] From *The Trumpet of the Swan*, by E. B. White, pp. 63–64, HarperCollins Publishers, Inc.

"It would depend on how tired he got after the first hour," replied Sam. The other pupils roared. Miss Snug rapped for order.

"Sam is quite right," she said. "I never looked at the problem that way before. I always supposed that man could walk twelve miles in four hours, but Sam may be right: that man may not feel so spunky after the first hour. He may drag his feet. He may slow up."

Albert Bigelow raised his hand. "My father knew a man who tried to walk twelve miles, and he died of heart failure," said Albert.

"Goodness!" said the teacher. "I suppose *that* could happen, too."

"Anything can happen in four hours," said Sam. "A man might develop a blister on his heel. Or he might find some berries growing along the road and stop to pick them. That would slow him up even if he wasn't tired or didn't have a blister."

"It would indeed," agreed the teacher. "Well, children, I think we have all learned a great deal about arithmetic this morning, thanks to Sam Beaver."

Everyone had learned how careful you have to be when dealing with figures.

Statistics involves the manipulation of numbers and the conclusions based on those manipulations (Miss Snug). Experimental design deals with all the things that influence the numbers you get (Sam and Albert). This text could have been a "pure" statistics book, from which you would learn to analyze numbers without knowing where they came from or what they referred to. You would learn about statistics, but such a book would be dull, dull, dull. On the other hand, to describe procedures for collecting numbers is to teach experimental design—and this book is for a statistics course. My solution to this conflict is generally to side with Miss Snug but to include some aspects of experimental design throughout the book. Perhaps the information on design that you pick up will encourage you to try your hand at your own data gathering. Here's a start on experimental design.

The overall task of an experimenter is to discover relationships among variables. Variables are things that vary, and researchers have studied personality, health, gender, anger, caffeine, memory, beliefs, age, skill. (I'm sure you get the picture—almost anything can be a variable.)

A simple experiment has two major variables, the **independent variable** and the **dependent variable**. In the simplest experiment, the researcher *selects* two values of the independent variable for investigation. Values of the independent variable are usually called **levels** and sometimes called *treatments*.

The values of the dependent variable are found by *measuring* or *observing* the participants in the investigation. The purpose of an experiment is to determine whether values of the dependent variable change when you switch from one level of the independent variable to a second level.

The basic idea is that the researcher finds or creates two groups of subjects that are similar except for the independent variable. These two groups are measured on the dependent variable. The question is whether the data will allow you to claim that the values on the dependent variable *depend* on the level of the independent variable.

In this simple example, the independent variable was varied by having two groups that are different for the independent variable. The two groups might have been selected because they were already different—in age, gender, personality, and so forth—or the experimenter might have produced the difference in the two groups by an experimental manipulation—degree of anxiety, amount of drug, and so forth. The dependent variable scores were obtained by measuring all the individuals in both groups.

An example might help. Suppose for a moment that as a budding gourmet cook you are trying to improve your spaghetti sauce, and one of your buddies has suggested adding marjoram. To investigate, you serve spaghetti sauce at two different gatherings. For one group the sauce is spiced with marjoram; for the other, it is not. At both gatherings, you count the number of favorable comments about the spaghetti sauce. *Stop reading; identify the independent and the dependent variables.*

independent →

dependent

The independent variable is marjoram. There were two levels of this variable, present and absent. The dependent variable is the number of favorable comments, which is a measure of the taste of the sauce.

One of the pitfalls in actually doing experiments is that every situation has other variables in addition to the independent variable that might possibly be responsible for the changes in the dependent variable. These other variables are called **extraneous variables**. In the story of Miss Snug, Sam and Albert noted several extraneous variables that could influence distance traveled.

Are there any extraneous variables in the spaghetti sauce example? Oh yes, there are many, and just one is enough to raise a suspicion about any conclusion that relates spaghetti sauce taste to marjoram. Extraneous variables in that situation include the amount and quality of the other ingredients in the sauce, the spaghetti itself, the "party moods" of the two groups, and how hungry everybody was. If any of these extraneous variables was actually operating, it weakens the claim that a difference in the comments about the sauce is the result of the presence or absence of marjoram.

The simplest way to remove an extraneous variable is to be sure that all subjects are equal on that variable. For example, the "party moods" variable can be controlled (equalized) by conducting the taste test in a laboratory. You can ensure that the sauces are the same except for marjoram by mixing up one pot of ingredients, dividing it into two batches, adding marjoram to one batch but not the other, and then cooking. Controlling extraneous variables is a complex topic that is covered in courses that focus on research or experiments.

In many experiments it is impossible or impractical to control all the extraneous variables. Sometimes researchers think they have controlled them all, only to find that they did not. The effect of an uncontrolled extraneous variable is to prevent a simple cause-and-effect conclusion. Even so, if the dependent variable changes when the independent variable changes, something is going on. In this case, researchers say that the two variables are *related*, but that other variables may play a part too.

At this point, you can test your understanding by generating answers to the question, What were the independent and dependent variables in the Zeigarnik experiment? How many levels of the independent variable were there? As for the question of how well Zeigarnik controlled extraneous variables, I'll give a partial answer. In her experiment each participant was tested at both levels of the independent variable. One of the advantages of this technique is that it naturally controls many extraneous variables. Thus, extraneous variables such as ability to remember, age, and motivation were exactly the same for tasks that were interrupted as for tasks that were not because each participant contributed data to *both* levels of the independent variable.

The general question that experiments try to answer is, What effect does the independent variable have on the dependent variable? At various places in the following chapters, I will explain experiments and statistical analyses of their results using the terms *independent*, *dependent*, and *extraneous* variables. These explanations will usually assume that all extraneous variables were well controlled; that is, you may assume that the experimenter knew how to design the experiment so that changes in the dependent variable could be correctly attributed to changes in the independent variable. However, I will present a few investigations (like the spaghetti sauce example) that I hope you recognize as being so poorly designed that conclusions about the relationship between the independent variable and dependent variable cannot be drawn. Be alert.

Here's my summary on statistics and experimental design. Researchers suspect that there is a relationship between two variables. They design and complete an experiment; that is, they *choose* the levels of the independent variable (treatments), control the extraneous variables, and then *measure* the subjects on the dependent variable. The measurements (data) are then analyzed using statistical procedures. Finally, the researcher tells a story that is consistent with the statistical analysis of the experiment.

STATISTICS AND PHILOSOPHY

The previous section directed your attention to the relationship between statistics and experimental design; this section will direct your thoughts to the place of statistics in the grand scheme of things.

The discipline of philosophy has the task of explaining the grand scheme of things and, as you know, there have been many schemes. For a scheme to be considered a grand one, it has to propose answers to questions of epistemology—that is, answers to questions about the nature of knowledge.

One of the big questions in epistemology is, How do we acquire knowledge? Answers such as *reason* and *experience* have gotten a lot of attention.[7] For those who emphasize the importance of reason, mathematics has been a continuing source of inspiration. In classical mathematics, you start with a set of axioms that are assumed to be true. Theorems are thought up and are then proved by giving axioms as reasons. Once a theorem is proved, it can be used as a reason in a proof of other theorems.

Statistics has its foundations in mathematics. Statistics, therefore, is based on reason. As you go about the task of memorizing definitions of terms such as σ^2 and ΣX, calculating their values from data, and telling the story of what they mean, know deep down that you are using a technique in logical reasoning. Logical reasoning is rationalism, which is one solution to the problem of epistemology. Experimental design is more complex—it is a combination of reasoning and experience (observation).

Let's move from this formal description of philosophy to a more informal one: A very common task of most human beings can be described as *trying to understand*. Statistics has helped many in their search for better understanding, and it is such people who have recommended (or demanded) that statistics be taught in school. A reasonable

---

[7] In philosophy, those who emphasize reason are rationalists and those who emphasize experience are empiricists.

expectation is that you, too, will find statistics useful in your future efforts to understand and persuade.

Speaking of persuasion, you have probably heard it said, "You can prove anything with statistics." The message conveyed is that a conclusion based on statistics is suspect because statistical methods are unreliable. Well, it just isn't true that statistical methods are unreliable—but it is true that people can misuse statistics (just as any tool can be misused). One of the great things about studying statistics is that you will get better at detecting when a conclusion is based on misused statistics.

STATISTICS: THEN AND NOW

Statistics began with counting. The beginning of counting, of course, was prehistory. The origin of the mean is almost as obscure. That statistic was in use by the early 1700s, but no one is credited with its discovery. Graphs, however, began when J. H. Lambert, a Swiss-German scientist and mathematician, and William Playfair, an English political economist, invented and improved graphs in the period 1765 to 1800 (Tufte, 1983).

In 1834 the Royal Statistical Society was formed in London. Just five years later, on November 27, 1839, at 15 Cornhill in Boston, a group of Americans founded the American Statistical Society. Less than three months later, for a reason that you can probably figure out even though over 150 years have passed, the group changed its name to the American Statistical Association.

According to Walker (1929), the first course in statistics given in a university in the United States was probably "Social Science and Statistics," taught at Columbia University in 1880. The professor was a political scientist and the course was offered in the economics department. In 1887, at the University of Pennsylvania, separate courses in statistics were offered by the departments of psychology and economics. By 1891, Clark University, the University of Michigan, and Yale had been added to the list of schools teaching statistics, and anthropology had been added to the list of departments. Biology was added in 1899 (Harvard) and education in 1900 (Columbia).

You might be interested to know when statistics was first taught at your school and in what department. College catalogs are probably the most accessible source for this information.

Today, statistics is a tool used by many people who are trying to understand or persuade or both. Kirk (1990) says that four groups of statistics users can be identified:

Category 1—those who are able to understand statistical presentations

Category 2—those who are able to select and apply statistical procedures to a particular problem

Category 3—applied statisticians who help others use statistics on a particular problem

Category 4—mathematical statisticians who develop new statistical techniques and discover new characteristics of old techniques[8]

---

[8] "Careers in Statistics" and "Statistics as a Career: Women at Work" are one-page flyers on careers as a statistician. Single copies are available free from the American Statistical Association, 1429 Duke Street, Alexandria, VA 22314-3402.

If the hopes of this book are realized, you will be well along the path of becoming a category 2 statistician by the end of your statistics course.

HOW TO USE THIS BOOK

At the beginning of this chapter, I mentioned the generative approach to learning. Your active participation is necessary if you are to learn statistics. I have tried to organize and write this book in a way that will encourage active participation. Here are some of the features you should find helpful.

■ Objectives

Each chapter begins with a list of skills the chapter is designed to help you acquire. Read this list of objectives first to find out what you are to learn to do. Then thumb through the chapter and read the headings. Next, study the chapter, working all the problems *as you come to them*. Finally, reread the objectives. Can you do each one? If so, put a check mark beside that objective.

■ Clues to the Future

Often I present an idea that will be used extensively in later chapters. You will find these ideas separated from the rest of the text in a box labeled "Clue to the Future." You have already seen two of these "Clues" in this chapter. Paying particularly close attention to these concepts will help you later in the course.

■ Computers, Calculators, and Statistics

Computers, calculators, and pencils and erasers are all tools used at one time or another by statisticians. Any or all of these devices may be part of the course you are taking. Regardless of the calculating aids that you use, however, your task remains the same: to read a problem, decide what statistical procedure to use, apply that procedure using the tools available to you, and write an interpretation of the results. In addition, you should be able to detect any gross errors by comparing the raw data, descriptive statistics, and inferential statistics.

Pencils, calculators, and computers represent, in ascending order, tools that are more and more error free. Most people who do statistics often use computers. You may or may not be at this point. Remember, though, whether you are using computers or not, your task remains as I described it in the previous paragraph.

■ Error Detection

I have boxed in, at various points in the book, some ways to detect errors. Some of these "Error Detection" tips will also help your understanding of statistical concepts.

Because many of these checks can be made early in a statistical problem, they can prevent the frustrating experience of getting an impossible answer when the error could have been caught in step 2.

■ Figure and Table References

When I think it would be worthwhile for you to examine a figure or a table, I have put the word "Figure " or "Table" and the number of that figure or table in boldface type. After your examination, it will be easy for you to find your place in the text again.

■ Problems and Answers

I have chosen a variety of problems from many disciplines for you to work. Because working problems and writing answers is necessary to learn statistics, I have used *interesting* problems and have sometimes written *interesting* answers for the problems; in some cases, I even put *new* information in the answers.

Many of the problems are conceptual questions that do not require any arithmetic. Think these through and write your answers. Being able to calculate a statistic is almost worthless if you cannot tell in English what it means. Writing tells you whether or not you really understand.

On several occasions, problems or data sets will be used more than once, either later in the chapter or even in another chapter. If you do not work the problem when it is first presented, you are likely to be frustrated when it appears again. To alert you, I have put an asterisk beside problems that will be used again.

Sometimes you may find a minor difference between your answer and mine (in the last decimal place, for example). This will probably be the result of rounding errors and does not deserve further attention.

■ Transition Pages

At various points in the book, there will be major differences between the material you have finished and that which you are about to begin. At these points I have provided a "Transition Page" that describes these major differences.

■ Glossaries

This book has three glossaries that you can use to jog your memory, if you encounter a term that was explained earlier. Two of them are in the back of the book and one is printed on the inside cover of the book.

1. *Glossary of words (Appendix C)*. This is a glossary of statistical words and phrases used in the text. When these items are introduced in the text, they appear in boldface type.
2. *Glossary of formulas (Appendix D)*. The formulas for all the statistical techniques discussed in the text are printed here, in alphabetical order according to the name of the technique.

3. *Glossary of symbols (inside cover of this book)*. Statistical symbols are defined here.

■ Arithmetic and Algebra Review

Appendix A gives you a review of arithmetic and algebra. It consists of a pretest (to see if you need to refresh your memory on any of the many techniques you will need for this text) and a review (to provide that refresher). I recommend that you use this appendix as soon as you finish Chapter 1.

CONCLUDING THOUGHTS FOR THIS INTRODUCTORY CHAPTER

Most students find that this book works well for them as a textbook in their statistics course. Those who keep this book after the course is over find that it becomes a very useful reference book. In future courses and after leaving school, they find themselves pulling out their copy to look up some forgotten definition or vaguely recalled procedure. I hope that you not only learn from this book but that you, too, join that group of students who count this book as part of their personal library.

This book is a fairly complete introduction to elementary statistics. There is more to the study of statistics—lots, lots more—but there is a limit to what you can do in one term. Some of you, however, will learn the material in this book and want to know more. If you are such a student, I have provided references in footnotes.

I also recommend that when you finish your course (but before any final examination), you study and work the problems in Chapter 14, the last chapter in the book. It is designed to be an overview/integrative chapter.

For me, the study of statistics has been very satisfying and useful. I hope you will find it worthwhile, too.

PROBLEMS

4. Name the four kinds of scales identified by S. S. Stevens.
5. Give the properties of each of the four kinds of scales.
6. Identify which kind of scale is being used in each of the following cases.
   a. Geologists have a "hardness scale" for identifying different rocks, called the Mohs' scale. The hardest rock (diamond) has a value of 10 and will scratch all others. The second hardest will scratch all but the diamond, and so on, down to a rock such as talc, with a value of 1, which can be scratched by any other rock. (A fingernail, a truly handy field-test instrument, has a value of between 2 and 3.)
   b. The volumes of three different cubes are 40, 64, and 65 cubic inches.
   c. Three different highways are identified by their numbers: 40, 64, and 65.
   d. Republicans, Democrats, Independents, and Others are identified on the voters' list with the numbers 1, 2, 3, and 4.

    e. The pages of a book are numbered 1 through 150.

    f. The winner of the Miss America contest was Miss California; runners-up were Miss Ohio and Miss Pennsylvania.[9]

    g. The prices on the three items were $3.00, $10.00, and $12.00.

    h. She earned three degrees: B.A., M.S., and PhD.

7. The three studies that follow were conducted by undergraduates. For each study, identify the independent and dependent variables and the number and names of the levels of the independent variable.

    a. Becca gave students in statistics a résumé, telling them that the person had applied for a position that included teaching statistics in their college. The students were asked to study the résumé and rate the candidate on a scale of 1 (not qualified) to 10 (extremely qualified). The résumés were identical except that the candidate's first name was Jane on half the résumés and John on the other half.

    b. Tracie showed participants a film clip that included a scene of a man spilling a box of oranges. In a discussion afterward, the fruit was identified as apples for some of the participants; for others, it was identified correctly as oranges; and for still others, no additional mention was made of the fruit. Two days later, the participants filled out a questionnaire that included their degree of confidence (on a scale of 1 to 7) that the fruit that was spilled was apples.

    c. Nancy measured the heights of 63 fifth grade students and then gave them the Coopersmith Self-Esteem test. Her next step was to divide the children into groups that were short, about average, or tall.

8. The three studies that follow were conducted by researchers who are now well known. For each study, identify the independent variable and the number and names of the levels of the independent variable. Name the dependent variable and answer items i and ii.

    a. Theodore Barber hypnotized 25 different people, giving each a series of suggestions. The suggestions included arm rigidity, hallucinations, color blindness, and enhanced memory. During hypnosis, Barber counted the number of suggestions that participants complied with (the mean was 4.8). For another 25 people, he simply asked them to achieve the best score they could (but no hypnosis was used). This second group was given the same suggestions and the number complied with was counted (the mean was 5.1). (See Barber, 1976.)

       i. Identify a nominal variable and a statistic.

       ii. In a sentence, describe what Barber's study shows.

    b. Elizabeth Loftus (1979) had participants view a film of a car accident. Afterward, some were asked, How fast was the car going? and others were asked, How fast was the car going when it passed the barn? (There was no barn in the film.) A week later, she asked the participants, Did you see a barn? If the barn had been mentioned earlier, 17 percent said yes; if it had not been mentioned, 3 percent said yes.

---

[9] Contest winners have come most frequently from these states, which have had 6, 6, and 5 winners, respectively.

      i. Identify a population and a parameter.

      ii. In a sentence, describe what Loftus's study shows.

  c. Stanley Schachter and Larry Gross (1968) gathered data from obese male students for about two hours in the afternoon. At the end of this time, a clock on the wall was correct (5:30 P.M.) for 20 participants, slow (4:30 P.M.) for 20 others, or fast (6:30 P.M.) for 20 more. The actual time, 5:30, was the usual dinnertime for these students. While participants filled out a final questionnaire, Wheat Thins were freely available. The weight of the crackers each student consumed was measured. The means were: 4:30 group—20 grams; 5:30 group—30 grams; 6:30 group—40 grams.

      i. Identify a ratio scale variable.

      ii. In a sentence, describe what this study shows.

9. There are uncontrolled extraneous variables in the study that follows. Name as many as you can.

    The investigator concluded that statistics Textbook A was better than Textbook B. The conclusion was based on a comparison of two statistics classes. The section 1 class, which met MWF at 10 A.M., used Textbook A, and was taught by Professor X. The section 2 statistics class, which met for 3 hours on Wednesday evening, used Textbook B, and was taught by Professor Y. At the end of the term, all students took the same comprehensive test. The mean score for the Textbook A students was higher than the mean score for Textbook B students.

10. Read the nine objectives at the beginning of this chapter. Responding to them will help you consolidate what you have learned.   ■

2

Organization of Data, Graphs, and Central Tendency

OBJECTIVES FOR CHAPTER 2

After studying the text and working the problems in this chapter, you should be able to:

1. Arrange data into simple and grouped frequency distributions

2. Describe the characteristics and uses of frequency polygons, histograms, bar graphs, and line graphs, and be able to interpret them

3. Name some distributions by looking at their shapes

4. Find the mean, median, and mode of a simple frequency distribution and of a grouped frequency distribution

5. Determine whether a central tendency measure is a statistic or a parameter and interpret it

6. Determine which central tendency measure is most appropriate for a set of data

7. Determine the direction of skew of a frequency distribution

8. Find the mean of a set of means

You have now invested some time, energy, and effort on the preliminaries of elementary statistics. Concepts have been introduced and you have an overview of the course. The three chapters that follow are about descriptive statistics. To quickly get an idea of the specific topics, read this chapter title and the next two chapter titles thoughtfully.

To begin your study of descriptive statistics, consider a group of **raw scores**. Raw scores, or raw data, can be obtained in many ways. For example, if you administer a questionnaire to a group of college students, the scores from the questionnaire are raw scores. I will illustrate several of the concepts in this chapter by using the raw scores from a questionnaire administered to 100 representative college students.

If you would like to fully engage in this exercise, take five minutes, complete the 10 questions in the following questionnaire, and calculate your score. (If you read the

TABLE 2.1 Scores of 100 college students on the Personal Control questionnaire

61	55	52	47	49	42	44	50	58	49
52	52	39	55	57	44	53	53	37	46
55	50	45	66	57	51	54	55	48	44
53	70	47	47	52	42	50	56	48	44
51	53	58	62	54	57	46	47	59	41
31	52	55	46	59	47	72	53	46	50
66	49	55	55	58	48	41	52	42	67
52	46	56	61	47	32	57	55	49	52
53	64	45	49	42	54	61	52	37	48
49	56	45	52	55	39	47	56	44	44

three chapter titles, I predict that you will answer the 10 questions. If you didn't read the chapter titles, it's not too late to read them. Participating in this exercise is a step toward engaging more fully in this course.) Answer the questions and score yourself now before you read further.

This particular scale was developed over a period of 10 years by Delroy Paulhus at the University of British Columbia. It is called the Personal Control (PC) scale, and it measures a person's perceived self-competence. (See Paulhus and Van Selst, 1990.)

Table 2.1 is an unorganized collection of the scores of our 100 representative college students on the PC scale.

QUESTIONNAIRE

For each item, indicate the extent to which the statement applies to you by using the following scale.

1 = Disagree strongly   5 = Agree slightly
2 = Disagree   6 = Agree
3 = Disagree slightly   7 = Agree strongly
4 = Neither agree nor disagree

1. I can usually achieve what I want when I work hard for it.
2. Once I make plans I am almost certain to make them work.
3. I prefer games involving some luck over games of pure skill.
4. I can learn almost anything if I set my mind to it.
5. My major accomplishments are entirely due to my hard work and ability.
6. I usually do not set goals because I have a hard time following through on them.
7. Bad luck has sometimes prevented me from achieving things.
8. Almost anything is possible for me if I really want it.
9. Most of what will happen in my career is beyond my control.
10. I find it pointless to keep working on something that is too difficult for me.

Scoring. Add the numbers beside items 1, 2, 4, 5, and 8. On items 3, 6, 7, 9, and 10, reverse the point value (that is, 1 = 7, 2 = 6, 3 = 5, 4 = 4, 5 = 3, 6 = 2, 7 = 1). Add these five values together. Finally, add your two sums together to get your final score.

In this chapter you will learn to (1) organize data like those of Table 2.1 into frequency distributions, (2) present the data graphically, and (3) calculate and interpret measures of central tendency and determine whether they are statistics or parameters. Incidentally, you will also be able to compare your PC score to those of your fellow college students.

SIMPLE FREQUENCY DISTRIBUTIONS

Table 2.1, which is just a jumble of numbers, is not very interesting or informative. (My guess is that you glanced at it and went quickly on.) A much more informative presentation of the 100 scores is an arrangement called a **simple frequency distribution**.

Table 2.2 is a simple frequency distribution—an ordered arrangement that shows the **frequency** of each score. (I would guess that you spent more time on Table 2.2 than on Table 2.1 and that you got more information from it.)

Look again at Table 2.2. The PC score column tells the name of the variable that is being measured. The generic name for any variable is X, which is the symbol used in (generic) formulas. The Frequency (f) column shows how frequently a score appeared. The tally marks are used when you construct your rough-draft version and are not usually included in the final form. N is the number of scores and is found by summing the f column.

You will have a lot of opportunities to construct simple frequency distributions, so here is a set of steps to follow:

1. Find the highest and lowest scores. In Table 2.1, the highest score is 72 and the lowest score is 31.
2. In column form, write in descending order all the numbers from 72 to 31. The heading for this column is the name of the variable being measured.
3. Start with the number in the upper left-hand corner of the unorganized scores (a score of 61 in Table 2.1), draw a line under it, and place a tally mark beside 61 in your frequency distribution.[1]
4. Continue this process through all the unorganized scores.
5. Count the number of tallies by each score and enter the count in a column headed f. Add the numbers in the f column. (If the sum is equal to N, you haven't left out any scores.)

A simple frequency distribution is a common way to present a set of data. It is common because it is useful, and it is useful because it allows you to pick up valuable information with just a glance. For example, the highest and lowest scores are readily

---

[1] If you use the unorganized scores again, you will discover that underlining is much better than crossing out.

T A B L E 2.2   **Simple frequency distribution of Personal Control scores for a representative sample of 100 college students (rough-draft format)**

PC score (X)	Tally marks	Frequency (f)	PC score (X)	Tally marks	Frequency (f)
72	|	1	51	||	222
71		0	50	||||	4
70	|	1	49	⊮ |	6
69		0	48	||||	4
68		0	47	⊮ ||	7
67	|	1	46	⊮	5
66	||	2	45	|||	3
65		0	44	⊮ |	6
64	|	1	43		0
63		0	42	||||	4
62	|	1	41	||	2
61	|||	3	40		0
60		0	39	||	2
59	||	2	38		0
58	|||	3	37	||	2
57	||||	4	36		0
56	||||	4	35		0
55	⊮ ||||	9	34		0
54	|||	3	33		0
53	⊮ |	6	32	|	1
52	⊮ ⊮	10	31	|	1
					$N = \overline{100}$

T A B L E 2.3   **Final form of a simple frequency distribution of Personal Control scores for a representative sample of 100 college students**

Personal Control scores (X)	f	Personal Control scores (X)	f
72	1	51	2
70	1	50	4
67	1	49	6
66	2	48	4
64	1	47	7
62	1	46	5
61	3	45	3
59	2	44	6
58	3	42	4
57	4	41	2
56	4	39	2
55	9	37	2
54	3	32	1
53	6	31	1
52	10		

apparent. In addition, after some practice with frequency distributions, you can ascertain the general shape or form of the distribution and can make an informed guess about central tendency values. Perhaps you have already had enough practice and can make these judgments for **Table 2.2**.

A formal presentation of a simple frequency distribution usually does not include the tally marks and sometimes does not include the zero-frequency scores. **Table 2.3** shows a formal presentation of the data of Table 2.1. You would use a final form like this to present data to others, such as colleagues, professors, supervisors, or editors.

PROBLEMS

*1. The following numbers are the heights in inches of two groups of 18- to 24-year-old Americans. Choose the more interesting group and organize the 50 numbers into a simple frequency distribution, using the rough-draft format. For the group you choose, your result will be fairly representative of the whole population of 18- to 24-year-olds (*Statistical Abstract of the United States: 1994*, 1994). (This is the first of many problems with an asterisk. See footnote.)

Women					Men				
64	67	63	65	66	68	65	72	68	70
63	66	62	65	65	69	73	71	69	67
60	72	64	61	65	77	67	72	73	68
69	64	64	66	61	68	64	72	69	69
65	67	65	62	69	70	71	71	70	75
62	65	63	64	61	72	68	62	68	74
66	64	59	63	65	66	70	72	66	71
63	63	68	64	65	69	71	68	73	69
66	67	62	62	63	71	69	69	65	76
64	70	64	67	65	71	73	65	70	70

*2. The frequency distribution you will construct for this problem is a little different. The "scores" are simply names, making this a frequency distribution of nominal data.

A political science student covered a voting precinct on election day morning, noting the yard signs for the five candidates. Her plan was to find the relationship between yard signs and actual votes. The five candidates were Attila (A), Gandhi (G), Lenin (L), Mao (M), and Bolivar (B). Construct an appropriate frequency distribution.

```
G  A M M M G G L A G B A A G G B L M M
A  G G B G L M A A L M G G M G L G A A
B  L G G A G A M L M G B A G L G M A
```

■

* Problems with an asterisk are multiple-use problems. Your answer and understanding of an asterisked problem will be helpful when the data set is presented again.

GROUPED FREQUENCY DISTRIBUTIONS

Most researchers would condense Table 2.3 even more. The result would be a **grouped frequency distribution**. **Table 2.4** is a rough-draft example of such a distribution.[2] It is often necessary to group data when you want to construct a graph or to present your results in the form of a table.

In a grouped frequency distribution, scores are grouped into equal-sized ranges called **class intervals**. In Table 2.4, the entire range of scores, from 72 to 31, has been reduced to 15 class intervals. Each interval covers three scores; the size of this interval (number of scores covered) is indicated by i. For Table 2.4, $i = 3$. The midpoint of each interval represents all scores in that interval. For example, ten students had scores of 42, 43, or 44. The midpoint of the class interval 42–44 is 43. All ten are represented by 43. There are no scores in the interval 33–35, but zero-frequency intervals are included in formal grouped frequency distributions if they are within the range of the distribution.

Class intervals have lower and upper limits, much like simple scores obtained by measuring a quantitative variable. A class interval of 42–44 has a lower limit of 41.5 and an upper limit of 44.5. You will use these limits later when you calculate the median.

The only difference between grouped frequency distributions and simple frequency distributions is class intervals, which is the next topic.

T A B L E 2.4 **A grouped frequency distribution of Personal Control scores with $i = 3$ (rough-draft format)**

PC scores (class interval)	Midpoint (X)	Tally marks	f
72–74	73	|	1
69–71	70	|	1
66–68	67	|||	3
63–65	64	|	1
60–62	61	||||	4
57–59	58	ⅢⅣ ||||	9
54–56	55	ⅢⅣ ⅢⅣ ⅢⅣ |	16
51–53	52	ⅢⅣ ⅢⅣ ⅢⅣ |||	18
48–50	49	ⅢⅣ ⅢⅣ ||||	14
45–47	46	ⅢⅣ ⅢⅣ ⅢⅣ	15
42–44	43	ⅢⅣ ⅢⅣ	10
39–41	40	||||	4
36–38	37	||	2
33–35	34		0
30–32	31	||	2
			$\Sigma = \overline{100}$

[2] As is the case with a simple frequency distribution, a grouped frequency distribution that is to be presented formally does not include tally marks.

■ Establishing Class Intervals

I use four conventions in establishing class intervals. Although many statisticians use these conventions, others do not. Thus, there is no clear-cut right or wrong way to set up class intervals.

1. *The number of class intervals should be 10 to 20.* One purpose of grouping scores is to provide a clearer picture of the data. For example, Table 2.4 shows that there are many scores near the center of the distribution but fewer and fewer as the ends of the distribution are approached. This characteristic is not as apparent if the data are grouped into fewer than 10 intervals. On the other hand, the use of more than 20 class intervals puts you back into the simple frequency distribution situation—it is difficult to get an overall picture of the form of the distribution.

2. *Choose a convenient size for the class interval (i).* What is convenient? Well, odd numbers are convenient because if i is odd, the midpoint will be a whole number and whole numbers look better on graphs than decimal numbers do. Three and five are often chosen as an interval size. If an i of 5 produces more than 20 class intervals, data groupers usually jump to an i of 10 or some multiple of 10. An interval size of 25 is popular. You will occasionally find that $i = 2$ is necessary to stay within 10 to 20 class intervals.

3. *Begin each class interval with a multiple of i.* For example, if the lowest score is 31 and $i = 3$ (as happened with the PC scores), the first class interval should be 30–32 because 30 is a multiple of 3. An exception to this convention that seems justified occurs when $i = 5$. When the interval size is 5, it may be more convenient to begin the interval such that multiples of 5 fall at the midpoint, because multiples of 5 make graphs easier to read. For example, an interval 23–27 has 25 as its midpoint, whereas an interval 25–29 has 27 as its midpoint.

4. *The largest scores go at the top of the distribution.* This is a convention that is not followed by some computer programs.

■ Converting Unorganized Scores into a Grouped Frequency Distribution

Now that you know the conventions for establishing class intervals, read through the steps for converting unorganized data like those in Table 2.1 into a grouped frequency distribution like that in Table 2.4:

1. Find the highest and lowest scores. In Table 2.1, the highest score is 72 and the lowest score is 31.

2. Find the range of the scores by subtracting the lower limit of the lowest score from the upper limit of the highest score (72.5 − 30.5 = 42). The use of lower and upper limits is necessary in order to include the full range of scores.[3]

3. Determine i by a trial-and-error procedure. Remember that there are to be 10 to 20 class intervals and that the interval size should be convenient (3, 5, 10, or a multiple of 10). Dividing the range by a potential i value tells the number of class intervals that

[3] For a more complete explanation, with a picture, see the section in Chapter 3, "The Range" (pp. 56–57).

will result. For example, dividing the range of 42 by 5 gives a quotient of 8.40. Thus, $i = 5$ produces 8.4 or 9 class intervals. That does not satisfy the rule calling for at least 10 intervals, but it is close and might be acceptable. In most such cases, however, it is better to use a smaller i and get a larger number of intervals. Dividing the range by 3 gives 14 class intervals. You'll notice that Table 2.4 has 15 intervals. This is because the lowest score (31) is not a multiple of i.

4. Begin the bottom interval with a multiple of i, which may or may not be a raw score. End the interval with a number that allows the number of scores in the interval to equal i. For Table 2.4 the first interval is 30–32. To start the next interval, add i to your beginning number. Notice that each interval begins with a number evenly divisible by 3.

5. The rest of the process is the same as for a simple frequency distribution. For each score in the unorganized data, put a tally mark beside its class interval and underline the score. Count the tally marks and put the number in the frequency column. Add up the frequency column to be sure that the sum equals N.

PROBLEMS

*3. You may have heard or read that the normal body temperature (oral) is 98.6°F. The numbers that follow are temperature readings from healthy adults, aged 18–40. Arrange the data into an appropriate rough-draft frequency distribution. (Based on Mackowiak, Wasserman, and Levine, 1992.)

98.1	97.5	97.8	96.4	96.9	98.9	99.5	98.6	98.2	98.3	97.9
98.0	97.2	99.1	98.4	98.5	97.4	98.0	97.9	98.3	98.8	100.0
98.7	97.9	97.7	99.2	98.0	98.1	98.3	97.0	99.4	98.9	97.9
97.4	97.8	98.6	98.7	97.9	98.6	98.1	98.8			

*4. A psychology instructor read 60 related statements to a General Psychology class. He then asked the students to indicate which of the next 20 statements they had heard among the first 60. Due to the relationships among the concepts in the sentences, many seemed familiar but, in fact, none of the 20 had been read before. The following scores represent the number (out of 20) that each student had "heard earlier." (See Bransford and Franks, 1971.) Arrange the scores into an appropriate rough-draft frequency distribution. Based on this description and your frequency distribution, write a sentence of interpretation.

14	11	10	8	12	13	11	10	16	11
11	9	9	7	14	12	9	10	11	6
13	8	11	11	9	8	13	16	10	11
9	9	8	12	11	10	9	7	10	

*5. A sociology professor who was trying to decide how much statistics to present in her Introduction to Sociology class developed a 65-item test of statistical knowledge that covered concepts such as the median, graphs, standard deviation, and correlation. She gave the test to one class of 50 students, and on the basis of the results, she planned a course syllabus for that class and the other four sections being taught that year. Arrange the data into an appropriate rough-draft frequency distribution.

20	56	48	13	30	39	25	41	52	44
27	36	54	46	59	42	17	63	50	24
31	19	38	10	43	31	34	32	15	47
40	36	5	31	53	24	31	41	49	21
26	35	28	37	25	33	27	38	34	22

GRAPHIC PRESENTATION OF DATA

You have no doubt heard the saying *A picture is worth a thousand words*. When it comes to numbers, a not-yet-well-known saying is *A graph is better than a thousand numbers*. Actually, as long as I am rewriting sayings, I would also like to say *The more numbers you have, the more necessary graphics are*.

Pictures that present statistical data are called graphics. The most common graphic is a graph, composed of a horizontal axis (variously called the baseline, X axis, or **abscissa**) and a vertical axis (called the Y axis or **ordinate**). To the right and upward are both positive directions; to the left and downward are both negative directions. Look at **Figure 2.1**.

In this section, you will read about two kinds of graphs. The first kind is used to present frequency distributions like those you have been constructing. **Frequency polygons**, **histograms**, and **bar graphs** are examples of this first kind of graph. The second kind is the **line graph**, which is used to show the relationship between two different variables. I will have more for you on graphs in later chapters.

■ Frequency Distributions

The kind of variable you have measured determines whether you present your data with a frequency polygon, a histogram, or a bar graph. A frequency polygon or histogram is

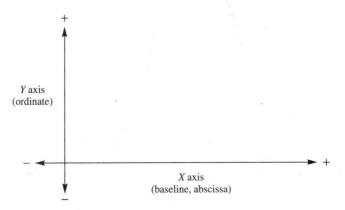

F I G U R E   2.1   The elements of a graph

used for quantitative data, and a bar graph is used for qualitative data. Although you sometimes see a bar graph of quantitative data, most researchers follow the rule just given. Qualitative data, however, *should not* be presented with a frequency polygon or a histogram. Think about the PC data in Table 2.1 and the yard sign data in problem 2. Which of these data sets is quantitative and which is qualitative?

The frequency polygon.   A frequency polygon is used to graph quantitative data. The PC scores in Table 2.4 are quantitative data, so a frequency polygon would be appropriate. The result is **Figure 2.2**. An explanation of Figure 2.2 will serve as your introduction to polygon construction.

On the X axis you will find the midpoints of the class intervals and the label, "Personal Control scores." Notice that the midpoints are spaced at equal intervals, with the smallest on the left and the largest on the right. The Y axis is labeled "Frequency" and is also marked off into equal intervals.

Excellent graphs "look right." The conventional advice is that graphs look better if the height of the figure is 60 percent to 75 percent of its width. Achieving this proportion usually requires a little juggling on your part as you are planning your graph. Huff (1954), Runyon (1981), and Tufte (1983) offer entertaining demonstrations of the misleading effects that occur when these proportions are drastically altered.

The intersection of the X and Y axes is often the zero point for both variables. For the Y axis in Figure 2.2, this is indeed the case. The distance on the Y axis is the same from 0 to 5 as from 5 to 10, and so on. On the X axis, however, that is not the case. Here the scale jumps from 0 to 28 and then is divided into equal units of 3. It is conventional to indicate a break in the measuring scale by breaking the axis with slash marks between 0 and the lowest score used, as you see in Figure 2.2. It is also conventional to close a polygon at both ends by connecting the curve to the X axis.

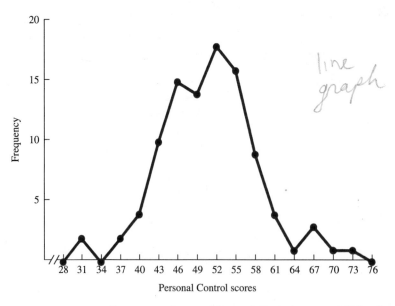

line graph

FIGURE 2.2   Frequency polygon of Personal Control scores of 100 college students

Each point of the frequency polygon represents two numbers: the class midpoint directly below it on the X axis and the frequency of that class directly across from it on the Y axis. By looking at the points in Figure 2.2, you can see that three students are represented by the midpoint 67; zero students by 34; and so on.

The histogram.   A histogram is another graphing technique that is appropriate for quantitative data such as the PC scores. **Figure 2.3** shows the data of Table 2.4 graphed as a histogram. A histogram is constructed by raising bars from the X axis to the appropriate frequency. The lines that separate the bars intersect the X axis at the lower and upper limits of the class interval. Notice that the bars of a histogram are separated by a line rather than a space.

Here are some considerations for deciding whether to use a frequency polygon or a histogram for quantitative data. If you are displaying two overlapping distributions on the same axes, a frequency polygon will do this more clearly than a histogram will. Also, frequency polygons (to my eye) are more graceful; histograms seem more cluttered to me. On the other hand, it is easier to read frequencies from a histogram, and histograms are the better choice when you are presenting discrete data. (*Discrete data* are quantitative data that do not have intermediate values—number of children in a family is an example.)

The bar graph.   A bar graph is used to present the frequencies of the categories of a qualitative variable. A conventional bar graph looks exactly like a histogram except that there are spaces between the bars. The space is the conventional signal that the variable being graphed is a qualitative one. If the variable is measured on an ordinal scale, then the order of the names on the X axis follows the order of the variable. If, however, the variable is a nominal one, *any* order on the X axis will be appropriate. Because you get

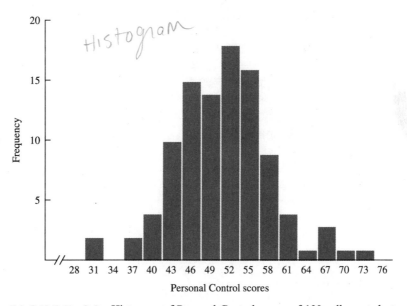

FIGURE 2.3 Histogram of Personal Control scores of 100 college students

to choose for nominal variables, you might decide that alphabetizing the names would be best. Other considerations may lead to some other order.

In thinking about a qualitative variable to describe with a bar graph, it occurred to me that data on college majors would be interesting to you. (Stop and think for a moment: Is a list of college majors a nominal or an ordinal measure?) I used data on 1991–92 college graduates in the United States to construct a conventional bar graph, with the names of the majors on the X axis (in alphabetical order) and the frequencies on the Y axis. It *looked* terrible. In addition, it was awkward to use because you had to turn the graph on its side to read the majors.

I returned to my drawing board, thinking of the admonition in Tufte's (1983) epilogue: "It is better to violate any principle than to place graceless or inelegant marks on paper." A progression of graphs followed; the final result was **Figure 2.4**. The variable being graphed (majors) is on the Y axis, allowing names to be written horizontally so they are easy to read. Frequency is on the X axis. I arranged the majors from most frequent to least frequent.[4] In addition, I wrote whole numbers at the end of each bar to indicate thousands of graduates. (In several early drafts, I used numbers with decimals at the end of the bar. The decimals, however, detracted from the goal of the graph, which is to present an overall picture of the quantitative relationships.) All in all, I was very pleased with the result.

Like histograms, bar graphs are not very satisfactory for presenting two or more variables. Imagine adding data from 10 years earlier to Figure 2.4. The result would be a cluttered graph. However, by using a little creative thought, the two sets of data can be combined, with the result that both variables are presented with elegance. **Figure 2.5** shows the results of combining two variables, the 1991–92 data of Figure 2.4 and the same majors for ten years earlier. The change from 1981–82 to 1991–92 in thousands is

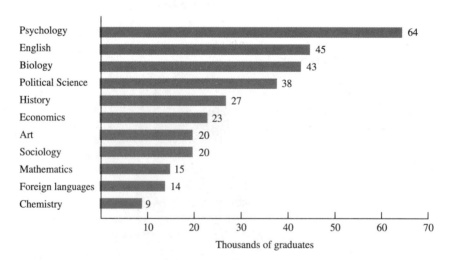

F I G U R E   2.4   College graduates with liberal arts majors for the academic year 1991–92

[4] Alphabetization is a boon when you are searching for one item among many. I decided that 11 was not many, leaving me free to choose some other order for this nominal variable.

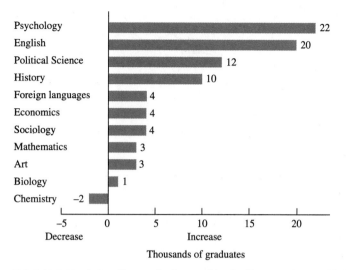

FIGURE 2.5 Change (in thousands) of college graduates with liberal arts majors, 1981–82 to 1991–92

plotted on the Y axis. The result is a graphic that tells you about change, a variable that seems to fascinate humans.

■ The Line Graph

Perhaps the graph most frequently used by scientists is the line graph. A line graph is a picture of the relationship between two variables.[5] A point on a line graph represents the value on the Y variable that goes with the corresponding value on the X variable. The point might represent two scores by one person or the mean score of a group of people.

Figure 2.6 is an example of a line graph. It shows the phenomenon called the *serial position effect*, which occurs when a set of ordered material is learned until it can be repeated without error. The most difficult part is just past the middle, with the first part of the material being the easiest. When you understand the serial position effect, you can figure out what material to study most.

Line graphs are used in many fields to convey the relationship between two variables. If you have a textbook for a course that investigates the relationship between quantitative variables, you might thumb through it and note the many relationships that are presented using line graphs. Should you decide to spend some time studying line graphs, I recommend Richard W. Bowen's 1992 book, *Graph It! How to Make, Read, and Interpret Graphs*.

---

[5] A frequency distribution is a special case of a line graph in which one of the variables is frequency.

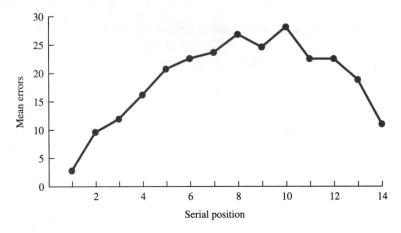

FIGURE 2.6 The serial position effect

■ Other Graphics

Many relationships can be effectively described using some sort of a graphic that is not a graph. In 1983 Edward Tufte published *The Visual Display of Quantitative Information*, a book that demonstrates and celebrates the power that graphics have. The book gives many examples of this power, including a reprint of "(perhaps) the best statistical graphic ever drawn."[6] Tufte says that, regardless of your field, when you construct a quality graphic, it improves your understanding of the phenomenon you are interested in. So if you find yourself somewhere on that path between confusion and understanding, you should try to construct a graphic. In the heartfelt words of a sophomore engineering student I know, "Graphs sure do help."

In addition to helping you understand, a graphic is a *powerful* way to educate and persuade others. After *you* understand, convincing others is usually the next step. To learn more about graphics, I especially recommend Tufte's 1983 book. If that works well for you, try Tufte's 1990 book, *Envisioning Information*.

DESCRIBING DISTRIBUTIONS

There are three ways to describe the form or shape of a distribution: with words, with pictures, and with mathematics. In this section you will use the first two of these. I will not cover mathematical methods that describe the form of a distribution except for one method, which appears in Chapter 12, the chapter on chi square.

[6] I'll give you just a hint about this graphic. It was drawn by a French engineer some time ago to convey a disastrous military campaign against Russia by Napoleon. Wainer (1984) was also impressed with this graphic, nominating it as the "World's Champion Graph."

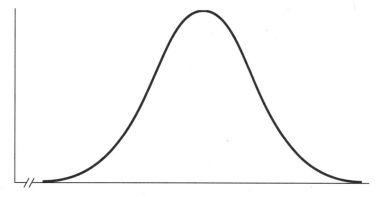

FIGURE 2.7 A normal distribution

■ Bell-Shaped Distributions

Look back at Table 2.4, graphed as Figure 2.2. Notice that the largest frequencies are in the middle of the distribution. The same thing is true in problem 1 of this chapter. These distributions are referred to as bell-shaped.

There is a special case of a bell-shaped distribution that you will soon come to know very well. It is called the **normal distribution**, or **normal curve**. **Figure 2.7** is an example of a normal distribution.

■ Skewed Distributions

In some distributions the scores with the greatest frequency are not found in the middle but near one end. These are called **skewed distributions**.

The word *skew* is similar to the word *skewer*, the name of the cooking implement used in making shish kebab. A skewer is thick at one end and pointed at the other (not symmetrical).

Figures 2.8 and 2.9 are illustrations of skewed distributions. **Figure 2.8** is *positively skewed*; the thin point is toward the high scores, and the most frequent scores are the low ones. Note that the point is to the right—the positive direction. **Figure 2.9** is *negatively skewed*; the thin point (skinny end) is toward the low scores, and the most frequent scores are the high ones. Note that the point is to the left—the negative direction.

■ Other Shapes

Curves of frequency distributions can, of course, take many different shapes. Some shapes are common enough to have names, and examples of these are presented in Figure 2.10.

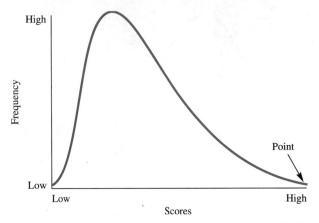

FIGURE 2.8 Example of a positively skewed distribution. Scores with the largest frequencies are concentrated among the low scores.

Rectangular distribution. **Figure 2.10a**, a **rectangular distribution** (also called a *uniform distribution*), occurs when the frequency of each value on the X axis is the same. You will see this distribution again in Chapter 5.

Bimodal distribution. A graph of a **bimodal distribution** has two humps in it. In some cases the humps are the same height, as they are in **Figure 2.10b**. To understand the name bimodal, you need to know that the mode of a distribution is the score that occurs most frequently. In the case of both distributions and graphs, if two high-frequency scores are separated by scores with lower frequencies, the name bimodal is appropriate.

J-curves. Finally, **Figure 2.10c** and **Figure 2.10d** are **J-curves**. If you look at Figure 2.10d and use your imagination, the name will make sense to you. The mode of a J-curve is very near one end of the distribution, resulting in a very skewed graph.

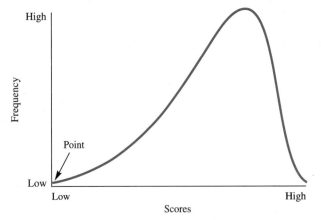

FIGURE 2.9 Example of a negatively skewed distribution. Scores with the largest frequencies are concentrated among the high scores.

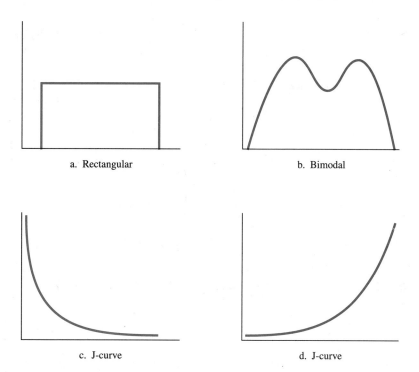

a. Rectangular

b. Bimodal

c. J-curve

d. J-curve

F I G U R E 2.10 Examples of rectangular, bimodal, and J-curves

DISTRIBUTIONS: A SUMMARY

Any set of measurements of a phenomenon can be arranged into a distribution. Scientists and others find these organized distributions to be most helpful, whether they are presented as a frequency distribution table or as a graph.

In the first chapter I said that by learning statistics you would be prepared to understand more and that you would be more persuasive. Frequency distributions and graphs (but especially graphs) will be valuable for many of the audiences you will face over the years.

PROBLEMS

6. Answer the following questions for Figure 2.2.
    a. What is the meaning of the number 55 on the X axis?
    b. What is the meaning of the number 10 on the Y axis?
    c. How many students scored in the class interval 48–50?
7. For problem 1 you constructed a frequency distribution. Graph it, being careful to include the score with a frequency of zero.

8. The average workweek for industrial workers varies from country to country. In 1992 the figures were: Australia: 37.9; Canada, 38.3; Chile, 44.4; Denmark, 31.5; Finland, 31.3; France, 38.7; Germany, 38.9; Japan, 38.8; Korea, 48.7; Mexico, 45.7; United Kingdom, 43.2; United States, 41.0. What kind of graph is appropriate for these data? Graph them after deciding on the best design for your graph.

*9. Decide whether the following distributions should be graphed as frequency polygons or as bar graphs. Graph both distributions.

a. Class interval	f		b. Class interval	f
50–54	1		54–56	3
45–49	1		51–53	7
40–44	2		48–50	15
35–39	4		45–47	14
30–34	5		42–44	11
25–29	7		39–41	8
20–24	10		36–38	7
15–19	12		33–35	4
10–14	6		30–32	5
5–9	2		27–29	2
			24–26	0
			21–23	0
			18–20	1

*10. Determine the direction of the skew for the two curves in problem 9 by examining the curves or the frequency distributions (or both).

11. Look at the simple frequency distribution that you constructed for problem 2. Which kind of graph should be used to display these data? Graph the distribution.

12. Without looking at Figures 2.7 or 2.10, sketch the form of the normal distribution, a rectangular distribution, a bimodal distribution, and two J-curves.

13. Is the point of a positively skewed distribution directed to the right or left?

14. Distinguish between a line graph and a frequency polygon.

15. For each frequency distribution listed, tell whether it would be positively skewed, negatively skewed, or approximately symmetrical (bell-shaped or rectangular).
    a. age of all people alive today
    b. age in months of all first graders
    c. number of children in families
    d. wages in a large manufacturing plant
    e. age at death of everyone who died in one year
    f. shoe size

MEASURES OF CENTRAL TENDENCY

So far in this chapter you have learned some ways to present scores using frequency distributions and graphs. In this section, you will learn other ways of summarizing data—the calculation of three measures of central tendency. Measures of **central tendency** (sometimes called measures of *central value*) give you one score or measure that represents, or is typical of, an entire group of scores.

Recall from Chapter 1 that statistics are the numbers you get from samples and parameters are the numbers you get from populations. In the case of all three measures of central tendency, the formula for calculating the statistic is the same as that for calculating the parameter.[7] The interpretations, however, are different.

I will use the *mean*, which you are already somewhat familiar with, as an example. A sample mean, $\bar{X}$ (pronounced "mean" or "ex-bar"), is only one of many possible sample means from a population. Suppose you have a sample from a population and you calculate the sample mean. Because other samples from that same population would probably produce somewhat different $\bar{X}$'s, a degree of uncertainty goes with the one $\bar{X}$ you have. However, if you were able to measure an entire population, its mean, μ (pronounced "mew"), would be the only one and would carry no uncertainty with it. Clearly, the parameter μ is more desirable than the statistic $\bar{X}$, but, unfortunately, it is often impossible to measure the entire population. Fortunately, $\bar{X}$ is the best estimator of μ.

The difference in interpretation, then, is that a statistic carries some uncertainty with it; a parameter does not.

■ The Mean

Here's a very simple example. Suppose a college freshman arrives at school in the fall with a promise of a monthly allowance for spending money. Sure enough, on the first of each month, there is money to spend. However, three months into the school term, our student discovers a recurring problem: There is too much month left at the end of the money.

In pondering this problem, it occurs to our student that lots of money escapes from his pocket at the Student Center. So, for a two-week period, he keeps a careful accounting of every cent he spends at the center on soft drinks, snacks, video games, coffee, and so forth. His data are presented in **Table 2.5**. You already know how to compute the mean of this set of scores, but before you do, eyeball the data and then write down your *estimate* of the mean. The formula for the **mean** is

$$\bar{X} = \frac{\Sigma X}{N}$$

where $\bar{X}$ = the mean[8]

Σ = an instruction to add (Σ is an uppercase Greek sigma)

X = the numbers symbolized by X; ΣX means to add all the X's

N = number; N is the number of X's

For the data in Table 2.5:

$$\bar{X} = \frac{\Sigma X}{N} = \frac{\$24.98}{14} = \$1.78$$

[7] This is not true for the standard deviation, which you will study in Chapter 3.

[8] Many scientific journals use a capital M to represent the mean.

TABLE 2.5 Amount of money spent per day at the Student Center over a two-week period

Day	Amount spent
1	$2.56
2	0.47
3	1.25
4	0.00
5	3.25
6	1.15
7	0.00
8	0.00
9	6.78
10	2.12
11	0.00
12	0.00
13	3.78
14	3.62
	$\Sigma = \$24.98$

These are data for a two-week period, but our freshman is interested in his expenditures for at least one month and, more likely, for many months. Thus, this is a sample mean and the $\bar{X}$ symbol is appropriate. The amount $1.78 is an estimate of the mean amount per day that our friend spends at the Student Center.

Now we come to an important part of any statistical analysis, which is to answer the question, So what? Calculating numbers or drawing graphs is a part of almost every statistical problem, but unless you can tell the story of what the numbers and pictures mean, you won't find statistics worthwhile.

The first interpretation you can make from Table 2.5 is to estimate monthly Student Center expenses. This is easy to do. Thirty days times $1.78 is $53.40.

Now, let's suppose that our student decides that this $53.40 is an important part of the "out of money before the end of the month" phenomenon. The student has three apparent options. The first is to get more money. The second is to spend less at the Student Center. The third is to justify leaving things as they are. For this third option, our student might perform an economic analysis to determine what he gets in return for his $50 + a month. His list might be pretty impressive: an improved social life, occasional information about classes and courses and professors, a borrowed book that was just super, hundreds of calories, and more.

The point of all this is that part of the attack on his money problem involved calculating a mean. However, an answer of $1.78 doesn't have much meaning by itself. Interpretation and comparisons are called for.

Characteristics of the mean. There are two characteristics of the mean you need to know. Both characteristics will come up again later.

First, if the mean of a distribution is subtracted from each score in that distribution and the differences are added, the sum will be zero; that is, $\Sigma(X - \bar{X}) = 0$. Each difference score is called a deviation and these will be explained more fully in Chapter 3. You might pick a few numbers to play with to *demonstrate* that $\Sigma(X - \bar{X}) = 0$. (1, 2, 3, 4, and 5 are easy to work with.) If you know the rules governing algebraic operations of summation (Σ) notation, you can *prove* the relationship that $\Sigma(X - \bar{X}) = 0$.

Second, the mean is defined as the point about which the sum of the squared deviations is minimized. (A deviation is the answer you get from $X - \bar{X}$.) If the mean is subtracted from each score and each deviation is squared and all squared deviations are added together, the resulting sum will be smaller than if any number other than the mean had been used; that is, $\Sigma(X - \bar{X})^2$ is a minimum. This "least squares" characteristic of the mean will come up in Chapters 3 and 4. Again, you can demonstrate this relationship by playing with some numbers or you can prove it to yourself (see Kirk, 1990, p. 104).

■ The Median

The **median** is the *point* that divides a distribution of scores into two parts that are equal in size. To find the median of the Student Center expense data, arrange the daily expenditures from highest to lowest, as shown in **Table 2.6**. Because there are 14 scores, the halfway point, or median, will have seven scores above it and seven scores below it. The seventh score from the bottom is $1.15. The seventh score from the top is $1.25. The median, then, is halfway between these two, or $1.20. Remember, the median is a *point* in the distribution; it may or may not be an actual score.

T A B L E 2.6 Data of Table 2.5 arranged in descending order

X	
$6.78	
3.78	
3.62	
3.25	7 scores
2.56	
2.12	
1.25	
	Median = $1.20
1.15	
0.47	
0	
0	7 scores
0	
0	
0	

What is the interpretation of a median of $1.20? The simplest is that on half the days our student spends less than $1.20 in the Student Center and on the other half he spends more.

What if there had been an odd number of days in our student's sample? Suppose he had chosen to sample half a month, or 15 days. Then the median would be the eighth score. This would leave seven scores above and seven below. For example, if an additional day had been included, during which $3.12 was spent, the median would be $1.25. If the additional day's expenditure was zero, the median would be $1.15.

■ The Mode

The third central tendency statistic is the **mode**. As mentioned earlier, the mode is the most frequently occurring score—the score with the greatest frequency.

For the Student Center expense data, the mode is $0.00. This mode can be seen most easily in **Table 2.6**. The zero amount occurred five times and all other amounts occurred only once.

When a mode is given, it is often accompanied by the percentage of times it occurred. You would probably agree that "The mode was $0.00, which occurred on 36 percent of the days" is much more informative than "The mode was $0.00."

FINDING CENTRAL TENDENCY OF SIMPLE FREQUENCY DISTRIBUTIONS

■ Mean

Table 2.7 is an expanded version of Table 2.3, the distribution of Personal Control (PC) scores. I will give you the steps for finding the mean from a simple frequency distribution, but first (looking only at the data and not the summary statistics at the bottom) *estimate* the mean of the PC scores and write it down.

The first step in calculating the mean from a simple frequency distribution is to multiply each score in the X column by its corresponding f value, so that all the people making a particular score are included. Sum the fX values and divide by N. (N is the sum of the f values.) This will give you the mean of a simple frequency distribution. In terms of a formula,

$$\mu \text{ or } \bar{X} = \frac{\Sigma fX}{N}$$

For the data in Table 2.7,

$$\mu \text{ or } \bar{X} = \frac{5100}{100} = 51.00$$

To answer the question of whether this 51.00 is an $\bar{X}$ or a μ, we would need more information. If the 100 scores were a population, we would have a μ, but if they represent a sample of some larger population, $\bar{X}$ would be the appropriate symbol.

TABLE 2.7 Calculating the mean of the simple
frequency distribution of the Personal Control scores

Personal Control scores (X)	f	fX
72	221	72
70	1	70
67	1	67
66	2	132
64	1	64
62	1	62
61	3	183
59	2	118
58	3	174
57	4	228
56	4	224
55	9	495
54	3	162
53	6	318
52	10	520
51	2	102
50	4	200
49	6	294
48	4	192
47	7	329
46	5	230
45	3	135
44	6	264
42	4	168
41	2	82
39	2	78
37	2	74
32	1	32
31	1	31
	$\Sigma = \overline{100}$	$\Sigma = \overline{5100}$

■ Median

With 100 PC scores, the median will be a point that has 50 scores above it and 50 below it. If you begin adding frequencies in Table 2.7 from the bottom $(1 + 1 + 2 + 2 + \cdots)$, you will find a total of 49 when you include the score of 51. Including 52 would make the total 59—more than you need. So the median is somewhere among those ten scores of 52.

Remember from Chapter 1 that any number actually stands for a range of numbers that has a lower and an upper limit. This number, 52, has a lower limit of 51.5 and an upper limit of 52.5. To find the exact median somewhere within the range of 51.5–52.5, use a procedure called **interpolation**. Here is the procedure and the reasoning that goes with it. Study it until you understand it. It will come up again (shortly).

There are 49 frequencies below a score of 51. To get to 50 frequencies, where the median is, you need one more score ($50 - 49 = 1$). Because there are ten scores of 52, you need $\frac{1}{10}$ of them to reach the median. Assume that those ten scores of 52 are distributed evenly throughout the interval of 51.5–52.5 and that, therefore, the median is $\frac{1}{10}$ of the way through the interval. Adding 0.1 to the lower limit of the interval, 51.5, gives 51.6, which is the median for these scores.

When the number of scores is odd, the same reasoning applies. However, the halfway point will be a number with 0.5 at the end of it. For example, if you drop the 72, the number of PC scores would be 99 and the median would be the point with 49.5 scores above it. The median now is $\frac{0.5}{10}$ of the way through the interval and is $51.5 + 0.05 = 51.55$. You will have a chance to practice this in the problems.

ERROR DETECTION

Calculating the median by starting from the top of the distribution will produce the same answer as calculating it by starting from the bottom.

◼ Mode

It is easy to find the mode from a simple frequency distribution. In **Table 2.7**, the score with the highest frequency, 10, is the mode. So 52 is the mode.

PROBLEMS

16. Find the median for the following sets of scores:
    a. 2, 5, 15, 3, 9
    b. 9, 13, 16, 20, 12, 11
    c. 8, 11, 11, 8, 11, 8
17. Which of the following distributions would be described as bimodal?
    a. 10, 12, 9, 11, 14, 9, 16, 9, 13, 20
    b. 21, 17, 6, 19, 23, 19, 12, 19, 16, 7
    c. 14, 18, 16, 28, 14, 14, 17, 18, 18, 6
18. For problem 4, find the mean, median, and mode. Write a general sentence of interpretation.
*19. For problem 2:
    a. Decide which of the measures of central tendency is appropriate and find it.
    b. Is this central tendency measure a statistic or a parameter?
    c. Write a sentence of interpretation.
20. Examine problems 3 and 5 to determine whether the numbers are samples or populations.

21. Starting from the top of the distribution, find the median of a and b.

a.

X	f
12	4
11	3
10	5
9	4
8	2
7	1

b. 9, 4, 3, 6, 5, 3, 7, 5, 2, 2, 3, 6, 5, 7, 4, 2, 5, 6, 4, 5

22. Write the two mathematical characteristics of the mean that were covered in this section. ■

FINDING CENTRAL TENDENCY OF GROUPED FREQUENCY DISTRIBUTIONS

As I mentioned earlier, the most common reason for constructing a grouped frequency distribution is to draw a graph or to present the data as a table. Sometimes, however, you need to find central tendency values from such presentations. This involves only a step or two more than you needed to find such values from a simple frequency distribution.

■ Mean

Compared to finding the mean of a simple frequency distribution, finding that of a grouped frequency distribution has one extra multiplication step, a step that involves the frequencies of each class interval. For this step, the midpoint of the interval takes the place of each score in the interval. Look at **Table 2.8**, which is an expansion of Table 2.4. Assume that the scores in each interval are evenly distributed throughout the interval. Thus, X is the mean for all scores within the interval. Multiply each X by its f value in order to include all scores in that interval. Place the product in the fX column. Summing the fX column gives ΣfX, which, when divided by N, yields the mean.

In terms of a formula,

$$\mu \text{ or } \bar{X} = \frac{\Sigma fX}{N}$$

For Table 2.8,

$$\mu \text{ or } \bar{X} = \frac{5092}{100} = 50.92$$

Note that the mean of the grouped data is 50.92, but that the mean of the simple frequency distribution is 51.00. The mean of grouped scores is often different, but seldom is this difference of any consequence.

TABLE 2.8 A grouped frequency distribution of
Personal Control scores with $i = 3$ (rough-draft format)

PC scores (class interval)	Midpoint (X)	f	fX
72–74	73	1	73
69–71	70	1	70
66–68	67	3	201
63–65	64	1	64
60–62	61	4	244
57–59	58	9	522
54–56	55	16	880
51–53	52	18	936
48–50	49	14	686
45–47	46	15	690
42–44	43	10	430
39–41	40	4	160
36–38	37	2	74
33–35	34	0	0
30–32	31	2	62
		$\Sigma = 100$	$\Sigma = 5092$

◼ Median

Finding the median of a grouped distribution usually requires interpolation within the
interval that contains the median. That is the case for **Table 2.8**. As before, you are
looking for the point that has 50 frequencies above it and 50 frequencies below it.
Adding frequencies from the bottom of the distribution, you find that there are 47
scores below the interval 51–53. You need three more frequencies ($50 - 47 = 3$) to
reach the median. Because 18 people scored in the interval, you need 3 of these 18.
*Again, assume that the 18 people in the interval are evenly distributed throughout the
interval.* The interval is 3 score points wide, and you need to go $\frac{3}{18}$ of the way into this
interval: $\frac{3}{18} \times 3 = 0.50$. The lower limit of the interval is 50.5, so you need to add 0.50
to 50.5 to get to the point that is $\frac{3}{18}$ of the way into the interval: $50.5 + 0.50 = 51.00$,
which is the median of the grouped PC scores. **Figure 2.11**, which will require *careful
study*, illustrates this procedure.

   In summary, the steps for finding the median in a grouped frequency distribution
are as follows:

1. Divide N by 2.
2. Starting at the bottom of the distribution, add the frequencies until you find the
   interval that contains the median.
3. Subtract from $N/2$ the total frequencies of all intervals below the interval that
   contains the median.
4. Divide the difference found in step 3 by the number of frequencies in the
   interval that contains the median.

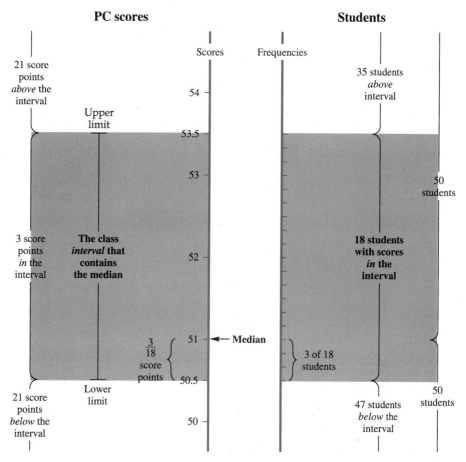

FIGURE 2.11 Finding the median within a class interval

5. **Multiply the proportion found in step 4 by i.**[9]
6. Add the product found in step 5 to the lower limit of the interval that contains the median. That sum is the median.

If you like the idea of creating your own formula, you might reduce these six steps to a formula for the median.

■ Mode

The mode is the midpoint of the interval that has the highest frequency. In Table 2.8, the interval 51–53 has the largest frequency count: 18. The midpoint of that interval, 52, is the mode.

---

[9] The most common error in calculating the median is failing to multiply by i.

You have already dealt with the situation of two large-frequency scores separated by small-frequency ones (the bimodal distribution). When the two or more largest-frequency scores are adjacent to one another, the mode is the mean of the midpoints of the intervals.

ERROR DETECTION

Eyeballing data is a valuable way to avoid big mistakes. Begin a problem by making a quick estimate of the answers. If the answers you calculate differ from what you had estimated, wisdom dictates that you reconcile the difference. You have either overlooked something when eyeballing or made a mistake in your computations.

THE MEAN, MEDIAN, AND MODE COMPARED

A common question is, Which measure of central tendency should I use? The general answer is, given a choice, use the mean. Sometimes, however, the data give you no choice. Here are three considerations that limit your choice.

■ Scale of Measurement

A mean is appropriate for ratio or interval scale data, but not for ordinal or nominal distributions. A median is appropriate for ratio, interval, and ordinal data, but not for nominal. The mode is appropriate for any of the four scales of measurement.

You have already thought through part of this in working problem 19. In it you found that the yard sign names (very literally, a nominal variable) could be characterized with a mode, but it would be impossible to try to add up the names and divide by N or to find the median of the names.

For an ordinal scale such as class standing in college, either a median or a mode would make sense. The median would probably be part of the way through the classification sophomore, and the mode would be freshman.

■ Skewed Distributions

Even if you have interval or ratio data, a mean is not recommended if the distribution is severely skewed. The following story demonstrates why the mean gives an erroneous impression for severely skewed distributions. (Note also that the story describes a population of data, not a sample; see if you can tell why.)

The developer of Swampy Acres Retirement Homesites is attempting, with a computer-selected mailing list, to sell the lots in a southern "paradise" to northern buyers. The marks express concern that flooding might occur. The developer reassures them by explaining that the average elevation of the lots is 78.5 feet and that the water has never exceeded 25 feet in that area. On the average, the developer has told the truth;

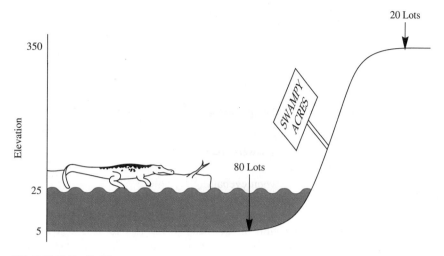

FIGURE 2.12   Elevation of Swampy Acres

TABLE 2.9   Frequency distribution of lot elevations at Swampy Acres

Elevation in feet (X)	Number of lots (f)	fX
348–352	20	7000
13–17	30	450
8–12	30	300
3–7	20	100
	100	7850

$$\mu = \frac{\Sigma fX}{N} = \frac{7850}{100} = 78.50 \text{ feet}$$

but this average truth is misleading. Look at the actual lay of the land in **Figure 2.12** and examine the frequency distribution in **Table 2.9**.

The mean elevation, as the developer said, is 78.5 feet; however, only 20 lots, all on a hill, are out of the flood zone. The other 80 lots are, on the average, under water. The mean for these data is misleading. The measure of central tendency that best describes these data is the median because it is unaffected by the size of the few extreme lots on the hill. The median elevation is 12.5 feet, well below the high-water mark. (Because our interest is only in this one development, the lot elevations constitute a population of data. There is no interest in generalizing from these data to some larger group.)

A number of books with engaging titles tell how statistics can be chosen to convey one particular idea rather than another. The grandparent of such books is probably Huff's *How to Lie with Statistics* (1954). More recent versions are Campbell's *Flaws and Fallacies in Statistical Thinking* (1974) and Runyon's *How Numbers Lie* (1981). All three books are delightfully written.

■ Open-Ended Categories

There is another instance that requires a median, even though you have symmetrically distributed interval or ratio data. This is when the class interval with the highest (or lowest) scores is not limited. In such a case, you do not have a midpoint and therefore *cannot* compute a mean. For example, age data are sometimes reported with the highest category as "75 and over"—the mean cannot be computed. Thus, when one or both of the extreme class intervals is not limited, the median is the appropriate measure of central tendency.

DETERMINING SKEWNESS FROM THE MEAN AND MEDIAN

Examining the relationship of the mean to the median is a way of determining the direction of skew in a distribution without having to draw a graph. When the *mean is smaller* than the median, there is some amount of negative skew. When the *mean is*

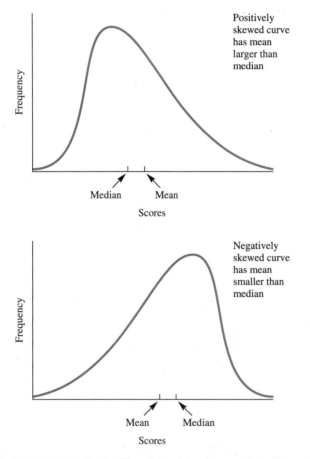

FIGURE 2.13   The effect of skewness on the relative position of the mean and median

larger than the median, there is positive skew. The reason for this is that the mean is affected by the size of the numbers and is pulled in the direction of the extreme scores. The median, however, is not influenced by the size of extreme scores. The relationship between the mean and the median is illustrated by **Figure 2.13**. The size of the difference between the mean and the median gives you an indication of how much the distribution is skewed. As you might suspect, there are mathematical ways to measure skewness. [See Kirk (1990), p. 140, for one.]

In summary, choose the mean if it is appropriate. (To follow this advice you must be able to recognize data for which the mean is *not* appropriate.)

THE MEAN OF A SET OF MEANS ($\bar{\bar{X}}$)

Occasions arise in which means are available from several samples taken from the same population. If these means are combined, the mean of the set of means ($\bar{\bar{X}}$) is the best estimator of the population parameter, μ. If every sample has the same N, you can compute $\bar{\bar{X}}$ by adding the means and dividing by the number of means. If the sample means are based on N's of different sizes, however, you cannot use this procedure. Here is a story that illustrates the right way and the wrong way to calculate the mean of a set of means.

At my undergraduate college, students with a cumulative grade-point average of 3.25 in the middle of their junior year were eligible to enter a program to "graduate with honors." [In those days (1960) usually fewer than 10 percent of a class had GPA's above 3.25.] Discovering this rule after my sophomore year, I decided to figure out if I had a chance to qualify. Calculating a cumulative GPA seemed easy enough to do: Four semesters had produced GPA's of 3.41, 3.63, 3.37, and 2.16. Given another GPA of 3.80, the sum of the five semesters would be 16.37 and dividing by 5 gave an average of 3.27, well above the required 3.25.

Graduating with honors seemed like a great ending for college, so I embarked on a goal-oriented semester: a B in German and an A in everything else—a GPA of 3.80. And, at the end of the semester the goal had been accomplished. Unfortunately, "graduating with honors" was not to be.

There was a flaw in my method of calculating my cumulative GPA. My method assumed that all of the semesters were equal in weight, that they had all been based on the same number of semester hours. The calculations based on this assumption are shown on the left side of **Table 2.10**.

However, all five semesters were not the same. Unfortunately, the semester with the GPA of 2.16 was based on 19 semester hours, rather than the usual 16 or so.[10] Thus, that semester had to be weighted more heavily than semesters in which only 16 hours were taken.

The right side of Table 2.10 shows the correct way to calculate cumulative GPA. Each semester's GPA is multiplied by its number of semester hours. These products are summed and that total is divided by the sum of the semester hours. As you can see from

[10] The semester was more educational than a GPA of 2.16 would indicate. A great deal of American literature was consumed that spring, but unfortunately I was not registered for any courses in American literature.

TABLE 2.10 Two methods of calculating a mean of a set of means from five semesters' GPA's; the method on the left is correct only if all semesters are based on the same number of credit hours

Flawed method	Correct method		
Semester GPA's	Semester GPA's	Hours credit	GPA's × hours
3.41	3.41	17	258
3.63	3.63	16	58
3.37	3.37	19	64
2.16	2.16	19	41
3.80	3.80	16	61
$\Sigma = 16.37$		$\Sigma = 87$	$\Sigma = 282$

$$\bar{\bar{X}} = \frac{16.37}{5} = 3.27 \qquad\qquad \bar{\bar{X}} = \frac{282}{87} = 3.24$$

the numbers on the right, the actual cumulative GPA was 3.24, not high enough to qualify for the honors program.

More generally, to find the mean of a set of means, multiply each separate mean by its N, add these products together, and divide by the sum of the N's. Thus, means of 2.0, 3.0, and 4.0 based on N's of 6, 3, and 2 produce an overall mean (mean of a set of means) of 2.64 (29 ÷ 11 = 2.64).

CLUE TO THE FUTURE

The distributions that you have been working with in this chapter are **empirical distributions** based on scores actually gathered in experiments. This chapter and the next two are about these empirical frequency distributions. Starting with Chapter 5 and throughout the rest of the book, you will also use **theoretical distributions**— distributions based on mathematical formulas and logic rather than on actual observations.

ESTIMATING ANSWERS

You may have noticed during this chapter that I have been (subtly?) working in some advice: *As your* first *step, estimate an answer.* Here's why I think you should start by estimating an answer: (1) Taking a few seconds to estimate keeps you from plunging into the numbers before you fully understand the problem. (2) An estimate is especially helpful if you have a calculator or a computer doing the bulk of your number crunching. Although these wonderful machines don't make errors, the people who enter the numbers or give instructions do, occasionally. Your at-the-beginning estimate will alert you to these mistakes. (3) An estimate that differs from a calculated value gives you a chance to correct an error before anyone else sees it.

It is rather exciting to look at an answer, whether it is on paper, a calculator display, or a computer screen, and say, "That *can't* be right!" and then to find out that, sure enough, the displayed answer is wrong. If you develop your ability to estimate answers, I promise that you will sometimes experience this excitement.

PROBLEMS

*23. Find the mean, median, and mode of the grouped frequency distribution you constructed for problem 3. You will need to add class midpoints to your previous answer before beginning your calculations.

24. Find the mean, median, and mode of the grouped frequency distribution you constructed for problem 5.

25. For the following situations, tell which measure of central tendency is appropriate and why.

   a. As part of a study on prestige, an investigator sat on a corner in a high-income residential area and classified passing automobiles according to producers: General Motors, Ford, Chrysler, and foreign.

   b. In a study of helping behavior, an investigator pretended to have locked himself out of his car. Passersby who stopped were classified on a scale of 1 to 5 as (1) very helpful, (2) helpful, (3) slightly helpful, (4) neutral, and (5) discourteous.

   c. In a study of family income in a city, the following income categories were established: $0–$10,000, $10,001–$20,000, $20,001–$30,000, $30,001–$40,000, $40,001–$50,000, $50,001–$60,000, and $60,001 and more.

   d. In a study of *per capita* income in a city, the following income categories were established: $0–$5000, $5001–$10,000, $10,001–$15,000, $15,001–$20,000, $20,001–$25,000, $25,001–$30,000, $30,001–$35,000, $35,001–$40,000, $40,001–$45,000, and $45,001–$50,000.

   e. First admissions to a state mental hospital for the previous five years were classified by disorder: schizophrenic, delusional, anxiety, dissociative, and other.

   f. A teacher gave her class an arithmetic test, and most of the children scored in the range 70–79. A few were above this, and a few were below.

26. A senior psychology major performed the same experiment on three groups and obtained means of 74, 69, and 75. The groups consisted of 12, 31, and 17 subjects, respectively. What is the overall mean for all subjects?

27. In problem 10, you looked over two distributions to determine the direction of the skew. Verify that judgment now with a comparison of the means and medians of the distributions.

28. A three-year veteran of the local baseball team was figuring out his lifetime batting average. During the first year he played for half the season and batted .350 (28 hits in 80 at-bats). The second year he had about twice the number of at-bats and his average was .325. The third year, although he played even more regularly, he was in a slump and batted only .275. By adding the three season averages, and dividing by 3, he found his lifetime batting average to be .317. Is this correct? Justify your answer.

29. Read and respond to the eight objectives at the beginning of the chapter. Generating responses will help you consolidate what you have learned.   ■

3

Variability

After studying the text and working the problems in this chapter, you should be able to:

1. Explain the concept of variability

2. Find the range of a distribution

3. Distinguish among the following: the standard deviation of a population, the standard deviation of a sample used to describe the sample, and the standard deviation of a sample used to estimate the population standard deviation

4. Calculate a standard deviation from grouped or ungrouped data and interpret its meaning

5. Add standard deviation measures to a line graph

6. Calculate the variance of a distribution

7. Convert raw scores to z scores

8. Use z scores to compare two scores in a distribution or a score in one distribution with a score in a second distribution

In Chapter 2 you began your study of descriptive statistics. *Frequency distributions* and *graphs* presented all the data. The *mean, median,* and *mode* each gave you one number that captured some central tendency characteristic of the distribution. Chapter 3 continues with more descriptive statistics. This time you get one number that expresses the **variability** of a distribution.

The value of knowing about variability is illustrated by the story of two brothers who went water skiing on Christmas Day (the temperature was about 35°F). On the *average*, each skier finished his turn 5 feet from the shoreline (where one may step off the ski into only 1 foot of very cold water). This bland average of the two actual stopping places, however, does not convey the excitement of the day.

The first brother, determined to avoid the cold water, held onto the towrope too long. Scraped and bruised, he finally stopped rolling at a spot 35 feet up on the rocky shore. The second brother, determined not to share the same fate, turned the towrope loose too soon. Although he swam the 45 feet to shore very quickly, his lips were very blue. No, the average stopping place of 5 feet from the shore doesn't capture the excitement of the day. To get the full story, you would also need to know about variability.

Here are some other situations in which knowing the average leaves you without enough information.

1. You are an elementary school teacher who has been assigned a class of fifth graders whose mean IQ is 115, well above the IQ of 100 that is the average for the general population. Because a child with an IQ of 115 can handle more complex, abstract material than the average child, you plan more sophisticated projects for the year.

If the variability of the IQs in the class is small, your projects will probably succeed. If the variability is large, however, the projects will be too complicated for some of the pupils, and for others, even these projects will not be challenging enough.

2. Your temperature, taken with a thermometer under your tongue, is 97.5°F. You begin to worry because this is below even the average of 98.2°F that you learned in the previous chapter (based on Mackowiak, Wasserman, and Levine, 1992).

There is variability around that mean of 98.2°F. Is 97.5°F below the mean by just a little or by a lot? Measuring variability is necessary if you are to answer this question.

3. Having graduated from college, you are considering two offers of employment, one in sales and the other in management. The pay is about the same for both. After checking out the statistics for salespersons and managers at the library, you find that those who have been working for 5 years also have similar averages. You conclude that the pay for the two occupations is equal.

Pay is more variable for those in sales than for those in management. Some in sales make much more than the average (and some make much less), whereas the pay of those in management is clustered together. Your feeling about this difference in variability might help you choose.

Clearly, variability is important. This chapter is about statistics and parameters that measure the variability of a distribution.

The range is the first measure I will describe. The second, the standard deviation, is the most important. Most of this chapter will be about the standard deviation. A third way to measure variability is with the variance. Finally, at the end of this chapter, I will describe z scores. The standard deviation is necessary in order to calculate z scores, which are used to compare the *relative standing of individual scores*. A z score is not a measure of the variability of a distribution.

THE RANGE

The **range** of a quantitative variable is the upper limit of the highest score minus the lower limit of the lowest score; that is,

$$\text{range} = X_H - X_L$$

where X_H = *upper limit* of the highest score in the distribution
X_L = *lower limit* of the lowest score in the distribution

As an example, the range from 5 to 10 is 6 (applying the formula: $10.5 - 4.5 = 6$). You can check this for yourself by counting the spaces in the following illustration:

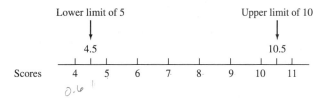

Applying this same logic to a distribution with a high score of 3.7 and a low score of 2.0, the range is 1.8 ($3.75 - 1.95$). Similarly, the range is 0.26 when the high score is 0.70 and the low score is 0.45. Often the range is reported with a statement like, "The scores ranged from 55 to 89."

As a matter of fact, you have already been exposed to the range as a measure of variability. You applied this idea in Chapter 2 when you were deciding how many class intervals to have in your grouped frequency distribution. (Remember? If not, review on page 27.)

In manufacturing, the range is used in some quality-control procedures. From a small sample of whatever is being manufactured, a range is calculated and compared to an expected figure. A large range means there is too much variability in the process, and adjustments are called for (see Chapter 6 of Berger and Hart, 1986).

Because the range depends on only the two *extreme* scores, you can probably imagine two very different distributions that have the same mean and the same range. (Go ahead, imagine them.) If you *are* able to dream up such distributions, you realize that having the same mean and range does not guarantee that the shape (form) of the two polygons will be identical.

THE STANDARD DEVIATION

When it comes to describing the variability of a distribution, the most widely used statistic is the **standard deviation**. It is popular because it is useful. Once you understand standard deviations, you will be able to express quantitatively the difference between the two distributions you imagined in the last section. (You did do the imagining, didn't you?) Also, you'll be able to measure the distances along the baseline of the normal curve, an important theoretical distribution that you will use throughout the rest of your statistical life (see Chapter 5). There are also many other uses for the standard deviation. Of course, in order to understand all this, you'll have to read all the material, work the problems, and do the interpretations. (Teachers take the position that this is not a very high price to pay for a lifetime of being able to understand the most popular yardstick of variability.)

Your principal task in this section is to learn the distinctions among three different standard deviations. For a particular situation or problem, the standard deviation you

T A B L E 3.1 **Symbols, purposes, and descriptions of three standard deviations**

Symbol	Purpose	Description
σ	Measure a population's variability	This lowercase Greek *sigma* is the symbol for the standard deviation of a population. σ is a parameter and is used to *describe* variability when a population of data is available.
$\hat{s}$	Estimate a population's variability	This lowercase $\hat{s}$ (*s*-hat) is an *estimate* of σ, in the same way that $\bar{X}$ is an estimate of μ. Parameters are rarely known, so often the best you can do is draw a sample from the population and use statistics such as $\bar{X}$ and $\hat{s}$ as estimates of parameters such as μ and σ. The variability statistic you will use most often in this book is $\hat{s}$.
S	Measure a sample's variability	There are occasions when you want to *describe* the variability of a sample but have no interest in estimating σ. In such cases, a capital S is the statistic to use. In this book you will encounter S in this chapter and in places where correlation is stressed (principally in Chapter 4).

use will be determined by your purpose. Table 3.1 lists purposes, symbols, and descriptions. It will be worth your time to study **Table 3.1** thoroughly now.

Distinguishing among these three standard deviations is sometimes a problem for beginning students. My advice is to study Table 3.1 and then be alert to situations in the text where a standard deviation is used. With each situation, you will acquire more understanding. I will first discuss the calculation of σ and S (which are computed in the same way except for the use of μ and $\bar{X}$, respectively, in the formulas) and then deal with $\hat{s}$.

THE STANDARD DEVIATION AS A DESCRIPTIVE INDEX OF VARIABILITY

Both σ and S are used to *describe* the variability of a set of data. The symbol σ is a parameter of a population; S is a statistic of a sample. The two are calculated with similar formulas. I will show you two ways to arrange the arithmetic of these formulas: the deviation-score formula and the raw-score formula. You can best learn what a standard deviation is actually measuring by working through the steps of the *deviation-score formula*. The *raw-score formula*, however, is quicker (and sometimes more accurate, depending on rounding). Algebraically, the two formulas are identical.

My suggestion is that you follow the arrangement in the next few pages. By the time you get to the end of the chapter, you should have both the understanding that the deviation-score formula produces and the efficiency that the raw-score formula gives you. From that point on, you may want to use a calculator or computer to do the arithmetic for you.

Before you can use the deviation-score method, you need to be introduced to a common statistic, the deviation score. Deviation scores will be used again in later chapters for other statistical situations.

■ Deviation Scores

A **deviation score** is a raw score minus the mean, either $X - \bar{X}$ or $X - \mu$. It is simply the difference between a score in the distribution and the mean of that distribution. Deviation scores are encountered so often that they have a special symbol, lowercase x. Note that a capital X is used for a score and a lowercase x is used for a deviation score. Not only must you know this, you must also be sure that you (and your instructor) can tell the difference between your written versions of X and x. Write them. Can your roommate tell the difference?

Because $x = X - \bar{X}$, raw scores that are larger than the mean will have positive deviation scores, raw scores that are smaller than the mean will have negative deviation scores, and raw scores that are equal to the mean will have a deviation score of zero.

Table 3.2 gives you a brief demonstration of how to compute deviation scores for a small population of data. In Table 3.2, I first computed the mean and then subtracted it from each score. This resulted in a set of deviation scores, which appear in the right-hand column.

A deviation score tells you the number of points by which a particular score deviates from, or differs from, the mean. In Table 3.2, the x value for Selene tells you that she scored six points above the mean, Caitlin scored at the mean, and Kelly was five points below the mean.

ERROR DETECTION

Notice that the sum of the deviation scores is always zero. Add the deviation scores; if the sum is not zero, you have made an error. You have studied this characteristic before. Remember that in the last chapter you learned that $\Sigma(X - \bar{X}) = 0$.

T A B L E 3.2   The computation of deviation scores from raw scores

Name	Score	$X - \mu$	x
Selene	14	14–8	6
Kevin	10	10–8	2
Caitlin	8	8–8	0
Sarah	5	5–8	−3
Kelly	3	3–8	−5
	$\Sigma X = 40$		$\Sigma x = 0$

$$\mu = \frac{\Sigma X}{N} = \frac{40}{5} = 8$$

PROBLEMS

For the following two distributions, find the range and the deviation scores. Check to see that $\Sigma x = 0$.

1. 17, 5, 1, 1
2. 0.45, 0.30, 0.30
3. Give the symbol and purpose of the three standard deviations.   ■

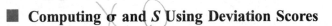

■ Computing σ and S Using Deviation Scores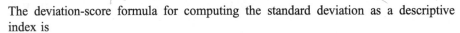

The deviation-score formula for computing the standard deviation as a descriptive index is

$$\sigma = \sqrt{\frac{\Sigma(X - \mu)^2}{N}} = \sqrt{\frac{\Sigma x^2}{N}} \quad \text{or} \quad S = \sqrt{\frac{\Sigma(X - \bar{X})^2}{N}} = \sqrt{\frac{\Sigma x^2}{N}}$$

where $\sigma = $ standard deviation of a population
        $S = $ standard deviation of a sample
        $N = $ number of deviations (same as the number of scores)
        $\Sigma x^2 = $ sum of the squared deviations[1]

Look at $\Sigma x^2/N$. It is not too different from $\Sigma X/N$. Indeed, $\Sigma X/N$ is an average of the scores (X), and $\Sigma x^2/N$ is an average of the squared deviations (x^2).

How can a standard deviation add to your understanding of a phenomenon? Let's take the phenomenon of Girl Scout cookies. Table 3.3 presents some imaginary (but true-to-life) data on the sales of boxes by six Girl Scouts. I'll use these data to illustrate the calculation of the standard deviation. Let's define these six Girl Scouts as a population; thus, the standard deviation will be σ.

The number of boxes sold (and collected for) is shown under the X column of **Table 3.3**. The rest of the arithmetic needed for calculating σ is also given. To compute σ by the deviation-score formula, first compute the mean.[2] Obtain a deviation score (x) for each raw score by subtracting the mean from the raw score. Square each x and sum the x^2 values to obtain Σx^2. Divide Σx^2 by N. Take the square root. Then $\sigma = 8.66$ boxes.

Now, what does $\sigma = 8.66$ boxes mean? How does it help your understanding? The 8.66 boxes is a measure of the variability in the number of boxes the six Girl Scouts sold. Had the σ been 0, you would know that each girl sold the same number of boxes. The closer σ is to 0, the more confidence you can have in predicting that the number of boxes any girl sold was equal to the mean of the group. Conversely, the farther σ is from 0, the less confidence you have. With $\sigma = 8.66$, you know that the girls varied a great deal in sales.

---

[1] Σx^2 is shorthand notation for $\Sigma(X - \bar{X})^2$ (or $\Sigma(X - \mu)^2$). In turn, $\Sigma(X - \bar{X})^2$ is shorthand notation that tells you to subtract the group mean from each score in the group (which produces N deviation scores), square each deviation score, and add all the squares together. One rule for working with summation notation is to perform the operations within the parentheses first.
[2] For convenience, I made up the data so the mean would be an integer.

TABLE 3.3 Using the deviation-score formula to compute σ for the cookie sales of a population of six Girl Scouts

Boxes of cookies X	Deviation scores x	x^2
28	18	324
11	1	1
10	0	0
5	−5	25
4	−6	36
2	−8	64
$\Sigma X = 60$	$\Sigma x = 0$	$\Sigma x^2 = 450$

$$\mu = \frac{\Sigma X}{N} = \frac{60}{6} = 10$$

$$\sigma = \sqrt{\frac{\Sigma (X - \mu)^2}{N}} = \sqrt{\frac{\Sigma x^2}{N}} = \sqrt{\frac{450}{6}} = \sqrt{75} = 8.66$$

[I realize that, so far, my interpretation of the standard deviation has not given you any more information than the range would. (A range of 0 means that each girl sold the same number of boxes, and so forth.) The range, however, has no additional information to give you—the standard deviation does. After Chapter 5 you will be able to give a much more complete description of the phenomenon of Girl Scout cookie sales by knowing the standard deviation.]

Now look again at **Table 3.3** and the formula for σ. Notice what is happening. The mean is being subtracted from each score. This difference, whether positive or negative, is squared and these squared differences are added together. This sum is divided by N and the square root is found. Every score in the distribution contributes to the final answer.

Notice the contribution made by a score like 28, which is far from the mean; its contribution is large. This makes sense because the standard deviation is a measure of variability and if there are scores far from the mean, they cause the standard deviation to be larger. Take a moment to think through the contribution to the standard deviation made by a score near the mean.[3]

ERROR DETECTION

All standard deviations are positive numbers. If you find yourself trying to take the square root of a negative number, you've made an error.

---

[3] If you play with a formula, you will become more comfortable with it and understand it better. Make up a small set of numbers and calculate a standard deviation. Change one of the numbers, add a number, or leave out a number. See what happens. Besides teaching yourself about standard deviations, you may learn more efficient ways to use your calculator.

PROBLEMS

Compute σ or S for each of the three distributions in problems 4–6, using the deviation-score method.

*4. 7, 6, 5, 2

5. 14, 11, 10, 8, 8

*6. 107, 106, 105, 102

7. Compare the standard deviation of problem 4 with that of problem 6. What conclusion can you draw about the effect of the size of the numbers on the standard deviation?

8. Does the size of the numbers in a distribution have any effect on the mean? *yes*

9. The temperatures in this problem are averages for the months of March, June, September, and December. Calculate the mean and standard deviation for each city. Summarize your results in a sentence.

San Francisco, CA	54°	59°	62°	52°
Albuquerque, NM	46°	75°	70°	36°

10. No computation is needed for this one; just eyeball the data. Determine whether the first distribution is more variable than the second or whether the two are equal.

   a. 1, 2, 4, 1, 3   and   9, 7, 3, 1, 0

   b. 9, 10, 12, 11   and   4, 5, 7, 6

   c. 1, 3, 9, 6, 7   and   14, 15, 14, 13, 14

   d. 114, 113, 114, 112, 113   and   14, 13, 14, 12, 13

   e. 8, 4, 6, 3, 5   and   4, 5, 7, 6, 15

■ Computing σ and S by the Raw-Score Method

The deviation-score formula helps you understand what is actually going on when you calculate a standard deviation. It is the formula to use until you do understand. Unfortunately, except for textbook examples, the deviation formula almost always has you working with many decimal values. The raw-score formula, which is algebraically equivalent, involves far fewer decimals. It also produces answers more quickly, especially if you are working with a calculator.

The raw-score formula is

$$\sigma \text{ or } S = \sqrt{\frac{\Sigma X^2 - \dfrac{(\Sigma X)^2}{N}}{N}}$$

where ΣX^2 = sum of the squared scores

      $(\Sigma X)^2$ = square of the sum of the raw scores

          N = number of scores

---

* Problems with an asterisk are multiple-use problems. Your answer and your understanding of an asterisked problem will be helpful when the data set is presented again.

Although the raw-score formula may at first glance appear forbidding, it is actually easier to use than the deviation-score formula because you don't have to compute deviation scores. The numbers you work with will be larger, but your calculator won't care.

Table 3.4 shows the steps for calculating σ or S by the raw-score formula. The data are those for boxes of cookies sold by the six Girl Scouts. Putting the arithmetic of **Table 3.4** into words, square the sum of the X column and divide by N. Subtract this quotient from the sum of the X^2 column. Divide this difference by N and extract the square root. The result is σ or S. Notice that the value of σ in Table 3.4 is the same as the one you calculated in Table 3.3. In this case, the mean is an integer, so the deviation scores introduced no rounding errors.[4]

ΣX^2 and $(\Sigma X)^2$: Did you notice the difference in these two terms when you were working with the data in Table 3.4? If so, congratulations. You cannot calculate a standard deviation correctly unless you understand the difference. Reexamine Table 3.4 if you aren't sure of the difference between ΣX^2 and $(\Sigma X)^2$. Be alert for these two sums in the problems that are coming up.

T A B L E 3.4 Using the raw-score formula to compute σ for the cookie sales of a population of six Girl Scouts

Boxes of cookies	
X	X^2
28	784
11	121
10	100
5	25
4	16
2	4
$\Sigma X = 60$	$\Sigma X^2 = 1050$

Note: $(\Sigma X)^2 = (60)^2 = 3600$

$$\sigma = \sqrt{\dfrac{\Sigma X^2 - \dfrac{(\Sigma X)^2}{N}}{N}} = \sqrt{\dfrac{1050 - \dfrac{(60)^2}{6}}{6}} = \sqrt{\dfrac{1050 - 600}{6}} = \sqrt{\dfrac{450}{6}} = \sqrt{75} = 8.66$$

---

[4] Some textbooks and statisticians prefer the algebraically equivalent formula

$$\sigma \text{ or } S = \sqrt{\dfrac{N\Sigma X^2 - (\Sigma X)^2}{N^2}}$$

I am using the formula in the text because the same form, or parts of it, will be used in other procedures.

Yet another formula is often used in the field of testing. This formula, which requires you to begin by calculating the mean, is

$$\sigma = \sqrt{\dfrac{\Sigma X^2}{N} - \mu^2} \quad \text{and} \quad S = \sqrt{\dfrac{\Sigma X^2}{N} - \bar{X}^2}$$

All three of these arrangements of the arithmetic are algebraically equivalent.

ERROR DETECTION

The range is usually two to five times larger than the standard deviation when $N = 100$ or less. The range (which can be calculated quickly) will tell you if you made any large errors in calculating a standard deviation.

CLUE TO THE FUTURE

You will be glad to know that in your efforts to calculate a standard deviation, you have produced two other useful statistics along the way. Each of these has a name and they will turn up again, in this chapter and in a future one. The number that you took the square root of is called the **variance** (also called a *mean square*), symbolized σ^2. The expression in the numerator, Σx^2, is called the **sum of squares**. You will see both of these terms in Chapters 9, 10, and 11, which are about the analysis of variance.

PROBLEMS

*11. Look at the following two sample distributions. Without any calculation (just look over the data), decide which one has the larger standard deviation. Guess the size of S for each distribution. (You may wish to calculate the range before you choose a number.) Finally, make a choice between the deviation-score and the raw-score formulas and compute S for each distribution. Compare your computation with your guess.
    a. 5, 4, 3, 2, 1, 0   **b.** 5, 5, 5, 0, 0, 0

12. For each of the distributions in problem 11, divide the range by the standard deviation. Is the result between 2 and 5?

13. By now you can look at the following two distributions and see that (a) is more variable than (b). The difference in the two distributions is in the lowest scores (2 and 6). Calculate σ for each distribution, using the raw-score formula. Notice the difference in σ produced by the change in just one score.
    a. 9, 8, 8, 7, 2   **b.** 9, 8, 8, 7, 6   ■

$\hat{s}$ AS AN ESTIMATE OF σ skip

Remember that $\hat{s}$ is the principal statistic you will learn in this chapter. It will be used again and again throughout the rest of this text.

As explained in Chapters 1 and 2, the purpose of a sample is usually to find out something about a population; that is, a statistic from a sample is used to estimate a parameter of the population. It is both obvious and unfortunate that samples from the same population are often slightly different, yielding different statistics. Which of the statistics is closest in value to the parameter? There is no way to know other than to measure the entire population, which is usually impossible. Statisticians say that the best you can do is to calculate the statistic in such a way that, *on the average*, its value is equal to the parameter. In the language of the mathematical statistician, you want a statistic that is an "unbiased estimator" of the corresponding population parameter.

If you have sample data and you want to calculate an estimate of σ, you should use the statistic, $\hat{s}$

$$\hat{s} = \sqrt{\frac{\Sigma(X - \bar{X})^2}{N - 1}}$$

Statisticians often add a "hat" to a symbol to indicate that something is being estimated. Note that the difference between $\hat{s}$ and σ is that $\hat{s}$ has $N - 1$ in the denominator, whereas σ has N.

This issue of dividing by N or by $N - 1$ has sometimes left students shrugging their shoulders and muttering, "OK, I'll memorize it and do it however you want." I would like to explain, however, why, when you have sample data and want to estimate σ, you use $N - 1$ in the denominator.

Because the formula for σ is

$$\sigma = \sqrt{\frac{\Sigma(X - \mu)^2}{N}}$$

it would seem logical just to calculate $\bar{X}$ on the sample data, substitute $\bar{X}$ for μ in the numerator of the formula, and calculate an answer. This solution will, more often than not, produce a numerator that is *too small* [as compared to the value of $\Sigma(X - \mu)^2$].

To explain this surprising state of affairs, remember from Chapter 2 (page 41) a characteristic of the mean: For any set of scores, the expression $\Sigma(X - \bar{X})^2$ is minimized. That is, for any set of scores, this sum will be smaller using $\bar{X}$ than it would be using any other number (either larger or smaller) in place of $\bar{X}$.

Thus, for a sample of scores, substituting $\bar{X}$ for μ gives you a numerator that is minimized. However, what you want is a value that is the same as $\Sigma(X - \mu)^2$. Now, if the value of $\bar{X}$ is at all different from μ, then the minimized value you get using $\Sigma(X - \bar{X})^2$ will be too small.

The solution that statisticians adopted when this underestimation problem became apparent was to use $\bar{X}$ and then divide the too-small numerator by a smaller denominator, namely $N - 1$. This results in a statistic that is a much better estimator of σ.[5]

---

[5] To illustrate this for yourself, use a small population of scores and do some calculations. For a population with scores of 1, 2, and 3, σ is $= 0.82$. With $N = 2$ there are three samples in the population. For each sample calculate the standard deviation using N in the denominator. Find the mean of these three standard deviations. Now, for each of the three samples, calculate the standard deviation using $N - 1$ in the denominator and find the mean of these three. Compare the two means to the σ that you want to estimate.

Here are two technical points that you should know if you plan to learn more about statistics after this course. First, even with $N - 1$ in the denominator, $\hat{s}$ is not an unbiased estimator of σ. Because the bias is not very serious, $\hat{s}$ is used as the best estimator of σ. Second, $\hat{s}^2$ (see the next section) *is* an unbiased estimator of σ^2.

You just finished four dense paragraphs—lots of ideas per square inch. You may understand it already, but if you don't, take 10 or 15 minutes to reread, do the exercise in footnote 5, and think.

Note also that as N gets larger, the subtraction of 1 from N has less and less effect on the size of the estimate of variability. This makes sense because the larger the sample size is, the closer on average $\bar{X}$ will be to μ.

One new task comes with the introduction of $\hat{s}$: It is the decision whether to calculate σ, S, or $\hat{s}$ for a given set of data. Your choice will be based on your purpose for the standard deviation. If your purpose is to describe, use σ or S, depending on whether you have population or sample data. If your purpose is to estimate a population σ from sample data (a common requirement in inferential statistics), use $\hat{s}$.

■ Calculating $\hat{s}$ *skip*

A raw-score formula is recommended for calculating $\hat{s}$:

$$\hat{s} = \sqrt{\frac{\Sigma X^2 - \dfrac{(\Sigma X)^2}{N}}{N - 1}}$$

This is practically the same as the formula for σ, so you will have no trouble calculating $\hat{s}$ (assuming you mastered the calculation of σ).[6]

This raw-score formula is the one you will probably use for your own data. Sometimes, though, you may be confronted with someone else's frequency distribution, for which you want to calculate a standard deviation. For a simple frequency distribution or a grouped frequency distribution, the formula is

$$\hat{s} = \sqrt{\frac{\Sigma f X^2 - \dfrac{(\Sigma f X)^2}{N}}{N - 1}}$$

where f is the number of scores in an interval.

Here are some data to illustrate the calculation of $\hat{s}$ both for ungrouped raw scores and for a frequency distribution. Consider puberty. As you know, females reach puberty earlier than males (about two years earlier on the average). Is there any difference between the sexes in the *variability* of reaching this developmental milestone? Comparing standard deviations will give you an answer.

If you have only a sample of ages for both sexes, and your interest is in all females and males, $\hat{s}$ is the appropriate standard deviation. **Table 3.5** shows the calculation of $\hat{s}$ for the females. Work through the calculations, noting that the standard deviation is 2.19 years.

Data for ages at which males reach puberty are presented in a simple frequency distribution in **Table 3.6**. Work through these calculations, noting that grouping causes two additional columns of calculations. For males, $\hat{s}$ is 1.44 years.

---

[6] Calculators that have a standard deviation function differ. Some use N in the denominator, some use $N - 1$, and some have both standard deviations. You will have to check yours to see how it is programmed.

TABLE 3.5 Calculation of $\hat{s}$ for ages at which females reach puberty (raw-score method)

Age (X)	X^2
17	289
15	225
13	169
12	144
12	144
11	121
11	121
11	121
$\Sigma X = 102$	$\Sigma X^2 = 1334$ $(\Sigma X)^2 = 10{,}404$

$$\hat{s} = \sqrt{\frac{\Sigma X^2 - \frac{(\Sigma X)^2}{N}}{N-1}} = \sqrt{\frac{1334 - \frac{(102)^2}{8}}{7}} = \sqrt{\frac{1334 - 1300.50}{7}} = 2.19$$

Thus, based on sample data, you can conclude that there is more variability among females in reaching puberty than there is among males. (Incidentally, you would be correct in your conclusion—I chose the numbers so they would produce results that are the same as those in the populations.)

Here are three final points about working with simple and grouped frequency distributions. Recognize that ΣfX^2 is found by squaring X, multiplying by f, and summing. $(\Sigma fX)^2$ is found by multiplying f by X, summing, and then squaring.

For grouped frequency distributions, use the midpoints of the class intervals as the X values, as you did in Chapter 2 when you worked with grouped frequency distributions. (Refer to Table 2.8 for a review.)

TABLE 3.6 Calculation of $\hat{s}$ for ages at which males reach puberty (simple frequency distribution)

Age (X)	f	fX	fX^2
18	1	18	324
17	1	17	289
16	2	32	512
15	4	60	900
14	5	70	980
13	3	39	507
	$N = 16$	$\Sigma X = 236$	$\Sigma X^2 = 3512$

$$\hat{s} = \sqrt{\frac{\Sigma fX^2 - \frac{(\Sigma fX)^2}{N}}{N-1}} = \sqrt{\frac{3512 - \frac{(236)^2}{16}}{15}} = \sqrt{\frac{3512 - 3481}{15}} = 1.44$$

For both grouped and simple frequency distributions, most calculators with memory will give you ΣfX and ΣfX^2 if you properly key in X and f for each line of the distribution. Procedures differ depending on the brand. The time you invest in learning (even if you have to read the instructions) will be repaid several times over in future chapters (not to mention the satisfying feeling you will get).

GRAPHING STANDARD DEVIATIONS

The information conveyed by standard deviations can often be added to line graphs. **Figure 3.1** is a double-duty graph of the puberty data, showing the standard deviations as well as the means. The dot shows the mean age (read from the Y axis), and the lines extend 1 standard deviation in either direction. From this graph you can see at a glance that the mean age of puberty is younger for females than for males and that for females there is more variability about the mean than there is for males.

THE VARIANCE

In the "Clue to the Future" box on page 64, you were told that you produce the variance as part of your calculation of the standard deviation. The variance is the number you take the square root of to get the standard deviation. The symbols for the variance are σ^2 (population variance) and $\hat{s}^2$ (sample variance used to estimate the population variance). By formula,

$$\sigma^2 = \frac{\Sigma(X - \mu)^2}{N} \quad \text{and} \quad \hat{s}^2 = \frac{\Sigma(X - \bar{X})^2}{N - 1} \quad \text{or} \quad \hat{s}^2 = \frac{\Sigma X^2 - \dfrac{(\Sigma X)^2}{N}}{N - 1}$$

F I G U R E 3.1   Puberty in females and males—means and standard deviations

The difference between σ^2 and $\hat{s}^2$ is the term in the denominator. The population variance uses N and the sample variance uses $N - 1$.[7]

The variance is not very useful as a *descriptive* statistic. It is, however, of enormous importance in inferential statistics. You will see more of the variance in Chapters 9, 10, and 11, which cover the **analysis of variance**.

A SUMMARY SECTION ON DESCRIPTIVE STATISTICS

For the first (but not the last) time, I want to call your attention to the subtitle of this book: Tales of Distributions. For two chapters you have been calculating statistics for a distribution and then interpreting them. You can find the value of the mean; if it is larger than the median, you can say that the distribution is positively skewed. You can calculate the range and the standard deviation. You can choose the proper method for graphing the distribution. In short, you are prepared to tell the tale of the distribution. Your story might be:

"It's a bimodal distribution with modes at 3 and 7."
"The errors are distributed symmetrically with a mean of 10 and a standard deviation of 3."
"It's a positively skewed distribution with a median at 6 and a standard deviation of 4."

What on earth would be the purpose of such stories? The purpose might be to better understand a phenomenon that you are interested in. The purpose might be to explain to your boss the changes that are taking place in your industry; perhaps it might be to convince your quantitative-minded customer to place a big order with you. Your purpose might be to convince a school board to buy (or not buy) a piece of property in your neighborhood.

In summary, you have a good start toward being able to tell the tale of a distribution of data. I am sure that these tales will be valuable to you.

PROBLEMS

14. Here is an interpretation problem. Remember the student who recorded the amount of money spent at the Student Center every day for 14 days? The mean was $1.78 per day. Suppose the student wanted to reduce Student Center spending. Write a sentence of advice for when $\hat{s} = \$0.02$, and a sentence for when $\hat{s} = \$2.00$.

15. Using words, describe the relationship between the variance and the standard deviation.

16. A researcher had a sample of scores from the freshman class on a test that measured attitudes toward authority. She wished to estimate the standard deviation

---

[7] Many calculators have a variance key. One kind uses $N - 1$ to calculate the standard deviation and N to calculate the variance. Given a square and a square root function key, this arrangement covers σ, σ^2, $\hat{s}$, and $\hat{s}^2$.

of the entire freshman class. (She had data from 20 years ago and she believed that current students were more homogeneous than students in the past.) Given the following summary statistics, calculate the proper standard deviation and variance.

$$N = 21 \qquad \Sigma X = 304 \qquad \Sigma X^2 = 5064$$

17. Here are those data on the heights of 18- to 24-year-old Americans that you graphed and found a mean for in the last chapter. Choose the more interesting group and find $\hat{s}$.

Females		Males	
Height (in.)	f	Height (in.)	f
72	1	77	1
70	1	76	1
69	2	75	1
68	1	74	1
67	4	73	4
66	5	72	5
65	10	71	7
64	9	70	6
63	7	69	8
62	5	68	7
61	3	67	2
60	1	66	2
59	1	65	3
		64	1
		62	1

18. In manufacturing, engineers strive for consistency. The following data are the errors in millimeters of giant doodads manufactured by two different processes. Choose S or $\hat{s}$ and determine which process produces the more consistent doodads.

Process A	0	1	−2	0	−2	3
Process B	1	−2	−1	1	−1	2

19. A high school English teacher measured the attitudes of 11th-grade students toward poetry. After a nine-week unit on poetry, she measured the students' attitudes again. She was disappointed to find that the mean change was 0. Following are some representative scores. (High scores indicate favorable attitudes.) Calculate $\hat{s}$ for both before and after, and write a conclusion based on the standard deviations.

Before	7	5	3	5	5	4	5	6
After	9	8	2	1	8	9	1	2

20. Estimate the population standard deviation for oral temperature using data you worked with in Chapter 2, problem 23. You may find ΣfX and ΣfX^2 either by working from the answer to problem 23, or, if you understand what you need to do

to find these values from that table, you can use $\Sigma fX = 4026.2$ and $\Sigma fX^2 = 395{,}393.36$. ∎

z SCORES

You have used measures of central tendency and measures of variability to describe a *distribution* of scores. The next statistic, z, is used to describe *a single score*.

Suppose one of your friends tells you he made a 95 on a math exam. What does that tell you about his mathematical ability? Due to your previous experience with tests, 95 may seem like a pretty good score. This conclusion, however, depends on a couple of assumptions. Without those assumptions, a score of 95 is *meaningless*. Let's return to the conversation with your friend.

After you say, "95! Congratulations," suppose he tells you that 200 points were possible. Now a score of 95 seems like something to hide. "My condolences," you say. But then he tells you that the highest score on that difficult exam was 101. Now 95 has regained respectability and you chortle, "Well, all right!" In response, he shakes his head and tells you that the mean score was 98. The 95 takes a nose dive. As a final blow, you find out that 95 was the lowest score, that nobody was worse than your friend. With your hand on his shoulder, your final remark is, "Come on, I'll buy you an ice cream cone."

This example illustrates that 95 acquires meaning only when it is compared with the rest of the test scores. In particular, a score gets its meaning from its relation to the mean and the variability of its fellow scores. A z **score** is a mathematical way to change a raw score so that it reflects its relationship to the mean and standard deviation of its fellow scores. The formula is

$$z = \frac{X - \bar{X}}{S} = \frac{x}{S}$$

Remember that x is an acquaintance of yours, the deviation score.

A z score describes the relation of an X to $\bar{X}$ with respect to the variability of the distribution. For example, if you know that a score (X) is five units from the mean $(X - \bar{X} = 5)$, you know only that the score is better than average, but you have no idea how far above average it is. If the distribution has a range of *10* units and $\bar{X} = 50$, then an X of 55 is a very high score. On the other hand, if the distribution has a range of *100* units and $\bar{X} = 50$, an X of 55 is barely above average. **Figure 3.2** is a picture of the ideas in this paragraph.

To find a score's position in a distribution, the variability of the distribution must be accounted for. The way to do this is to divide $X - \bar{X}$ by a unit that measures variability, the standard deviation. This results in a deviation per unit of standard deviation.[8] When an X is converted to a z score, the z represents the number of standard deviations the score is above or below the mean.

---

[8] This is the same concept as a percentage (per centum, or per hundred). Thus, 24 out of 50 and 12 out of 25 may appear to be different, but you recognize them as the same when you convert both figures to per hundred (by multiplying numerators and denominators by the number necessary to make the denominator 100).

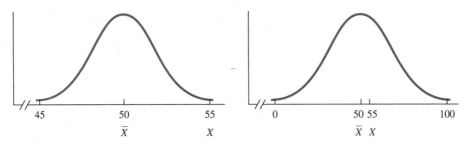

FIGURE 3.2 A comparison of $X - \bar{X} = 5$ in two distributions with different standard deviations

A *z* score is sometimes referred to as a **standard score** because it is a deviation score expressed in standard deviation units. Any distribution of raw scores can be converted to a distribution of *z* scores; for each raw score, there is a *z* score. Raw scores above the mean will have positive *z* scores; those below the mean will have negative *z* scores.

Converting a raw score to a *z* score gives you a number that indicates the score's relative position in the distribution. If *two* raw scores are converted to *z* scores, you will know their positions relative to *each other* as well as to the distribution. Finally, *z* scores are also used to compare two scores from different distributions, even when the scores are measuring different things. (If this seems like trying to compare apples and oranges, see problem 25.) Table 3.7 and its description will illustrate these uses of *z* scores.

In the General Psychology course I took years ago, the professor returned tests with a *z* score rather than a percentage score. This *z* score was *the* key to figuring your grade. A *z* score of +1.50 or higher was an A and −1.50 or lower was an F. (*z* scores between +.50 and +1.50 received B's. If you assume the professor practiced symmetry, you can figure out the rest of the grading scale.)

Table 3.7 shows the raw scores (percentage correct) and *z* scores for four of the many students who took two of the tests in that class. Consider the first student, Kris, who scored 76 on both tests. The two 76s appear to be the same, but the *z* scores show that they are not. The first 76 was a high A and the second 76 was a high F. The second student, Robin, appears to have improved on the second test, going from 54 to 86, but in fact, the scores were the test averages both times (a grade of C). Robin stayed the same. Marty also appears to have improved if you examine only the raw scores. However, the *z* scores reveal that Marty did *worse* on the second test. Finally, comparing Terry's and Robin's raw scores, it appears that Terry was 4 points higher than Robin on each test, but the *z* scores show that Terry's edge was greater on the second test.

The reason for these surprising comparisons is that the means and standard deviations were so different for the two tests. Perhaps the second test was easier, the material may have been more motivating to students, the students may have studied more, or maybe the teacher was better. Perhaps all of these reasons may have been true.

To summarize, *z* scores give you a way to compare raw scores. The basis of the comparison is the distribution itself rather than some external standard (such as a grading scale of 90–80–70–60 percent for A's, B's, and so on).

TABLE 3.7 The z scores of selected students on two
100-point tests in General Psychology

Student	Test 1 Learning and memory		Test 2 Psychopathology	
	Raw score	z Score	Raw score	z Score
—	—	—	—	—
—	—	—	—	—
Kris	76	+2.20	76	−1.67
Robin	54	.00	86	.00
—	—	—	—	—
Marty	58	+.40	82	−.67
Terry	58	+.40	90	+.67
—		—		—
	Test 1: $\bar{X} = 54$		Test 2: $\bar{X} = 86$	
		$S = 10$		$S = 6$

When used as a *descriptive* measure of one distribution of 100 or so scores, z scores range from approximately −3 to +3. Thus, a z score of −2.5 is at or near the bottom of the distribution. For some distributions, however, the highest z score may be as low as 1, and the lowest z score may be −1. When a z score is used as an *inferential* statistic, scores much larger than 3 can occur.

CLUE TO THE FUTURE

z scores will turn up often as you study statistics. They will be prominent in this book in Chapters 5, 6, and 13.

PROBLEMS

21. For any distribution, the mean will have a z score equal to what?
22. What conclusion can you reach about Σz?
23. Under what conditions would you prefer that your z score be negative rather than positive?
24. Hattie and Missy, twin sisters, were intense competitors, but they never competed against each other. Hattie specialized in long-distance running and Missy was an excellent sprint swimmer. As you can see from the distributions in the table, each was the best in her event. Take the analysis one step farther and use z scores to determine who is the more outstanding twin. You might start by looking at the data and making an estimate.

10K runners	Time (min)	50M swimmers	Time (sec)
Hattie	37	Missy	24
Dott	39	Nona	26
Liz	40	Nancy	27
Marette	42	Betty	28

25. Tobe grows apples and Zeke grows oranges. In the local orchards, the mean weight of apples is 5 ounces, with $S = 1.0$ ounces. For oranges the mean weight is 6 ounces, with $S = 1.2$ ounces. At harvest time, each enters his largest specimen in the Warwick County Fair. Tobe's apple was 9 ounces and Zeke's orange was 10 ounces. This particular year Tobe fell ill on the day of judgment, so he sent his friend Hamlet to inquire who had won. Adopt the role of judge and use z scores to determine the winner. Hamlet's query to you would be, "Tobe, or not Tobe; that is the question."

26. Milquetoast's anthropology professor announced that he would drop the poorest exam grade for each student. Milquetoast scored 79 on the first anthropology exam. The mean was 67 and the standard deviation 4. On the second exam, he made 125. The class mean was 105 and the standard deviation 15. On the third exam, the mean was 45 and the standard deviation 3. Milquetoast made 51. On which test was Milquetoast's performance poorest?

27. Return to the objectives at the beginning of the chapter. Can you do each one of them? ■

Thus far in this exposition of the wonders of statistics, each example and each problem has been about only one variable. The variable might have been Personal Control scores, heights, money, or apples, but your description was just for that one variable. For such *univariate* (one-variable) data, you have studied and practiced some techniques that help you describe the variable.

Another kind of statistical analysis tells you about the relationship between *two* variables. In this next chapter you will learn about two techniques that allow you to *describe* the relationship that exists between two variables.

What is the relationship between a person's verbal ability and mathematical ability? You will find out in this chapter. The following are other pairs of variables that might be related. By the time you finish this next chapter you will know whether or not they are related and *to what degree*.

Height of daughters and height of their fathers
Amount of motivation and quality of performance
Church membership and homicide
Inches of rainfall and bushels of wheat per acre
Income level and probability of being diagnosed as psychotic

This next chapter, "Correlation and Regression," explains a method that is used to determine the degree of relationship between two variables (correlation) and a method that is used to make predictions about one variable when you have measurements on another variable (regression).

4

Correlation and Regression

OBJECTIVES FOR CHAPTER 4

After studying the text and working the problems in this chapter, you should be able to:

1. Explain the difference between univariate and bivariate distributions

2. Explain the concept of correlation

3. Draw scatterplots

4. Explain the difference between positive and negative correlation

5. Compute a Pearson product-moment correlation coefficient (*r*)

6. Interpret correlation coefficients using the terms *reliability, causation,* and *common variance*

7. Identify situations in which the Pearson *r* will not accurately reflect the degree of relationship

8. Name and explain the elements of the regression equation

9. Compute regression coefficients and fit a regression line to a set of data

10. Interpret the appearance of a regression line

11. Make predictions for one variable from measurements of another variable

Correlation and regression: My guess is that you have some understanding of the concept of correlation and that you are less comfortable with the word *regression.* Speculation aside, correlation is simpler. Correlation is a statistical technique for measuring the *degree* of relationship between two variables.

Regression is a more complex set of ideas. In this chapter you will learn to use the regression technique for two tasks—*drawing* the line that best fits the data and *predicting* a person's score on one variable when you know that person's score on a second, correlated variable. Regression has other, more sophisticated uses, but you will have to put those off until you study more advanced statistics.

The ideas that we identify by the terms *correlation* and *regression* were developed by Sir Francis Galton in England a little over 100 years ago. Galton was a genius (he could read at age 3) who had an amazing variety of interests, many of which he actively pursued during his 89 years. He once listed his occupation as "private gentleman," which meant that he had inherited money and did not have to work at a job. Lazy, however, he was not. Galton wrote 16 books and more than 200 articles.

From an early age, Galton was enchanted with counting and quantification. Among the many things he tried to quantify were weather, individuals, beauty, characteristics of criminals, boringness of lectures, and the effectiveness of prayers. He was successful in many of his attempts. For example, it was Galton who discovered that atmospheric pressure highs produce clockwise winds around a calm center, and one of his efforts at quantifying individuals resulted in a book on fingerprints. Because it worked so well for him, Galton actively promoted the philosophy of **quantification**, the idea that you can understand a phenomenon much better if you can translate its essential parts into numbers.

Now, as you well know, simply measuring some interesting phenomenon such as a breeze and getting a number like 12 doesn't leave you sighing the sigh of the newly enlightened. According to researchers, such satisfying sighs come when your measurements confirm your expectation that two variables are related. Francis Galton was a master at identifying variables that were related.

Many of the variables that interested Galton were in the field of heredity. Although it was common in the 19th century to comment on physical similarities within a family (height and facial characteristics, for example), Galton thought that psychological characteristics, too, tended to run in families. Specifically, he thought that characteristics such as genius, musical talent, sensory acuity, and quickness had a hereditary basis. [A clue to the position that Galton took on this matter is that he and his illustrious cousin, Charles Darwin, shared an illustrious grandfather, Erasmus Darwin (although they had no grandmother in common).] Galton's 1869 book, *Hereditary Genius*, listed many families and their famous members.

Galton wasn't satisfied with the list in that early book; he wanted to express the relationships in quantitative terms. To get quantitative data, he established an anthropometric (people-measuring) laboratory at a health exposition (a fair) and later at the South Kensington Museum in London. Approximately 17,000 people who stopped at a booth paid three pence to be measured. They left with self-knowledge; Galton left with quantitative data and a pocketful of coins. In today's research the investigator usually pays the participants. For us researchers, the old days were better in some ways. (For one summary of Galton's results, see Johnson et al., 1985.)

Galton's most important legacy is probably his invention of the concepts of correlation and regression. Using correlation, we can express the degree of relationship between *any* two paired variables. (The relationship between the height of fathers and the height of their adult sons was Galton's classic example.)

Galton was not enough of a mathematician to work out the theory and formulas for his concepts. This task fell to Galton's young friend and protégé, Karl Pearson, Professor of Applied Mathematics and Mechanics at University College in London.[1] Pearson's 1896 *product-moment correlation coefficient* and other **correlation coefficients** that he and his students developed were quickly adopted by researchers in many fields and are widely used today in psychology, sociology, education, political science, the biological sciences, and other areas.

---

[1] I have some biographical information on Pearson in Chapter 11, the chapter on chi square. Chi square is another statistical invention of Professor Pearson.

Finally, although Galton and Pearson became famous for their data gathering and statistics, this was not what they had set out to do. For both, the principal goal was to develop recommendations that would improve the human condition. Developing recommendations required a better understanding of heredity and evolution, and statistics was simply the best way to arrive at this better understanding.

In 1889 Galton described how valuable statistics are (and also let us in on his emotional feelings about statistics):

> Some people hate the very name of statistics, but I find them full of beauty and interest.... Their power of dealing with complicated phenomena is extraordinary. They are the only tools by which an opening can be cut through the formidable thicket of difficulties that bars the path of those who pursue the Science of [Human Beings].[2]

My plan in this chapter is for you to read about bivariate distributions (necessary for both correlation and regression), to learn to compute and interpret Pearson product-moment correlation coefficients, and to use the regression technique to draw a best-fitting straight line and predict outcomes.

BIVARIATE DISTRIBUTIONS

In the chapters on central tendency and variability, you worked with one variable at a time (**univariate distributions**). Height, time, test scores, and errors all received your attention. If you look back at those problems, you'll find a string of numbers under one heading (see, for example, page 41). Compare those distributions with the one in **Table 4.1**. In Table 4.1 there is a set of test scores under Humor Test and another set of scores under the variable Intelligence Test. You could find the mean and standard deviation of either of these variables. The characteristic of the data in this table that makes it a **bivariate distribution** is that the scores on the two variables are *paired.* The 50 and the 8 go together, and the 20 and 4 go together. They are paired, of course, because the same person made the two scores. As you will see, there are also other reasons for pairing scores. All in all, bivariate distributions are fairly common.

The essential idea of a bivariate distribution (which is required for correlation and regression techniques) is that there are two variables with values that are paired for

TABLE 4.1 A bivariate distribution of scores on two tests taken by the same individuals

	Humor test X variable	Intelligence test Y variable
Larry	50	8
Shep	40	9
Curly	30	5
Moe	20	4

[2] For a short biography of Galton, I recommend David (1968).

some logical reason. A bivariate distribution may show positive correlation, negative correlation, or zero correlation, which, as you might suspect, are the next three headings in this chapter.

POSITIVE CORRELATION

In the case of a *positive correlation* between two variables, high measurements on one variable tend to be associated with high measurements on the other variable, and low measurements on one variable with low measurements on the other. This is the case for the manufactured data in **Table 4.2**. Tall fathers tend to have sons who grow up to be tall men. Short fathers tend to have sons who grow up to be short men. If the true relationship were so undeviating that every son grew to be exactly his father's height (as in the manufactured data), the correlation would be perfect, and the correlation coefficient would be 1.00. A graph plotting this relationship would look like **Figure 4.1**. If such were the case (which, of course, is ridiculous; mothers and environments have their say, too), then it would be possible to predict without error the adult height of an unborn son simply by measuring his father.

Examine Figure 4.1 carefully. Each point represents a pair of scores—the height of a father and the height of his son. Such an array of points is called a **scatterplot**, or scattergram. These scatterplots will be used throughout this chapter. Incidentally, it was when Galton cast his data as a scatterplot graph that the idea of a co-relationship began to become clear to him.

The line that runs through the points in Figure 4.1 (and in Figures 4.2, 4.3, 4.4, and 4.5) is called a **regression line**. It is a "line of best fit." When there is perfect correlation ($r = 1.00$), all points fall exactly on the regression line. It is from Galton's use of the term *regression* that we get the symbol r for correlation.[3]

In the past generation or so, with changes in nongenetic factors such as nutrition and medical care, sons tend to be somewhat taller than their fathers, except for extremely tall fathers. If every son grew to be exactly 2 inches taller than his father (or 1 inch or 6 inches, or 5 inches shorter), the correlation would still be perfect, and the coefficient would again be 1.00. **Figure 4.2** demonstrates this point. That is, you can have a perfect correlation even if the paired numbers aren't the same. The only requirement for perfect correlation is that the *differences* between pairs of scores all be the same. If they are the same, all points of a scatterplot lie on the regression line, correlation is perfect, and an exact prediction can be made.

---

[3] The term *regression* can lead to confusion because it has two separate meanings. As you already know, it is a *statistical method* that allows you to draw the line of best fit and to make predictions with bivariate data.

Regression also refers to a *phenomenon* that occurs when a select group is tested a second time. Suppose that, from a large sample of scores, you select a subsample of extreme scores for retesting (those who did either very well or very poorly the first time). On the retest, the mean of the subsample tends to be closer to the larger group's mean than it was on the first test. Galton found that the mean height of sons of extremely tall men was shorter than the mean height of their fathers and also that the mean height of the sons of extremely short men was greater than the mean height of their fathers. In both cases the sons' mean is closer to the population mean. Because the mean of an extreme group, when measured a second time, tended to regress toward the population mean, Galton named this phenomenon *regression*. Unfortunately, he then used the same word for one of the statistical techniques he was developing.

TABLE 4.2 Manufactured data on two variables: heights of fathers and their sons*

Father	Height (in.) X	Son	Height (in.) Y
Michael Smith	74	Mike, Jr.	74
Christopher Johnson	72	Chris, Jr.	72
Matthew Williams	70	Matt, Jr.	70
Joshua Brown	68	Josh, Jr.	68
Andrew Jones	66	Andy, Jr.	66
James Miller	64	Jim, Jr.	64

* The first names are, in order, the six most common in the United States. Rounding out the top ten are John, Nicholas, Justin, and David. The surnames are also the six most common. Completing this top ten are Davis, Wilson, Anderson, and Taylor (Dunkling, 1993).

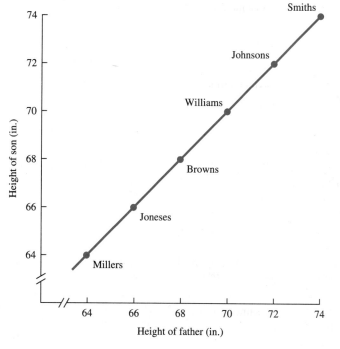

FIGURE 4.1 A scatterplot with a perfect positive correlation ($r = 1.00$)

Of course, people cannot predict their sons' heights precisely. The correlation is not perfect and the points do not all fall on the regression line. As Galton found, however, there is some positive relationship; the correlation coefficient is about .50. The points do tend to cluster around the regression line.

Here is a new problem. In your academic career you have taken an untold number of aptitude and achievement tests. For several of these tests, separate scores have been

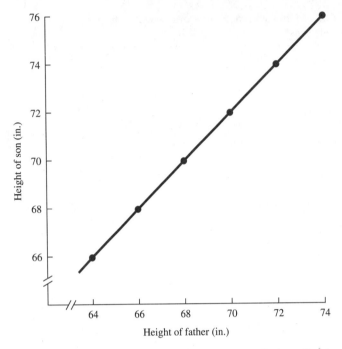

FIGURE 4.2 A scatterplot with every son 2 inches taller than his father ($r = 1.00$)

computed for verbal aptitude and mathematics aptitude. In general, what is the relationship between verbal aptitude and math aptitude? That is, are people who are good in one also good in the other, or are they poor in the other, or is there no relationship? Stop for a moment and compose your own answer to that question.

As you may have suspected, the next graph shows a scatterplot of data that will begin to answer my question. **Figure 4.3** shows the scores of eight high school seniors who took the Scholastic Aptitude Test (SAT). The SAT verbal scores are on the X axis, and the SAT math scores are on the Y axis. As you can see in Figure 4.3, there is a positive relationship, though not a perfect one. Verbal and mathematics aptitude scores tend to vary together; if the score on one is high, the other tends to be high, and if one is low, the other tends to be low. Because there is a relationship, you can predict students' math scores if you know their verbal scores. Later in this chapter you'll learn to calculate the precise *degree* of relationship and to use a formula for making precise *predictions*. (Examining Figure 4.3, you might complain that the graph is oddly shaped; it violates the 60 to 75 percent rule for the Y axis and all the data points are stuck up in one corner. I agree; it looks ungainly, but I was in a dilemma, which I'll explain later in the chapter.)

Here is another bivariate distribution for you to think about. For a particular wheat field over a period of several years, what is the relationship between yield and rainfall? Do you suppose that it is positive like that of verbal and math aptitude scores, that there is no relationship, or that the relationship is negative?

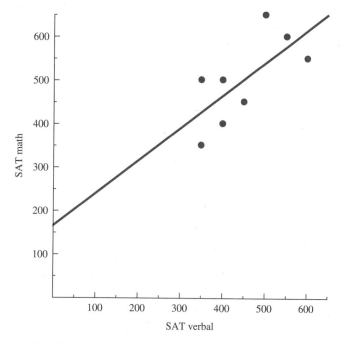

FIGURE 4.3 Scatterplot and regression line for SAT verbal and SAT math scores for eight high school students ($r = .69$)

NEGATIVE CORRELATION

As you have probably figured out from the heading, the answer to the preceding question is, Negative. **Figure 4.4** is a scatterplot of yield and rainfall for a wheat field over a 35-year period. These data came from the first edition (1925) of a book by R. A. Fisher, *Statistical Methods for Research Workers*. This book became a bible for research workers in many fields, with 14 editions published between 1925 and 1973.

As you can see in Figure 4.4, the relationship between rainfall and wheat yield is that as rainfall goes up, yield goes down. Any relationship in which increases in one variable go with decreases in the other produces a negative correlation coefficient. When the correlation is negative, the regression line goes from the upper left corner of the graph to the lower right corner. As you may recall from algebra, such lines have a negative slope.

Some other examples of negative correlation are

1. highway driving speed and miles per gallon,
2. daily rain and daily sunshine, and
3. grouchiness and friendships.

As was the case with perfect positive correlation, there is such a thing as perfect negative correlation. In cases of perfect negative correlation also, all the data points of the scatterplot fall on the regression line.

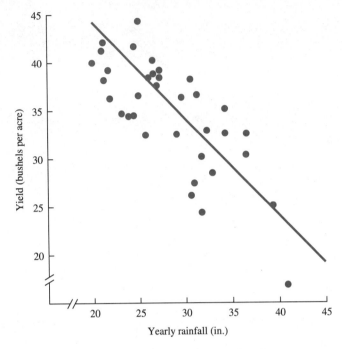

FIGURE 4.4 Wheat yield and rainfall for 35 years

A positive correlation coefficient is not more valuable than a negative correlation coefficient. The algebraic sign simply tells you the direction of the relationship (which is important when you are describing how the variables are related). The absolute size of *r*, however, tells you the *degree* of the relationship. A strong relationship (either positive or negative) is usually more valuable than a weaker one.

ZERO CORRELATION

A *zero correlation* means there is no relationship between the two variables. High and low scores on the two variables are not associated in any predictable manner.

The 50 American states have different murder rates per 100,000. (Recent data show Louisiana with the highest and South Dakota with the lowest.) The states also differ in per capita church membership. (Utah has the highest and Nevada has the lowest.) For one year, the correlation between murder and church membership was −.04, which is practically zero.

Figure 4.5 shows a scatterplot that produces a zero correlation coefficient. When $r = 0$, the regression line is horizontal with a Y value equal to $\bar{Y}$. This makes sense; if $r = 0$, then your best estimate of Y for *any* value of X is $\bar{Y}$.

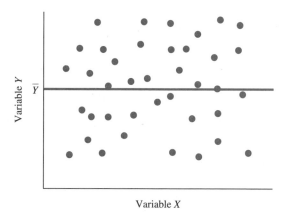

Variable X

FIGURE 4.5   Scatterplot that produces a zero correlation

CLUE TO THE FUTURE

Correlation will come up again in future chapters. If there is correlation between two sets of measurements, and you want to compare the *mean* of one set with the mean of the other set, you can make a more accurate comparison if you take the correlation into account (see Chapters 8, 11, and 13).

PROBLEMS

1. What are the characteristics of a bivariate distribution?
2. What is meant by the statement "Variable X and variable Y are correlated"?
3. Tell how X and Y vary in positive correlation. In negative correlation.
4. Can the following variables be correlated and, if so, would you expect the correlation to be positive or negative?
    a. Heights and weights of a group of adults
    b. Height of oak trees and height of pine trees
    c. Average daily temperature and the cost of heating a home
    d. IQ and reading comprehension
    e. The quiz 1 scores and the quiz 2 scores of students in section 1 and section 2 of General Biology
    f. The section 1 scores and the section 2 scores of students in General Biology on quiz 1

THE CORRELATION COEFFICIENT

A correlation coefficient provides a quantitative way to express the *degree* of relationship that exists between two variables. In terms of a definitional formula:

$$r = \frac{\Sigma(z_X z_Y)}{N}$$

where r = Pearson product-moment correlation coefficient
   z_X = a z score for variable X
   z_Y = the corresponding z score for variable Y
   N = number of pairs of X and Y values

Think through this formula and discover what happens when high scores on one variable are paired with high scores on the other variable (positive correlation). The large positive z scores are paired and the large negative z scores are paired. In both cases, multiplying them produces large positive products which, when added together, make a large positive numerator. The result is a large positive value of r. Think through for yourself what happens in the formula when there is a negative correlation or a zero correlation.

Though valuable for definitional purposes, the z-score formula is not easy to use when you are *calculating* a value for r. Fortunately, though, there are many equivalent formulas for r; I'll describe two and give you some guidance on when to use each.

Computational Formulas

One useful formula for computing r is the *blanched* (partially cooked) formula. The second formula has you enter the scores directly into the formula and is referred to as the *raw-score formula*. The blanched formula requires that you calculate means and standard deviations from the data before you calculate r. If your calculator produces these statistics (but not r itself), the blanched formula will be quicker.

Also, the blanched formula gives you a better feel for the data because you know means and standard deviations. The raw-score formula, on the other hand, does not require any intermediate steps between the summary data and r. If all you want is r, the raw-score formula is the quicker method.

Blanched formula.   The blanched formula requires you to "cook out" the means and standard deviations of both X and Y before computing r. Researchers often use means and standard deviations when telling the story of the data, so this formula is used by many:

$$r = \frac{\dfrac{\Sigma XY}{N} - (\bar{X})(\bar{Y})}{(S_X)(S_Y)}$$

where X and Y are paired observations
   XY = product of each X value multiplied by its paired Y value
   $\bar{X}$ = mean of variable X
   $\bar{Y}$ = mean of variable Y

S_X = standard deviation of variable X
S_Y = standard deviation of variable Y
N = number of pairs of observations

The new expression in this formula, ΣXY, instructs you to multiply each of the X values by its paired Y value and then sum those products. Do *not* try to obtain ΣXY by multiplying $(\Sigma X)(\Sigma Y)$. *It won't work.* The following table demonstrates this fact. Work through it.

	X	Y	XY
	2	5	10
	3	4	12
Sum	5	9	22

$$\Sigma XY = 22$$
$$(\Sigma X)(\Sigma Y) = (5)(9) = 45$$

The term ΣXY is called "the sum of the cross-products." Some form of the sum of the cross-products is necessary for computing Pearson r. One other caution about formulas for a correlation coefficient: Remember that N is the number of *pairs* of observations.

Table 4.3 demonstrates the steps you use to compute r by the blanched procedure. The data are the ones used to draw Figure 4.3, the scatterplot of the SAT verbal and SAT math scores for the eight students. I made up the numbers so that they would produce the same correlation coefficient as reported by Wallace (1972). Work through the numbers in Table 4.3, paying careful attention to the calculation of ΣXY.

Raw-score formula.   With the raw-score formula, you calculate r from the raw scores without computing means and standard deviations. The formula is

$$r = \frac{N\Sigma XY - (\Sigma X)(\Sigma Y)}{\sqrt{[N\Sigma X^2 - (\Sigma X)^2][N\Sigma Y^2 - (\Sigma Y)^2]}}$$

You have already learned what all the terms of this formula mean. Remember that N is the number of *pairs* of values.

All calculators with two or more memory storage registers—and some with one—allow you to accumulate simultaneously the values for ΣX and ΣX^2. After doing so, you may compute the values for ΣY and ΣY^2 simultaneously. This leaves only ΣXY for you to compute.

Many calculators have a built-in function for r; by entering X and Y values and pressing the r key, the coefficient is displayed. If you have such a calculator, I recommend that you use this labor-saving device after you have used the computation formulas a number of times. Working directly with terms like ΣXY leads to an understanding of what goes into r.

If you follow my advice and calculate sums, your calculator may switch into scientific notation when sums reach above the millions. A display such as 3.234234 08 might appear. To convert this number back to familiar notation, just move the decimal to he right the number of places indicated by the number on the right. Thus, 3.234234 08 becomes 323,423,400. The display 1.23456789 12 becomes 1,234,567,890,000.

TABLE 4.3   Calculation of r for SAT verbal and SAT math aptitude scores by the blanched formula

Student	SAT verbal X	SAT math Y	X^2	Y^2	XY
1	350	350	122,500	122,500	122,500
2	350	500	122,500	250,000	175,000
3	400	400	160,000	160,000	160,000
4	400	500	160,000	250,000	200,000
5	450	450	202,500	202,500	202,500
6	500	650	250,000	422,500	325,000
7	550	600	302,500	360,000	330,000
8	600	550	360,000	302,500	330,000
	3600	4000	1,680,000	2,070,000	1,845,000

$$\bar{X} = \frac{\Sigma X}{N} = \frac{3600}{8} = 450 \qquad \bar{Y} = \frac{\Sigma Y}{N} = \frac{4000}{8} = 500$$

$$S_X = \sqrt{\frac{\Sigma X^2 - \frac{(\Sigma X)^2}{N}}{N}} = \sqrt{\frac{1,680,000 - \frac{(3600)^2}{8}}{8}} = 86.60$$

$$S_Y = \sqrt{\frac{\Sigma Y^2 - \frac{(\Sigma Y)^2}{N}}{N}} = \sqrt{\frac{2,070,000 - \frac{(4000)^2}{8}}{8}} = 93.54$$

$$r = \frac{\frac{\Sigma XY}{N} - (\bar{X})(\bar{Y})}{S_X S_Y} = \frac{\frac{1,845,000}{8} - (450)(500)}{(86.60)(93.54)} = .69$$

Table 4.4 illustrates the use of the raw-score procedure for computing r. The data are the same as those used in Table 4.3, demonstrating that the value of r is the same for both methods.

The final step in any statistics problem is interpretation. What story goes with a correlation coefficient of .69 between SAT verbal scores and SAT math scores? An r of .69 is a fairly substantial correlation coefficient. Students who have high SAT verbal scores *tend* to have high SAT math scores. Note, however, that if the correlation had been near zero, you could say that the two abilities are unrelated. If the coefficient had been strong and negative, you could say, Good in one, poor in the other.

Correlation coefficients should be based on an "adequate" number of pairs of observations. As a general rule of thumb, "adequate" means 30 or more. My SAT example, however, had an N of 8, and I will ask you to work problems with fewer than 30 pairs. Small-N problems are a textbook device that allows you to spend your time on interpretation and understanding rather than on "number crunching." In Chapter 7 you will learn the reasoning behind my admonition that N be adequate.

You may have noticed that I am describing r as a sample statistic; there are no σ_X, σ_Y, μ_X, or μ_Y symbols in this chapter. Of course, if a population of data is available that

TABLE 4.4 Calculation of r for SAT verbal and SAT math aptitude scores by the raw-score formula using data from Table 4.3

$\Sigma X = 3600 \quad \Sigma Y = 4000 \quad \Sigma X^2 = 1,680,000 \quad \Sigma Y^2 = 2,070,000 \quad \Sigma XY = 1,845,000$

$$r = \frac{N\Sigma XY - (\Sigma X)(\Sigma Y)}{\sqrt{[N\Sigma X^2 - (\Sigma X)^2][N\Sigma Y^2 - (\Sigma Y)^2]}}$$

$$= \frac{(8)(1,845,000) - (3600)(4000)}{\sqrt{[(8)(1,680,000) - (3600)^2][(8)(2,070,000) - (4000)^2]}}$$

$$= \frac{360,000}{518,459} = .69$$

---

can be correlated, the parameter may be computed. The procedure is the same as for sample data. Most recent statistical texts use ρ (the Greek letter rho) as the symbol for this parameter, although you must be cautious in your outside reading; ρ has also been used to symbolize the Spearman correlation coefficient, which you will learn about in Chapter 13.

ERROR DETECTION

The Pearson correlation coefficient ranges between -1.00 and $+1.00$. Values smaller than -1.00 or larger than $+1.00$ indicate that you have made an error.

Now it is time for you to try your hand at computing r. In problem 5, use both the raw-score formula and the blanched formula. The answer is given for both formulas.

PROBLEMS

*5. This problem is based on data published in 1903 by Karl Pearson and Alice Lee. In the original article, 1376 pairs of father–daughter heights were analyzed. The scores here produce the same means and the same correlation coefficient that Pearson and Lee obtained. For these data, draw a scatterplot and calculate r.

| Father's height, X (in.) | 69 | 68 | 67 | 65 | 63 | 73 |
| Daughter's height, Y (in.) | 62 | 65 | 64 | 63 | 58 | 63 |

*6. This problem is based on actual data also. By now you know how to obtain the summary values for ΣX, ΣX^2, ΣY, ΣY^2, and ΣXY, so I am providing them for you. The raw data are scores on two different tests that measure self-esteem. Subjects were seventh-grade students. Summary values: $N = 38$, $\Sigma X = 1755$, $\Sigma Y = 1140$, $\Sigma X^2 = 87,373$, $\Sigma Y^2 = 37,592$, $\Sigma XY = 55,300$. Compute r.

*7.   The *X* variable in this problem is the population of each of the states in the United States in millions. The *Y* variable is expenditure per pupil in public schools for each of the states. Calculate *r* using either method. $N = 50$, $\Sigma X = 202$, $\Sigma Y = 41{,}048$, $\Sigma X^2 = 1740$, $\Sigma Y^2 = 35{,}451{,}830$, $\Sigma XY = 175{,}711$.   ■

SCATTERPLOTS

You already know something about scatterplots–what their elements are and what they look like when $r = 1.00$, $r = .00$, and $r = -1.00$. In this section, I will illustrate some intermediate cases and reiterate my philosophy about the value of pictures.

Figure 4.6 shows scatterplots of data that have correlation coefficients of .20, .40, $-.60$, .80, .90, and $-.95$. If you draw an envelope around the points in a scatterplot, the

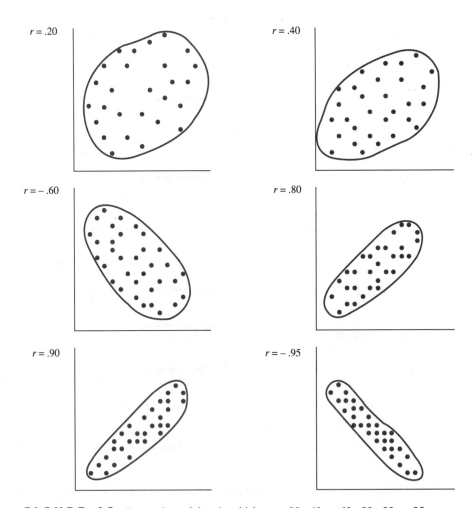

FIGURE 4.6   Scatterplots of data in which $r = .20$, .40, $-.60$, .80, .90, $-.95$

picture that the data present becomes clearer. The thinner the envelope, the higher the correlation. To say this in more mathematical language, the closer the points are clustered around the regression line, the higher the correlation.

Pictures help you understand. Scatterplots are easy to construct. Although they require some time, the benefits are worth it. If I had given you the raw data on the populations of states and their school expenditures (see problem 7) and you had drawn a scatterplot, you would have plotted one point in the upper left-hand corner of the graph (a place where data points contribute to a *negative r*). Very likely you would have raised your eyebrows at the unusual point and wondered what state it was.[4] So this is a paragraph that encourages you to construct scatterplots—they're worth it.

THE USE AND INTERPRETATION OF *r*

The basic and simple interpretation of *r* is probably familiar to you at this point. A correlation coefficient measures the degree of linear relationship between the two variables of a bivariate distribution. Fortunately for those eager to understand, correlation coefficients can produce even more information. However, the interpretation of *r* can be a tricky business; there are several common errors. Over the next few pages you will learn additional uses and interpretations of correlation coefficients and also how to detect common errors that people make in interpreting correlation coefficients.

■ Reliability

One important use of correlation coefficients is to assess the **reliability** of devices that measure things. A device (a test is an example) is reliable if the score you get is not subject to chance variation.

One way to assess chance variation is to measure a number of individuals and then measure them a second time. If the measurement you get the first time has not been influenced by chance, then you would expect to get the same measurement the second time. If the second measurement is *exactly* the same as the first for each individual, it is easy to conclude that the method of measurement is perfectly reliable—that chance has nothing to do with the score you get. However, if the measurements are not exactly the same, you experience some uncertainty. Fortunately, a correlation coefficient tells you the degree of agreement between the test and the retest scores. High correlations mean lots of agreement and therefore high reliability, and low correlations mean lots of disagreement and therefore low reliability. For those of you who are asking, What size coefficient indicates reliability? a rule of thumb is .80 or better for social science measurements.[5]

For example, in Galton's data, for 435 adults whose height was measured twice, the correlation was .98. Not surprisingly, Galton's method of measuring height was very reliable. The correlation, however, for "highest audible tone," a measure of pitch

---

[4] Alaska

[5] This section is concerned with the reliability of a *measuring instrument*. The reliability of a *relationship* is different and will be covered in material on hypothesis testing. When you ask whether a relationship is reliable, the .80 rule of thumb does not apply.

perception, was only .28 for the 349 who were tested a second time within a year (Johnson et al., 1985). One of two interpretations is possible. Either people's ability to hear high sounds changes up and down during a year, or the test was not reliable. In this case, the test lacked reliability. Two possible explanations for this lack of reliability are (1) the test environment was not as quiet from one time to the next and (2) the instruments the researchers used were not calibrated exactly the same on each test.

■ Correlation as Evidence for Causation

A high correlation coefficient does *not* give you the kind of evidence that allows you to make cause-and-effect statements. Therefore, don't do it. Ever.

Jumping to a cause-and-effect conclusion is easy to do; it may even be tempting at times. For example, Shedler and Block (1990) found that among a sample of 18-year-olds whose marijuana use ranged from abstinence to once-a-month, there was a positive correlation between use and psychological health. As Shedler and Block point out, this is *not* evidence that occasional drug use promotes psychological health. Because their study had followed the sample from age 3 on, they knew the quality of the parenting the 18-year-olds had received. Not surprisingly, parents who were responsive, accepting, patient, and who valued originality had children who were psychologically healthy. In addition, these same children as 18-year-olds had used marijuana on occasion. Thus, two variables—drug use and parenting style—were each correlated with psychological health. Shedler and Block concluded that psychological health and adolescent drug use were both traceable to quality of parenting. (This research also included a sample of frequent users, who were not psychologically healthy and who had been raised with a parenting style not characterized by the adjectives above.)

Of course, if you have a sizable correlation coefficient, it *may* be the result of a cause-and-effect relationship between the two variables. For example, the early statements about cigarette smoking causing lung cancer were based on simple correlational data. Persons with cancer were often heavy smokers. Also, national comparisons indicated a relationship (see problem 13). However, as careful thinkers (and the cigarette companies) pointed out, both cancer and smoking might have been caused by a third variable; stress was often suggested. That is, stress caused cancer and stress also caused people to smoke. Thus, cancer rates and smoking rates were related (a high correlation), but one did not cause the other; both were caused by a third variable. What is required to establish a cause-and-effect relationship are data from controlled experiments, not correlational data. Experimental data, complete with control groups, finally established the cause-and-effect relationship between smoking and lung cancer. (Chapter 8, "Hypothesis Testing and Effect Size: Two-Sample Tests," includes a discussion of experimental data.)

■ Coefficient of Determination

The correlation coefficient is also the basis of the **coefficient of determination**, which tells you the proportion of variance that two variables in a bivariate distribution have in common. The coefficient of determination is calculated by squaring r and is always a positive value between 0 and 1.

 coefficient of determination $= r^2$

Look back at Table 4.2 (page 81), the height data I manufactured that produced $r = 1.00$. There is variation among the fathers' heights, as well as among the sons' heights. How much of the variation among the sons' heights is associated with the variation in the fathers' heights? All of it! That is, the variation among the sons' heights (74, 72, 70, and so on) is exactly the same variation that is seen among their fathers' heights (74, 72, 70, and so on). In the same way, the variation among the sons' heights in Figure 4.2 (page 82) (76, 74, 72, and so on) is the same *variation* that is seen among their fathers' heights (74, 72, 70, and so on). For Table 4.2 and Figure 4.2, $r = 1.00$ and $r^2 = 1.00$.

Now look at Table 4.3 (page 88), the SAT verbal and SAT math aptitude scores. There is variation among the SAT verbal scores, as well as among the SAT math scores. How much of the variation among the SAT math scores is associated with the variation among the SAT verbal scores? Some of it. That is, the variation among the SAT math scores (350, 500, 400, and so on) is only partly reflected in the SAT verbal scores (350, 350, 400, and so on). The *proportion* of variance in the SAT math scores that is associated with the SAT verbal scores is r^2. In this case, $(.69)^2 = .48$.

Think for a moment about the many factors that influence the scores in Table 4.3. Some factors influence both scores—factors such as motivation, mental sharpness on test day, and, of course, the big one: general intellectual ability. Other factors influence one test but not the other one—factors such as anxiety about math tests, chance successes and errors, and, of course, the big ones: specific verbal knowledge and specific math knowledge.

What a coefficient of determination of .48 tells you is that 48 percent of the variance of the two tests is **common variance**. However, 52 percent of the variance is independent variance—that is, variance in one test that is not associated with variance in the other test.

Here is another example. The correlation of first-term college grade-point averages (GPAs) with academic aptitude test scores is about .50. The coefficient of determination is .25. This means that of all that variation in GPAs (from flunking out to 4.0), 25 percent is associated with the aptitude scores. The rest of the variance (75 percent) is related to other factors. As examples of other factors that influence GPA (for good or for ill) consider health, roommates, new relationships, and financial situation. Academic aptitude tests cannot predict the variation that these factors will produce.

Common variance is often illustrated by two overlapping circles, each of which represents the total variance of one variable. The overlapping portion is the amount of common variance. The left half of **Figure 4.7**, for example, shows these circles for the GPA–college aptitude test scores and the right half shows the SAT verbal–SAT math data.

Note what a difference there is between a correlation of .69 and one of .50 when they are interpreted in terms of common variance. Although .69 and .50 seem fairly close, an r of .69 predicts almost twice the amount of variance that an r of .50 predicts: 48 percent to 25 percent.

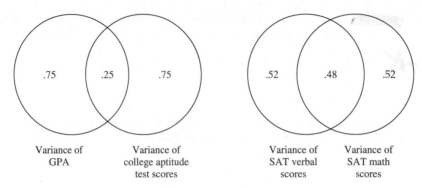

Variance of   Variance of   Variance of   Variance of
GPA   college aptitude   SAT verbal   SAT math
                test scores   scores   scores

F I G U R E   4.7   Two separate illustrations of common variance

By the way, this method of interpreting correlation coefficients is the one usually used by professional statisticians. Common variance is important to them.[6]

PROBLEMS

8. Estimate the correlation coefficients for these scatterplots.

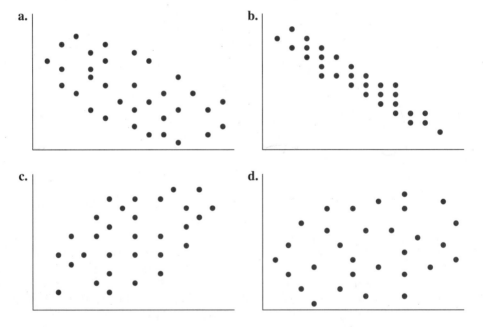

9. You found a correlation of .57 between two measures of self-esteem. What is the coefficient of determination, and what does it mean?

[6] If you are moderately skilled in algebra and would like to understand more about common variance, I'd recommend that you finish the material in this chapter on regression and then work through the presentation of Minium, King, and Bear (1993, pp. 206–208).

10. You found a correlation of .25 between the population of American states and their expenditure per pupil in public schools. Calculate the coefficient of determination and write an interpretation.

11. Here are some summary statistics that you have seen before. Can you find the correlation coefficient?

	Height of women (inches)		Height of men (inches)
ΣX	2,572	7170736	2,788
ΣX^2	165,623		194,675
N		40 pairs	

12. What percent of variance in common do two variables have if their correlation is .10? What if the correlation is increased to .40?

13. Eleven countries are listed in the table. To the right of each is the cigarette consumption per capita in 1930 and the male death rate from lung cancer 20 years later in 1950 (Doll, 1955; reprinted in Tufte, 1983). Calculate a Pearson r and write a statement telling what the data show.

② Draw scatter plot of the data
③ Compute + draw regression line for the same data (Find b+a)

Country	Per capita cigarette consumption			Male death rate (per million)	
Iceland	12803	217	47089	59	3481
Norway	22750	250	62500	91	8281
Sweden	34804	308	94864	113	12769
Denmark	61790	370	136900	167	27889
Australia	78260	455	207025	172	29584
Holland	111294	458	209764	243	59049
Canada	75750	505	255025	150	22500
Switzerland	135500	542	293764	250	62500
Finland	391424	1112	1236544	352	123904
Great Britain	535649	1147	1315609	467	218089
United States	245053	1283	1646089	191	36481

1705077 6647   2255
5505173   54 8277

14. Interpret the following statements:
   a. The correlation between vocational-interest scores at age 20 and at age 40 for the same subjects was found to be .70.
   b. The correlation between intelligence test scores of identical twins raised together is .86.
   c. The correlation between IQ and family size is about −.30.
   d. $r = .22$ between height and IQ for 20-year-old men.
   e. $r = −.83$ between income level and probability of psychosis.
   f. $r = .72$ between scores on a test of intolerance of ambiguity and scores on a test of authoritarianism.

−0.7   0·7

STRONG RELATIONSHIPS BUT LOW CORRELATIONS

One good thing about understanding something is that you come to know what's going on beneath the surface. Knowing the inner workings, you can judge whether the surface appearance is to be trusted or not. You are about to learn of two of the "inner workings" of correlation. These will help you evaluate the meaning of low correlations. Low correlations do not always mean there is no relationship between two variables.[7]

▨ Nonlinearity

For *r* to be a meaningful statistic, the best-fitting line through the scatterplot of points must be a *straight line*. If a curved regression line fits the data better than a straight line, *r* will be low, not reflecting the true degree of relationship between the two variables.

Figure 4.8 is an example of a situation in which *r* is inappropriate because the best-fitting line is curved. The *X* variable is arousal and the *Y* variable is efficiency of performance. At low levels of arousal (sleepy, for example), performance is not very good. Likewise, at very high levels of arousal (agitation, for example), people don't perform well. However, in the middle range, there is a degree of arousal that is optimum; performance is best at moderate levels of arousal.

There is obviously a strong relationship between arousal and performance, but *r* for the distribution in Figure 4.8 is −.10, a value that indicates a very weak relationship. The product-moment correlation coefficient is just not useful for measuring the strength of curved relationships. Special nonlinear correlation techniques for such relationships do exist and are described in texts such as Howell (1992) and Rosenthal and Rosnow (1991).

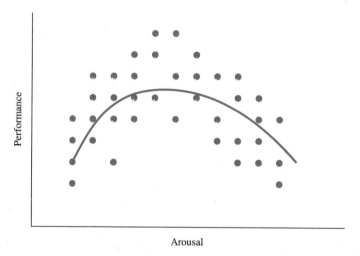

F I G U R E 4.8 Generalized relationship between arousal and efficiency of performance

[7] Correlations that do not reflect the true degree of relationship are said to be *spuriously* low or high.

■ Truncated Range

Besides nonlinearity, a second situation can give you a low Pearson coefficient even though there is a strong relationship between the two variables. Spuriously low r values can occur when the sample range is much smaller than the population range (a **truncated range**).

Suppose you wanted to know the relationship between IQ and self-esteem. To get data you used students from an Introduction to Psychology class and got the picture shown in **Figure 4.9**. These data look like a snowstorm; the conclusion would be that there is no relationship between IQ and self-esteem.

However, college students have IQ scores higher than those of the general population, so the study does not include those with lower IQs. What effect does this restriction of the range have? You can get an answer to this question by looking at **Figure 4.10**, which shows a hypothetical scatterplot of scores for the *population*, with the scores of the psychology students again shown as open circles. This scatterplot shows a moderate relationship in the population. So, unless you recognized that your sample of college students truncated the population range, you would not find the actual relationship that exists.

OTHER KINDS OF CORRELATION COEFFICIENTS

The kind of correlation coefficient you have been learning about—the Pearson product-moment correlation coefficient—is appropriate for measuring the degree of relationship between two linearly related, continuous variables. Much of the time, this is the kind of data you have.

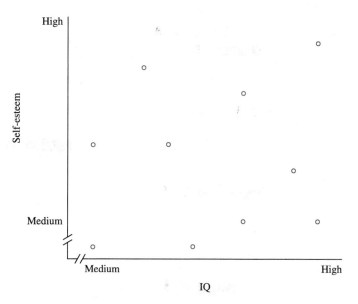

F I G U R E 4.9 Hypothetical scatterplot of IQ and self-esteem scores for Introduction to Psychology students

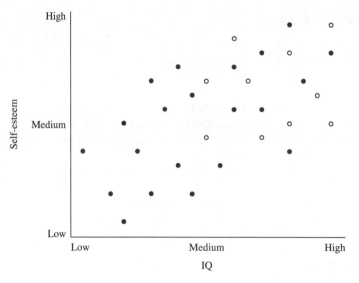

F I G U R E   4.10   Hypothetical scatterplot of IQ and self-esteem scores for the population

Sometimes, however, the data do not consist of two linearly related, continuous variables. What follows is a description of five other situations. In each case, you can express the degree of relationship in the data with a correlation coefficient—but not a Pearson product-moment correlation coefficient. Fortunately, these correlation coefficients can be interpreted much like Pearson product-moment coefficients.

1. If one of the variables is **dichotomous** (has only two values), then a **biserial** correlation (r_b) or a **point-biserial** correlation (r_{pb}) is appropriate. Variables such as height (measured simply as tall or short) and gender (male or female) are examples of dichotomous variables.

2. If the relationship between two variables is curved rather than linear, the correlation ratio *eta* (η) gives the degree of association.

3. Several variables can be combined, and the resulting combination can be correlated with one variable. With this technique, called **multiple correlation**, a more precise prediction can be made. Performance in school can be predicted better by using several measures of a person rather than one.

4. A technique called **partial correlation** allows you to separate or partial out the effects of one variable from the correlation of two variables. For example, if you want to know the true correlation between achievement test scores in two school subjects, it will probably be necessary to partial out the effects of intelligence, because IQ and achievement are correlated.

5. When the data are ranks, rather than scores from a continuous variable, the proper coefficient is the **Spearman r_s,** which is covered in Chapter 13.

These and other correlational techniques are covered in intermediate-level textbooks. (See Howell, 1992; Rosenthal and Rosnow, 1991; and Edwards, 1984.)

know when it is appropriate to use these

PROBLEMS

15. The correlation between number of older siblings and degree of acceptance of personal responsibility for one's own successes and failures is −.37.
  a. How would you interpret this correlation?
  b. What can you say about the cause of this correlation?
  c. What practical significance might it have?

16. A correlation of .97 has been found between anxiety and neuroticism.
  a. Interpret the meaning of this r.
  b. Are you justified in assuming that being neurotic is responsible in large part for manifestations of anxiety?

17. Examine the following data, make a scatterplot, and compute r if appropriate.

Serial position	1	2	3	4	5	6	7	8
Errors	2	5	6	9	13	10	6	4

CORRELATION AND REGRESSION

You are now prepared to learn how to make predictions. The technique you will use is called *linear regression*. As you will see, what you have learned about correlation (the degree of relationship between variables) will be most helpful when you interpret the results of a regression analysis (a formula that allows predictions).

A few sections ago I said that the correlation between college entrance examination scores and first-semester grade-point averages is about .50. Thus, you can predict that those who score high on the entrance examination are more likely to succeed as freshmen than those who score low. All this is rather general, though. Usually you want to predict a *specific* grade point for a *specific* applicant. For example, if you were in charge of admissions at Collegiate U., you would want to know the entrance examination score that predicts a GPA of 2.00, the minimum required for graduation. To make specific predictions, you must do a regression analysis.

Regression analysis is a technique that uses the data to write an equation for a straight line. This equation is then used to make predictions. I'll begin this with some background material on making predictions from equations.

MAKING PREDICTIONS FROM A LINEAR EQUATION

You are used to making predictions; some you make with a great deal of confidence. "If I get my average up to 80 percent, I'll get a B in this course." "If I spend $15 plus tax on this compact disc, I won't have enough left from my $20 to go to a movie."

Often predictions are based on an assumption that the relationship between two variables is linear, that a straight line will tell the story exactly. Frequently, this assumption is quite justified. For the short-term economics problem above, imagine a set of axes with "amount spent" on the X axis ($0 to $20) and "amount left" on the Y axis ($0 to $20). A straight line that connects the two $20 marks on the axes tells the whole story. (This is a line with a negative slope that is inclined 45 degrees from horizontal.) Draw this picture. Can you also write the equation for this graph?

Part of your education in algebra was about straight lines. You may recall that the slope-intercept formula for a straight line is

$$Y = mX + b$$

where Y and X are variables representing scores on the Y and X axes
        m = slope of the line (a constant)
        b = intercept (intersection) of the line with the Y axis (a constant)

Here are some reminders about the *slopes* of lines. If the highest point on the line is to the right of the lowest point, its slope is a positive number; if the highest point is to the left of the lowest point, its slope is negative. Horizontal lines have a slope equal to zero. Lines that are almost vertical have slopes that are either large positive numbers or large negative numbers.

Now, back to the slope-intercept formula. If $m = 3$ and $b = 6$, the formula becomes $Y = 3X + 6$. If you are given the value of X, you can easily find the value of Y. If $X = 4$, then $Y = \_$?

The unsolved problem, of course, is how to go from the general formula $Y = mX + b$ to a specific formula like $Y = 3X + 6$—that is, how to find the values for m and b.

A common solution involves a rule and a little algebra. The rule is that if a point lies on the line, the point satisfies the equation of the line. Thus, the point where $X = 5$ and $Y = 8$, represented as (5, 8), produces $8 = 5m + b$ when substituted into the general equation. If you are given a second point that lies on the line, you will get another equation with m and b as unknowns. Now you have two equations with two unknowns and you can solve for each in turn, giving you values for m and b. If you would like to check your understanding of this on a simple problem, you might figure out the formula for the line that tells the story of the $20 problem. For simplicity, use the two points where the line crosses the axes, (0, 20) and (20, 0).

There are many situations in which a practical prediction about Y is needed. For example, suppose a truck's springs are half compressed by its load (say, 10,000 pounds of pulp wood that is to be used to make paper for statistics books). How much more can the truck hold?

If you added 2000 pounds and then 2000 pounds more and found compression percentages of 60 percent and 70 percent, you would have enough information to leap to the conclusion that there is a linear relationship between the compression of a spring and its load.[8] And sure enough, if you predict that the truck can carry 20,000 pounds (enough for an average year's supply—6000 copies—of this book), you would be correct.

---

[8] This relationship is known as Hooke's law of elasticity, and it applies to all solid bodies such as metal and wood. Robert Hooke articulated this law in 1678, some years after he observed and chose the name "cells" for the fundamental stuff studied by biologists.

Once you have assumed or have satisfied yourself that the relationship is linear, *any* two points will allow you to draw the line that tells the story of the relationship between spring compression and load.

THE REGRESSION EQUATION—A LINE OF BEST FIT

With this background in place, you are in a position to appreciate the problem Karl Pearson faced at the end of the 19th century when he looked at father–daughter height data like those in problem 5. Look at your scatterplot of those data (or at Figure 4.3 or 4.4). The assumption that the relationship is linear is reasonable, but you quickly realize that the two-point solution will not work. The equation would depend on which two particular points you chose to plug into the formula. The two-point solution produces dozens of different lines. The question is how to find the best line. What solutions come to mind?

Perhaps one of your solutions was to find the mean Y value for each X value, connect those means with straight lines, and then choose two points that make the line fall within this narrower range of the means. Such a line (a "moving average") might be thought of as a best line. For every X value there would be a predicted Y value (symbolized by Y' and pronounced "Y prime" or "Y predicted"). The Y values (your observations) would vary around the Y' values (your predictions).

Pearson's solution, which statistics has embraced, is to use the **least squares** method, which is a more mathematically sophisticated version of the solution you looked at in the previous paragraph. Look at **Figure 4.11**, which shows the heights of two of the daughters from Pearson and Lee's 1903 data, both of whom had fathers who were 5′9″ tall. Based on all the data, I drew a regression line calculated by the least squares method, and, as you can see, the height predicted for a daughter of a father 5′9″ is 5′4″. Although both fathers were 5′9″ tall, Ann was 5′6″ and Beth was 5′3″. Thus, the prediction was 2 inches in error for Ann and 1 inch in error for Beth. You can imagine

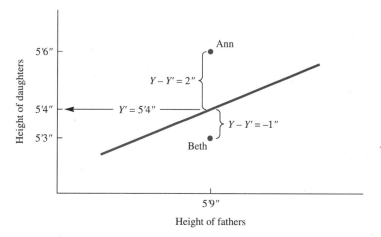

FIGURE 4.11   Hypothetical case of two daughters with a father 5′9″ tall

that if the other daughters' heights were on the graph, most of them would not fall on the regression line either. With all this error, why use the least squares method for drawing the line?

The answer is that a least squares regression line *minimizes* error in a particular way. For example, you can calculate an error for each person using the formula

$$\text{error} = Y - Y'$$

Using inches for the two daughters in our example, you get

Ann: error $= Y - Y' = 66 - 64 = 2$
Beth: error $= Y - Y' = 63 - 64 = -1$

In a similar way, there is an error for each person on the scatterplot (though for some the error would be zero). The least squares method, then, places a straight line such that the *sum of the squares* of the errors is a minimum. In symbol form, $\Sigma(Y - Y')^2$ is a minimum for a straight line calculated by the least squares method.

As a result of applying the least squares method to a set of bivariate data, you get *the slope and the intercept*, which are values you need for the equation for a particular straight line. With a slope and an intercept, you can write the equation for a line and this will be a line that *best* fits the data.

To summarize, the problem was to find a way to express mathematically what is clear in the scatterplot of the Pearson and Lee data (and in Figure 4.3), which is that the scores vary together. Assuming that a straight line will be adequate leads you to the slope-intercept formula. To actually write a specific formula, use the least squares method to find the slope and the intercept (because a two-point solution will not work).

One more transitional point is necessary. In the language of algebra, the idea of a straight line is expressed as $Y = mX + b$. In the language of statistics, exactly the same idea is expressed as $Y = a + bX$. Y and X are used the same way in the two formulas, but different letters are used for the slope and the intercept. Unfortunately, the terminology is well established in both fields. Fortunately, the translation is easy and doesn't cause many problems. Thus, in statistics the letter b stands for the slope of the line and a is the intercept of the line with the Y axis.

In statistics, the **regression equation** is

$$Y' = a + bX$$

where $Y' = Y$ value predicted from a particular X value
   $a =$ point at which the regression line intersects the Y axis
   $b =$ slope of the regression line
   $X = X$ value for which you wish to predict a Y value

In a correlation problem, the symbol Y can be assigned to either variable, but in a regression equation, Y is assigned to the variable you wish to predict.

■ The Regression Coefficients

To use the equation, $Y' = a + bX$, you need the values for a and b, which are called **regression coefficients**. Values for a and b can be calculated from any bivariate set of data. The arithmetic for these calculations comes from the least squares method of line fitting.

If you have already computed r and the standard deviations for both X and Y, b can be obtained by the formula

$$b = r\frac{S_Y}{S_X}$$

where r = correlation coefficient for X and Y
S_Y = standard deviation of the Y variable
S_X = standard deviation of the X variable

Is it clear to you that for positive correlation, b will be a positive number? For negative correlation, b will be negative.[9]

You can compute a from the formula

$$a = \bar{Y} - b\bar{X}$$

where $\bar{Y}$ = mean of the Y scores
b = regression coefficient computed previously
$\bar{X}$ = mean of the X scores

The regression coefficients are illustrated in **Figure 4.12**. The coefficient a is the point where the regression line crosses the Y axis (the Y intercept, which is 4.00 in Figure 4.12). The coefficient b is the numerical value of the slope of the regression line. To determine the slope of a line from a graph, divide the vertical distance the line rises by the horizontal distance the line covers. In Figure 4.12 the line DE (vertical rise of the line FD) is half the length of FE (the horizontal distance of FD). Thus, the slope of the regression line is 0.50 ($DE/FE = b = 0.50$). Put another way, the value of Y increases

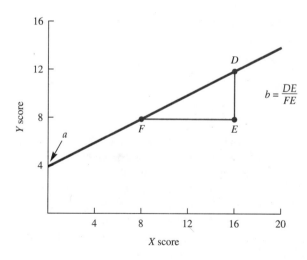

FIGURE 4.12   The regression coefficients a and b

---

[9] The value of b can sometimes be obtained more easily by the formula

$$b = \frac{N\Sigma XY - (\Sigma X)(\Sigma Y)}{N\Sigma X^2 - (\Sigma X)^2}$$

one-half point for every one-point increase in X. Expressed in a third way, the regression line is "going out" horizontally twice as fast as it is "going up" vertically.

■ Writing an SAT Regression Equation

To illustrate the construction of a regression equation, I'll use the SAT data, choosing to make predictions about math aptitude scores (Y) from verbal aptitude scores (X). The descriptive statistics needed from the data are

$$r = .69 \qquad S_X = 86.60 \qquad S_Y = 93.54 \qquad \bar{X} = 450 \qquad \bar{Y} = 500$$

Using the formula

$$b = r\frac{S_Y}{S_X}$$

you get

$$b = (.69)\frac{93.54}{86.60} = (.69)(1.08) = .75$$

Using the formula

$$a = \bar{Y} - b\bar{X}$$

you get

$$a = 500 - (.75)(450) = 500 - 337.50 = 162.50$$

The b coefficient tells you that when verbal aptitude increases by one point, math aptitude increases by three-fourths of a point. The a coefficient tells you that the regression line will intersect the Y axis at a value of 162.50. Thus, the formula for the regression line that relates math achievement and verbal achievement is

$$Y' = 162.50 + .75X$$

■ Drawing an SAT Regression Line

To draw a regression line on a scatterplot, you need a straightedge and two points that are on the line. Any two points will do. I'll demonstrate with **Figure 4.13**, which is a redrawing of the SAT data. One point that is always on the regression line for any bivariate data is $(\bar{X}, \bar{Y})$. Thus, for the SAT data, the two means (450, 500) identify a point, which is marked on Figure 4.13 with a circle. A second point can be found by substituting an X value of your choosing into the formula in the previous section. By choosing X as 400, the arithmetic can be done almost at a glance:

$$Y' = 162.50 + .75X = 162.50 + 300 = 462.50$$

The second point is (400, 462.50), which is marked on Figure 4.13 with an **X**. Finally, simply line up the straightedge on the two points and extend the line in both directions. Notice that the line crosses the Y axis at 400, which may surprise you because $a = 162.5$.

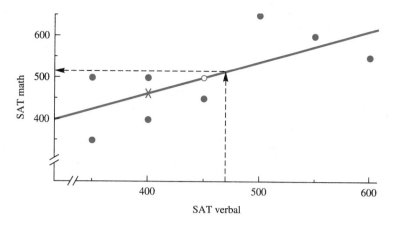

FIGURE 4.13   Scatterplot and regression line for the verbal and math aptitude data (Table 4.3)

■ The Appearance of Regression Lines

A word of caution is in order about appearances. The *appearance* of the slope of a line drawn on a scatterplot or other graph depends heavily on the units chosen for the X and Y axes. Look at **Figure 4.14**. Although the two lines appear different, $b = 1.00$ for both. They appear different because the space allotted to each Y unit in the right graph is half that allotted to each Y unit in the left graph.

I can now explain the dilemma I faced when I composed the "ungainly" **Figure 4.3** for you. The graph is ungainly because it is square (100 X units is the same length as 100 Y units) and because both axes start at 0 (even though the lowest score is 350). I composed it the way I did because I wanted the regression line to cross the Y axis at a (note that it does) and because I wanted its slope to *appear* equal to b (note that it does).

The more attractive **Figure 4.13**, of course, is a scatterplot of the same data as that in Figure 4.3. The difference is that Figure 4.13 has breaks in the axes and 100 Y units

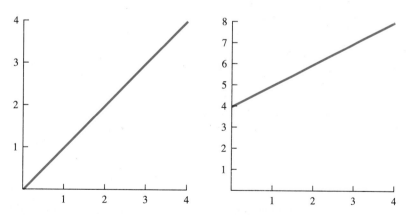

FIGURE 4.14   Two regression lines with the same slope ($b = 1.00$) but with different appearances. The difference is caused by the units chosen for the Y axis.

is about one-third the length of 100 X units. I'm sure by this point you have the message—you cannot necessarily determine a and b from a graph.

■ Predicting a Y Score—Finding Y'

There are two methods you might use to find Y' for a particular X score. For a rough estimate, you can use an accurately drawn regression line. From the X score, draw a vertical line up to the regression line. Then draw a horizontal line to the vertical axis. The Y score there is Y'. This graphical method is demonstrated by the broken lines in **Figure 4.13** for a verbal aptitude score of 470. The projection to the Y axis shows a predicted math aptitude score of just over 500.

The precise way to find Y' is to use the regression equation. To predict a math aptitude score for a verbal aptitude score of 470, the regression equation is

$$Y' = a + bX = 162.50 + (.75)(470)$$
$$= 162.50 + 352.50$$
$$= 515$$

Thus, given an X score of 470, the predicted Y score is 515.

You can find Y' values from summary data without calculating a and b by using the following formula:

$$Y' = r\frac{S_Y}{S_X}(X - \bar{X}) + \bar{Y}$$

For the verbal aptitude score of 470, the math aptitude score is

$$Y' = (.69)(1.08)(20) + 500 = 14.90 + 500 = 515$$

This is the same value as before.

Now you know how to make predictions. Predictions, however, are cheap; anyone can make them. Respect accrues only when the predictions come true. So far, I have dealt with accuracy by simply pointing out that when r is high, accuracy is high, and when r is low, you cannot put much faith in your predicted values of Y'. The other variable that influences the accuracy of Y predictions is the standard deviation of Y. The smaller the standard deviation of Y, the greater the accuracy of prediction.

Listing the factors that influence the accuracy of regression analysis predictions does not get you very far toward "true understanding." However, that list is as far as I am going in this introductory book. To go further, to *measure* the accuracy of predictions made from a regression analysis, you need to know about the **standard error of estimate**. This statistic is discussed in most intermediate-level textbooks and in textbooks on testing. After you have covered the material in Chapters 5 and 6 of this book, you will have the background necessary to understand the standard error of estimate.

■ There Are Two Regression Lines

There are *two* regression lines for a bivariate distribution. One is a regression line for Y on X, the regression line you have been computing in this chapter. The other is a regression line for X on Y. When using the formulas in this chapter to make predictions,

assign Y to the variable you want to predict and X to the variable that provides the "given."

PROBLEMS

18. In problem 5, the father–daughter height data, you found $r = .51$.
    a. Compute the regression coefficients a and b.
    b. Use your scatterplot from problem 5 and draw the regression line.
19. In problem 6, the data on self-esteem tests, you computed an r.
    a. Compute a and b.
    b. What Y score would you predict for a child who scores 42 on X?
20. Regression is a technique that economists and businesspeople rely on heavily. Think about the relationship between advertising expenditures and sales. For the data in the table, which are based on national statistics, (a) write the regression equation, (b) plot the regression line on the scatterplot, and (c) predict sales that would result if $10,000 were spent on advertising. (d) Write a sentence that tells whether any confidence at all can be put in this prediction.

Advertising, X ($ thousands)	Sales, Y ($ thousands)
3	70
4	120
3	110
5	100
6	140
5	120
4	100

21. The correlation between Stanford-Binet IQ scores and Wechsler Adult Intelligence Scale (WAIS) IQs is about .80. Both tests have a mean of 100. The standard deviation of the Stanford-Binet is 16. For the WAIS, $S = 15$. What WAIS IQ would you predict for a person scoring 65 on the Stanford-Binet? (An IQ score of 70 has been used by some schools as a cutoff point between regular classes and special education classes.)
22. From 1977 through 1982, more than 900,000 people a year graduated from college with a baccalaureate degree. Using the following data from the *Statistical Abstract of the United States*, predict the year when 1 million will graduate with a bachelor's degree. Carefully choose which variable to call X and which to call Y.

Year	'77	'78	'79	'80	'81	'82
Graduates (100,000s)	9.2	9.2	9.2	9.3	9.4	9.5

23. Once again, look over the objectives at the beginning of the chapter. Can you do them?
24. Now it is time for *integrative* work on the descriptive statistics you have studied in Chapters 2–4. Choose one of the two options that follow.

a. Write an essay on descriptive statistics. Start by jotting down from memory things you could include. Review the three chapters, adding to your list additional facts or other considerations. Draft the essay. Rest. Revise it.

b. Construct a table that summarizes the descriptive statistics in Chapters 2–4. List the techniques in the first column. Across the top of the table list topics such as purpose, formula, and so forth that distinguish among the techniques. Fill in the table.

Whether you choose option a or b, save your answer for that time in the future when you are reviewing what you are learning in this course (final exam time?).

■

WHAT WOULD YOU RECOMMEND? Chapters 1–4

At this point in the text (and at two later points), there is a set of *What would you recommend?* problems. Serving a review/integration function, these problems ask you to choose an appropriate statistic from among several you have learned in the previous four chapters. For each problem that follows, recommend a statistic that will answer the question and note why you recommend that statistic.

A. Registration figures for the American Kennel Club show which breeds are common and which are uncommon. For a frequency distribution for all breeds, what central tendency statistic would be appropriate?

B. Each year there is a contest for the biggest watermelon and one for the biggest pumpkin. How could an overall winner be chosen, given the fact that pumpkins grow larger than watermelons (over 800 pounds for pumpkins to less than 300 pounds for watermelons)?

C. Tuition here at Almamater U. has gone up each of the past five years. How can I predict what it will be in 20 years when my child enrolls?

D. Each of the American states has a certain number of miles of ocean coastline (ranging from 0 to 6640 miles). Consider a frequency distribution of these 50 scores. What central tendency statistic is appropriate for this distribution (and why)? What measure of variability would you choose to recommend?

E. The two applicants completed 30 judgments; the average error for each was zero. Are the two equivalent or might another statistic distinguish between the two and be interpretable?

F. Suppose you study some new, relatively meaningless material until you know it all. If you are tested 40 minutes later, you recall 85 percent; four hours later, 70 percent; four days later, 55 percent; and four weeks later, 40 percent. How can you express the relationship between time and memory?

G. What was the age of the average voter for the year 1996? The table from which an answer can be derived shows age categories that start with "18–20" and end with "65 and over."

H. For a class of 40 students, the study time for the first test ranged from 30 minutes to six hours. The grades ranged from a low of 48 to a high of 98. What statistic would describe how the variable *study time* is related to the variable *grades*?

You are now through with the part of the book that is clearly descriptive statistics. By now, you should be able to describe a set of data by using a graph, a few choice words, and numbers such as a mean, standard deviation, and (if appropriate) a correlation coefficient.

The next chapter serves as a transition between descriptive and inferential statistics. All the problems you will work in the following chapter will give you answers that *describe* something about a person, score, or group of people or scores. However, the ideas about probability and theoretical distributions that you will use to work these problems are essential elements of inferential statistics.

So, the transition this time is to concepts that will prepare you to plunge into material on inferential statistics. As you will see rather quickly, most of the descriptive statistics that you have been learning will be very useful for inferential statistics.

Theoretical Distributions Including the Normal Distribution

OBJECTIVES FOR CHAPTER 5

After studying the text and working the problems in this chapter, you should be able to:

1. Distinguish between a theoretical and an empirical distribution

2. Distinguish between theoretical and empirical probability

3. Predict the probability of certain events from your knowledge of the theoretical distribution of those events

4. List the characteristics of the normal distribution

5. Find the proportion of a normal distribution that lies between two scores

6. Find the scores between which a certain proportion of a normal distribution falls

7. Find the number of scores associated with a particular proportion of a normal distribution

This chapter has more figures than any other chapter, almost one per page. The reason for all these figures is that they are the best way I know to convey to you ideas about theoretical distributions and probability. So, please examine these figures carefully, making sure you understand what each part means. When you are working problems, drawing your own pictures helps.

I want to begin by distinguishing between empirical distributions and theoretical distributions. In Chapter 2, you learned to arrange scores into frequency distributions. The scores you worked with were selected because they were representative of scores from actual research. Distributions of such observed scores are **empirical distributions.**

This chapter is about theoretical distributions. Like the empirical distributions in Chapter 2, a theoretical distribution is a presentation of all the scores, usually presented as a graph. **Theoretical distributions**, however, are based on mathematical formulas and logic rather than on empirical observations. Statisticians find many theoretical distributions useful. This chapter covers three of them: rectangular, binomial, and normal.

In statistics, theoretical distributions are used to determine probabilities. When there is a correspondence between an empirical distribution and a theoretical distribution, you can use the theoretical distribution to arrive at *probabilities* about future empirical events. Probabilities, as you know, are valuable aids in reaching decisions.

Probability is the first topic in this chapter. Rectangular and binomial distributions will be used to illustrate probability more fully and to establish some points that are true for all theoretical distributions. The third distribution, the normal distribution, will occupy the bulk of your time and attention in this chapter.

PROBABILITY

The concept of probability is already somewhat familiar to you. You know, for example, that probability values range from .00 (there is no possibility that an event will occur) to 1.00 (the event is certain to happen).

In statistics, events are sometimes referred to as "successes" or "failures," and calculating the actual probability of a success takes one of two courses. With the *theoretical* approach, you first enumerate all the ways a *success* can occur. Then you enumerate all the *events* that can occur (whether successes or failures). Finally, you form a ratio with successes on top (the numerator) and total events on the bottom (the denominator). This fraction, changed to a decimal, is the theoretical probability of the event.

In coin flipping, the theoretical probability of "head" is .50. A head is a success and it can occur in only one way. The total number of possible outcomes is two (head and tail), and the ratio $\frac{1}{2}$ is .50. In a similar way, the probability of a six on a die is $\frac{1}{6} = .167$. For playing cards, the probability of a jack is $\frac{4}{52} = .077$.

With the *empirical* approach to finding probability, you *observe* actual events, some of which are successes and some of which are failures. The ratio of successes divided by the total events produces a probability, a decimal number between .00 and 1.00. To find an empirical probability, you use observations rather than logic to get the numbers.

What is the probability of choosing a college graduate at random and getting someone with a particular major? Remember the data in Figure 2.4, showing college graduate majors? The probability question can be answered by processing numbers from that figure. Here's how. Choose the major that you are interested in and label the frequency of that major as "number of successes." Divide that number by 1,137,000, which is the total number of baccalaureate degrees granted in 1991–92. The figure you get will answer the probability question. If the major in question is sociology, then 20,000/1,137,000 = .02 is the answer. For English, 45,000/1,137,000 = .04 is the answer.[1] Now, here's a question for you to answer for yourself. Were these probabilities

---

[1] If I missed doing the arithmetic for the major *you* are interested in, I hope you'll do it for yourself.

determined theoretically or empirically? Finally, you should know that another name for the empirical probability approach is the relative frequency approach.

The rest of this chapter emphasizes theoretical distributions and theoretical probability. You will work with coins and cards next, but before you are finished, I promise you a much wider variety of applications.

A RECTANGULAR DISTRIBUTION

To show you the relationship between theoretical distributions and theoretical probabilities, I'll use an example of a theoretical distribution that you may be familiar with. **Figure 5.1** is a histogram that shows the distribution of types of cards in an ordinary deck of playing cards. There are 13 kinds of cards, and the frequency of each card is 4. This theoretical curve is a **rectangular distribution**. (The line that encloses a frequency polygon is called a curve, even if it is straight.) The number in the area above each card is the probability of obtaining that card in a chance draw from the deck. That theoretical probability (.077) was obtained by dividing the number of cards that represent the event (4) by the total number of cards (52).

Probabilities are often stated as "chances in a hundred." The expression $p = .077$ means that there are 7.7 chances in 100 of the event in question occurring. Thus, from Figure 5.1 you can tell at a glance that there are 7.7 chances in 100 of drawing an ace from a deck of cards. This knowledge might be helpful in some card games.

With this theoretical distribution, you can determine other probabilities. Suppose you wanted to know your chances of drawing a face card or a 10. These are the shaded events in Figure 5.1. Simply add the probabilities associated with a 10, jack, queen, and king. Thus, $.077 + .077 + .077 + .077 = .308$. This knowledge might be helpful in

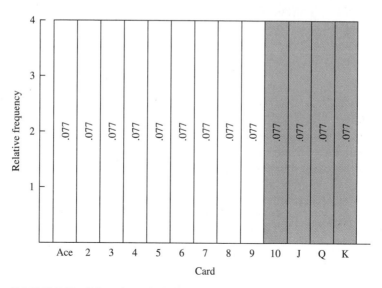

FIGURE 5.1   Theoretical distribution of 52 draws from a deck of playing cards

a game of blackjack, in which a face card or a 10 is an important event (and may even signal "success").

In Figure 5.1, there are 13 kinds of events, each with a probability of .077. Not surprisingly, when you add up all the events [(13)(.077)], the result is 1.00. In addition to the probabilities adding up to 1.00, *the areas add up to 1.00*. That is, by conventional agreement, the *area* under the curve is taken to be 1.00. With this arrangement, any statement about area is also a statement about probability. (If you like to verify things for yourself, you'll find that each slender rectangle has an *area* that is .077 of the area under the curve.) Of the total area under the curve, the proportion that signifies ace is .077, and that is also the probability of drawing an ace from the deck.[2]

CLUE TO THE FUTURE

The probability of an event or a group of events corresponds to the *area* of the theoretical distribution associated with the event or group of events. This idea will be used throughout this book.

PROBLEMS

1. What is the probability of drawing a card that falls between 3 and jack, excluding both?
2. If you drew a card at random, recorded the result, and replaced it, how many 7s would you expect in 52 draws?
3. What is the probability of drawing a card that is higher than a jack *or* lower than a 3?
4. If you made 78 draws from a deck, replacing each card, how many 5s and 6s would you expect?

A BINOMIAL DISTRIBUTION

The **binomial distribution** is another example of a theoretical distribution. Suppose you took three new quarters and tossed them into the air. What is the probability that all three will come up heads? As you may know, the answer is found by multiplying together the probabilities of each of the independent events. For each coin, the probability of a head is $\frac{1}{2}$, so the probability that all three will be heads is $(\frac{1}{2})(\frac{1}{2})(\frac{1}{2}) = \frac{1}{8} = .1250$.

[2] In gambling situations, uncertainty is commonly expressed as odds. The expression "odds of 5 : 1" means that there are five ways to fail and one way to succeed; 3 : 2 means three ways to fail and two ways to succeed. The odds of drawing an ace are 12 : 1. To convert odds to a probability of success, divide the second number by the sum of the two numbers.

T A B L E 5.1   **All possible outcomes when three coins are tossed**

Outcomes	Number of heads	Probability of outcome
Heads, heads, heads	3	.1250
Heads, heads, tails	2	.1250
Heads, tails, heads	2	.1250
Tails, heads, heads	2	.1250
Heads, tails, tails	1	.1250
Tails, heads, tails	1	.1250
Tails, tails, heads	1	.1250
Tails, tails, tails	0	.1250

Here are two other questions about tossing those three coins. What is the probability of two heads? What is the probability of one head or zero heads? You could answer these questions easily if you had a theoretical distribution of the probabilities, so I will show you how to construct one. Start by listing, as in **Table 5.1**, the eight possible outcomes of tossing the three quarters into the air. Each of these eight outcomes is equally likely, so the probability for any one of them is $\frac{1}{8} = .1250$. There are three outcomes in which two heads appear, so the probability of two heads is $.1250 + .1250 + .1250 = .3750$. This is the answer to the first question. Based on Table 5.1, I constructed **Figure 5.2**, which is the theoretical distribution of probabilities we need. You can use it to answer problems 5 and 6, which follow.[3]

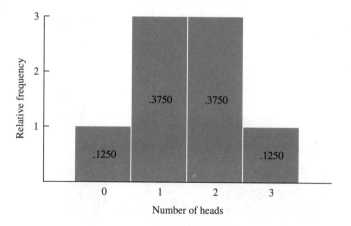

F I G U R E 5.2   A binomial distribution showing the number of heads when three coins are tossed

---

[3] The binomial distribution is discussed more fully by Moore and McCabe (1993), Howell (1992), and Loftus and Loftus (1988).

PROBLEMS

5. If you toss three coins into the air, what is the probability of a success if success is (a) either one head or two heads? (b) all heads or all tails?

6. If you threw the three coins into the air 16 times, how many times would you expect to find zero heads? ■

COMPARISON OF THEORETICAL AND EMPIRICAL DISTRIBUTIONS

I have carefully called Figures 5.1 and 5.2 *theoretical* distributions. A theoretical distribution may not reflect *exactly* what would happen if you drew cards from an actual deck of playing cards or tossed quarters into the air. Actual results could be influenced by lost or sticky cards, sleight of hand, unbalanced coins, or chance deviations. Now let's turn to the empirical question of what a frequency distribution of actual draws from a deck of playing cards looks like. **Figure 5.3** is a histogram based on 52 draws from a used deck shuffled once before each draw.

As you can see, Figure 5.3 is not exactly like Figure 5.1. In this case, the differences between the two distributions are due to chance or worn cards and not to lost cards or sleight of hand (at least not conscious sleight of hand). Of course, if I made 52 more draws from the deck and constructed a new histogram, the picture would probably be different from both Figures 5.3 and 5.1. However, if I continued, drawing 520 or 5200 or 52,000 times,[4] and only chance was at work, the curve would be practically flat on the top; that is, the empirical curve would look like the theoretical curve.

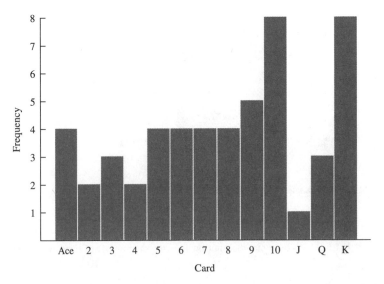

FIGURE 5.3 Empirical frequency distribution of 52 draws from a deck of playing cards

[4] Statisticians describe this extensive sampling as "the long run."

The major point here is that a theoretical curve represents the "best estimate" of how the events would actually occur. As with all estimates, a theoretical curve may produce predictions that vary from actual observations, but in the world of real events, it is better than any other estimate.

In summary, then, a theoretical distribution is one based on logic and mathematics rather than on observations. It shows you the probability of each event that is part of the distribution. When it is similar to an empirical distribution, the probability figures obtained from the theoretical distribution are accurate predictors of actual events.

THE NORMAL DISTRIBUTION

One theoretical distribution that has proved to be extremely valuable is the **normal distribution**, a distribution that, among other things, describes how chance operates. In about 1800, Carl Friedrich Gauss (1777–1855) worked out the mathematics of the curve and promoted it as a way to represent random error in astronomy observations (Stewart, 1977). Because the "Gaussian" curve was such an accurate picture of the effects of random variation, early writers referred to the curve as the *law* of error.[5] At the end of the 19th century, Karl Pearson, as part of his efforts to standardize the terminology in statistics, coined the name *normal distribution*.[6] Pearson probably chose the word "normal" because a wide variety of measurements were distributed in this bell-shaped fashion.

One of the early promoters of the normal curve was Adolphe Quetelet (Ka´-tle) (1796–1874), a Belgian who showed that many social and biological measurements are distributed normally. Quetelet, who knew about the "law of error" from his work as an astronomer, presented tables showing the correspondence between measurements such as height and chest size and the normal curve. During the 19th century, Quetelet promoted statistics and was widely influential (T. M. Porter, 1986). For example, Florence Nightingale, a pioneer in using statistical analyses to improve health care, said that Quetelet was the founder of "the most important science in the whole earth" (Cohen, 1984). Also, Quetelet's work suggested to Francis Galton that genius could be treated mathematically, an idea that led to the concept of correlation.[7]

Although many measurements are distributed approximately normally, it is not the case that data "should" be distributed normally. This unwarranted conclusion has been reached by some scientists in the past.

Finally, the theoretical normal curve has an important place in statistical theory. This importance is quite separate from the fact that empirical frequency distributions often correspond closely to the normal curve.

---

[5] In this context, *error* means random variation. Used in this manner, the term *error* is an important part of today's statistical terminology.

[6] Pearson also chose the term *standard deviation* for the statistic that you know by that name. (See T. M. Porter, 1986.)

[7] Quetelet qualifies as a famous person: A statue was erected in his honor in Brussels, he was the first foreign member of the American Statistical Association, and the Belgian government commemorated the centennial of his death with a postage stamp (1974). You can read a short intellectual biography of Quetelet by Landau and Lazarfeld (1968) in the *International Encyclopedia of the Social Sciences*.

■ Description of the Normal Distribution

Figure 5.4 is a normal distribution. It is a bell-shaped, symmetrical, theoretical distribution based on a mathematical formula rather than on any empirical observations. (Even so, if you peek ahead to Figures 5.7, 5.8, and 5.9, you will see that empirical curves often look similar to this theoretical distribution.) When the theoretical curve is drawn, the Y axis is sometimes omitted. On the X axis, z scores are used as the unit of measurement for the standardized normal curve, where

$$z = \frac{X - \mu}{\sigma}$$

where X = a raw score
μ = the mean of the distribution
σ = the standard deviation of the distribution

I will discuss the standardized curve first and later add the raw scores to the X axis.

There are several other things to note about the normal distribution. The mean, the median, and the mode are the same score—the score on the X axis at which the curve is at its peak. If a line is drawn from the peak to the mean score on the X axis, the area under the curve to the left of the line will be half the total area—50 percent—leaving half the area to the right of the line. The tails of the curve are **asymptotic** to the X axis; that is, they never actually cross the axis but continue in both directions indefinitely, with the distance between the curve and the X axis becoming less and less. Although, in theory, the curve never ends, it is convenient to think of (and to draw) the curve as extending from -3σ to $+3\sigma$. (The *table* for the normal curve in Appendix B, however, covers the area from -4σ to $+4\sigma$.)

Another point about the normal distribution is that the two inflection points in the curve are at exactly -1σ and $+1\sigma$. The **inflection points** are where the curve is the steepest—that is, from where the curve changes from bending upward to bending over. (See the points above -1σ and $+1\sigma$ on Figure 5.4 and think of walking up, over, and down a bell-shaped hill.)

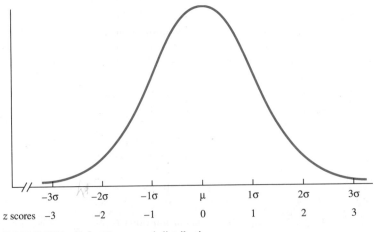

z scores   −3   −2   −1   0   1   2   3

FIGURE 5.4 The normal distribution

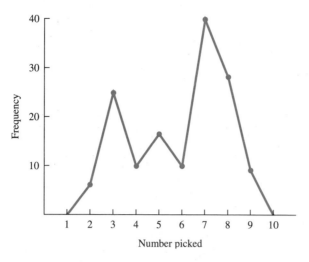

FIGURE 5.5 Frequency distribution of choices of numbers between 1 and 10

I want to end this introductory section with a caution about the word *normal*. The antonym for normal is abnormal. *Curves that are not normal distributions, however, are definitely not abnormal.* There is nothing uniquely desirable about the normal distribution. Many nonnormal distributions are also useful to statisticians. Figure 5.1 is an example. It isn't a normal distribution, but it can be very useful. Figure 5.5 shows what numbers were picked when an instructor asked introductory psychology students to pick a number between 1 and 10. **Figure 5.5** is a bimodal distribution with modes at 3 and 7. It isn't a normal distribution, but it will prove useful later in this book.

■ The Normal Distribution Table

The theoretical normal distribution is used to determine the probability of an event, just as Figure 5.1 was. **Figure 5.6** is a picture of the normal curve, showing the probabilities associated with certain areas. The figure shows that the probability of an event with a z score between 0 and 1.00 is .3413. For events with z scores of 1.00 or larger, the probability is .1587. These probability figures were obtained from Table C in Appendix B. Look at **Table C** now. It is arranged so that you can begin with a z score (column A) and find the following:

1. The area between the mean and the z score (column B)
2. The area beyond the z score (column C)

In column A find the z score of 1.00. The proportion of the curve between the mean and a z score of 1.00 is .3413. The proportion beyond the z score of 1.00 is .1587. Because the normal curve is symmetrical and because the area under the entire curve is 1.00, you will not be surprised when you add .3413 and .1587 together. Also, because the curve is symmetrical, these same proportions hold for $z = -1.00$. Thus, all the proportions in Figure 5.6 were derived by finding the proportions associated with a z

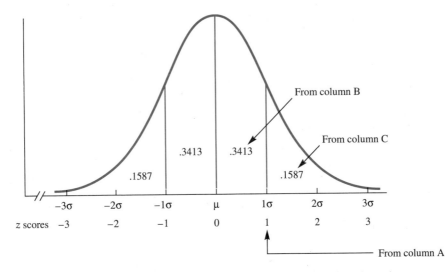

FIGURE 5.6   The normal distribution showing the probabilities of certain z scores

value of 1.00 in Table C. Don't just read this paragraph; do it. Understanding the normal curve *now* will pay you dividends throughout the book.

Notice that the proportions in Table C are carried to four decimal places and that I used all of them. This is customary practice in dealing with the normal curve, because you often want two decimal places when a proportion is converted to a percentage.

PROBLEMS

7. In Chapter 3 you read of a professor who gave A's to those with z scores of $+1.50$ or higher.
   a. What proportion of a class would be expected to make A's?
   b. What assumption must you make in order to arrive at the proportion you found in part a?
8. What proportion of the normal distribution is found in the following areas:
   a. Between the mean and $z = .21$
   b. Beyond $z = .55$
   c. Between the mean and $z = -2.01$
9. Is the distribution in Figure 5.5 theoretical or empirical?

As I've already mentioned, many empirical distributions are approximately normally distributed. **Figure 5.7** shows a set of 261 IQ scores, **Figure 5.8** shows the diameter of 199 ponderosa pine trees, and **Figure 5.9** shows the hourly wage rates of 185,822 union truck drivers in 1944. As you can see, these distributions from diverse fields are similar to Figure 5.4, the theoretical normal distribution. Please note that all of these empirical distributions are based on a "large" number of observations. Usually more than 100 observations are required for the curve to fill out nicely.

So far in this section, I have made two statistical points: first, that Table C can be

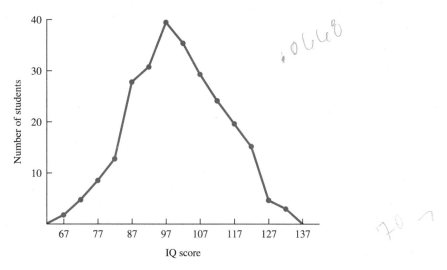

FIGURE 5.7 Frequency distributions of IQ scores of 261 fifth-grade students (unpublished data from J. O. Johnston)

used to determine areas (proportions) of a normal distribution, and, second, that many empirical distributions are approximately normally distributed. My final point is that any normally distributed empirical distribution can be made to correspond to the theoretical distribution in Table C by using z scores. Converting the raw scores of *any* empirical normal distribution to z scores will give the distribution *a mean equal to zero* and *a standard deviation equal to 1.00,* and that is exactly the scale used in the theoretical normal distribution (also called the standardized normal distribution). With this correspondence established, the theoretical normal distribution can be used to determine the probabilities of empirical events, whether they are IQ scores, tree diameters, or hourly wages.

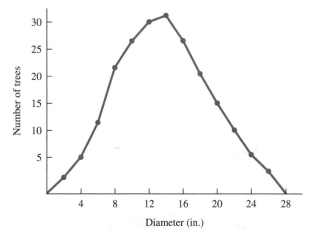

FIGURE 5.8 Frequency distribution of diameters of 100-year-old ponderosa pine trees on one acre, $N = 199$ (Forbs and Meyer, 1955)

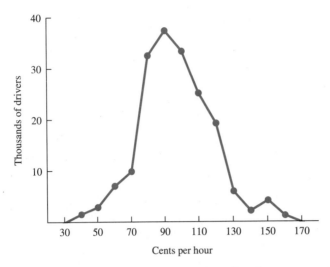

FIGURE 5.9 Frequency distribution of hourly wage rates of union truck drivers on July 1, 1944, $N = 185,822$ (U.S. Bureau of Labor Statistics, December 1944)

In the following examples, the normal curve will be used with IQ scores. IQ scores are distributed normally (Figure 5.7 is evidence for that) with a theoretical mean of 100 and a standard deviation of 15.[8]

PROBLEM

10. Calculate the z scores for IQ scores of 55, 110, 103, and 100. ■

■ Finding the Proportion of a Population That Has Scores of a Particular Size or Greater

Suppose you were faced with finding out what proportion of the population has an IQ of 120 or higher. Begin by sketching a normal curve (either in the margin or on a separate paper). Note on the baseline the positions of IQs of 100 and 120. What is your estimate of the proportion with IQs of 120 or higher?

Look at **Figure 5.10**. It is a more formal version of your sketch, giving additional IQ scores on the X axis. The proportion of the population with IQs of 120 or higher is shaded. The z score that corresponds with an IQ of 120 is

$$z = \frac{120 - 100}{15} = \frac{20}{15} = 1.33$$

[8] This is true for the Wechsler intelligence scales (WAIS, WISC, and WPPSI). The Stanford-Binet has a mean of 100 also, but the standard deviation is 16. Flynn (1987) suggests that the actual population mean IQ is well *above* 100 in many countries.

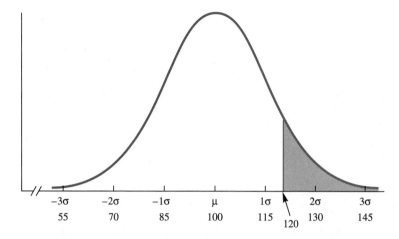

FIGURE 5.10 Theoretical distribution of IQ scores

Table C shows that the proportion beyond $z = 1.33$ is .0918. Thus, you would expect a proportion of .0918 or 9.18 percent of the population to have an IQ of 120 or higher. *Because the size of an area under the curve is also a probability statement about the events in that area, there are 9.18 chances in 100 that any randomly selected person will have an IQ of 120 or above.* **Figure 5.11** shows the proportions just determined.

Table C gives the proportions of the normal curve for positive z scores only. However, because the distribution is symmetrical, knowing that .0918 of the population has an IQ of 120 or higher tells you that .0918 has an IQ of 80 or lower. An IQ of 80 has a z score of -1.33.

You can answer questions of "how many" as well as questions of proportions using the normal distribution. Suppose there were 500 first graders entering school. How many would be expected to have IQs of 120 or higher? You just found that 9.18 percent of the population would have IQs of 120 or higher. If the population is 500, then calculating 9.18 percent of 500 would give you the number of children. Thus,

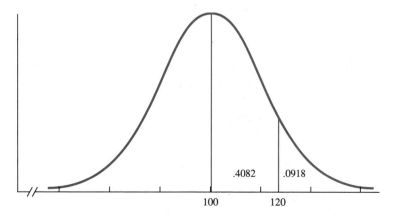

FIGURE 5.11 Proportion of the population with an IQ of 120 or higher

$(.0918)(500) = 45.9$. So 46 of the 500 first graders would be expected to have an IQ of 120 or higher.

There are 18 more normal curve problems for you to do in the rest of this chapter. Would you like to maximize your chances of working every one of them correctly the first time? Here's how. Start by sketching a normal curve. Read the problem and write in the givens and the unknowns on your curve. Estimate the answer. Apply the z-score formula. Compare your answer with your estimate; if they don't agree, decide which is in error and make any changes that are appropriate. Confirm your answer by checking the answer in back of the book. If you decide to go for 18 out of 18, good luck!

ERROR DETECTION

Sketching a normal distribution is the best way to understand a problem and avoid errors. Draw vertical lines at the scores you are interested in. Write in proportions.

PROBLEMS

11. Many school districts place children with IQs of 70 or lower in special education classes. What proportion of the general population would be expected in these classes?

12. In a school district of 4000 students, how many would be expected to be in special education?

13. What proportion of the population would be expected to have IQs of 110 or higher?

14. Answer the following questions for 250 first-grade students:
    a. How many would you expect to have IQs of 110 or higher?
    b. How many would you expect to have IQs lower than 110?
    c. How many would you expect to have IQs lower than 100?

■ Finding the Score That Separates the Population into Two Proportions

Instead of starting with an IQ score and calculating proportions, you can also work backward and answer questions about scores if you are given proportions. For example, what IQ score is required to be in the top 10 percent of the population?

My picture of this problem is shown as **Figure 5.12**. I began by sketching a more or less bell-shaped curve and writing in the mean (100). Next, I separated the "top 10 percent" portion with a vertical line. Because I need to find a score, I put a question mark on the score axis. The next step is to look in Table C under the column "area beyond z" for .1000. It is not there. You have a choice between .0985 and .1003.

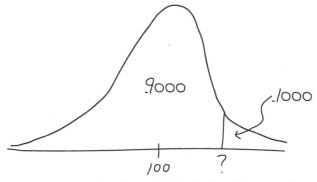

FIGURE 5.12   Theoretical distribution of IQ scores divided into an upper 10 percent and a lower 90 percent

Because .1003 is closer to the desired .1000, use it.[9] The z score that corresponds to a proportion of .1003 is 1.28. Now you have all the information you need to solve for X. To begin, solve the basic z-score formula for X.

$$z = \frac{X - \mu}{\sigma}$$

Multiplying both sides by σ produces

$$(z)(\sigma) = X - \mu$$

Adding μ to both sides isolates X. Thus, when you need to find a score (X) associated with a particular proportion of the normal curve, the formula is

$$X = \mu + (z)(\sigma)$$

Returning to the 10 percent problem and substituting numbers for the mean, the z score, and the standard deviation,

$$X = 100 + (1.28)(15)$$
$$= 100 + 19.20$$
$$= 119.2$$
$$= 119 \quad \text{(IQs are usually expressed as whole numbers.)}$$

Therefore, the IQ score required to be in the top 10 percent of the population is 119.

Here is a similar problem. Suppose a mathematics department wants to restrict the remedial math course to those who really need it. The department has the scores on the math achievement exam taken by entering freshmen for the past 10 years. The scores on this exam are distributed in an approximately normal fashion, with $\mu = 58$ and $\sigma = 12$. The department wants to make the remedial course available to those students whose mathematical achievement places them in the bottom third of the freshman class.

---

[9] You might use interpolation to determine a more accurate z score for a proportion of .1000. This extra precision (and labor) is unnecessary because the final result is rounded to the nearest whole number. For IQ scores, the extra precision does not make any difference in the final answer.

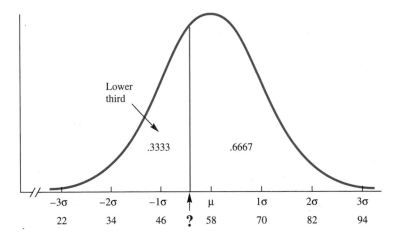

F I G U R E 5.13 Distribution of scores on a math achievement exam

The question is, What score will divide the lower third from the upper two-thirds? **Figure 5.13** should help to clarify the question.

The first step is to look in column C of Table C to find .3333. Again, such a proportion is not listed. The nearest proportion is .3336, which has a z value of −.43. (This time you are dealing with a z score below the mean, where all z scores are negative.) Applying z = −.43,

$$X = \mu + (z)(\sigma)$$
$$= 58 + (-.43)(12)$$
$$= 58 - 5.16 = 52.84 = 53$$

Using the theoretical normal curve to establish a cutoff score is efficient. All you need is the mean, the standard deviation, and confidence in your assumption that the scores are distributed normally. The empirical alternative for the mathematics department is to sort physically through all scores for the past 10 years, arrange them in a frequency distribution, and calculate the score that separates the bottom one-third.

PROBLEMS

15. Mensa is an organization of people who have high IQs. To be eligible for membership, a person must have an IQ "higher than 98 percent of the population." What IQ is required to qualify?

16. The mean height of American women aged 18–24 is 64.5 inches, with a standard deviation of 2.5 inches (*Statistical Abstract of the United States: 1994*, 1994).
 a. What height divides the tallest 5 percent of the population from the rest?
 b. The minimum height required for women to join the U.S. Army is 58 inches. What proportion of the population will be excluded?

*17. The mean height of American men aged 18–24 is 69.7 inches, with a standard deviation of 3.0 inches (*Statistical Abstract of the United States: 1994*, 1994).

a. The minimum height required for men to join the U.S. Army is 60 inches. What proportion of the population will be excluded?

b. What proportion of the population is taller than Napoleon Bonaparte, who was 5′2″?

18. The weight of many manufactured items is approximately normally distributed. For new U.S. pennies, the mean is 3.11 grams and the standard deviation is .05 gram (Youden, 1962).

a. What proportion of all new pennies would you expect to weigh more than 3.20 grams?

b. What weights separate the middle 80 percent of the pennies from the lightest 10 percent and the heaviest 10 percent? ∎

∎ Finding the Proportion of the Population Between Two Scores

Table C in Appendix B can also be used to determine the proportion of the population between two scores. For example, IQ scores that fall in the range from 90 to 110 are often called "average." What proportion of the population falls in this range? **Figure 5.14** is a picture of the problem.

In this problem, you must add an area on the left of the mean to an area on the right of the mean. First, you need z scores that correspond to the IQ scores of 90 and 110:

$$z = \frac{90 - 100}{15} = \frac{-10}{15} = -.67$$

$$z = \frac{110 - 100}{15} = \frac{10}{15} = .67$$

The proportion of the distribution between the mean and $z = .67$ is .2486, and, of course, the same proportion is between the mean and $z = -.67$. Therefore, $(2)(.2486) = .4972$ or 49.72 percent. So approximately 50 percent of the population is classified as "average," using the "IQ = 90 to 110" definition.

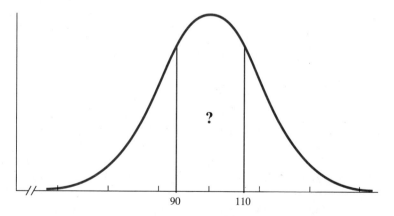

FIGURE 5.14 The normal distribution showing the IQ scores that define the "average" range

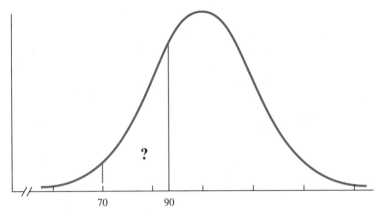

70   90

FIGURE 5.15 The normal distribution illustrating the area bounded by IQ scores of 90 and 70

What percent of the population would be expected to have IQs between 70 and 90? **Figure 5.15** illustrates this question. There are two approaches to this problem. One is to find the area from 100 to 70 and then subtract the area from 90 to 100. The other way is to find the area beyond 90 and subtract from it the area beyond 70. I'll illustrate with the second approach. The corresponding z scores are

$$z = \frac{90 - 100}{15} = -.67 \quad \text{and} \quad z = \frac{70 - 100}{15} = -2.00$$

The area beyond $z = -.67$ is .2514, and the area beyond $z = -2.00$ is .0228. Subtracting the second proportion from the first, you find that .2286 of the population has an IQ in the range of 70 to 90.

PROBLEMS

*19. The distribution of 800 test scores in an Introduction to Psychology course was approximately normal, with $\mu = 35$ and $\sigma = 6$.
   a. What proportion of the students had scores between 30 and 40?
   b. What is the probability that a randomly selected student would score between 30 and 40?
20. Now that you know the proportion of students with scores between 30 and 40, would you expect to find the same proportion between scores of 20 and 30? If so, why? If not, why not?
21. Calculate the proportion of scores between 20 and 30. Be careful with this one; drawing a picture is especially advised.

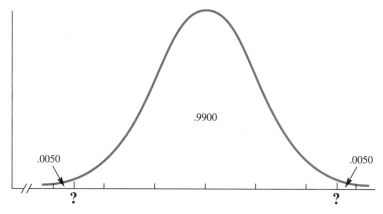

FIGURE 5.16 Distribution of IQs showing 1 percent of the scores divided between the two extremes

22. How many of the 800 students would be expected to have scores between 20 and 30? ■

■ Finding the Extreme Scores in a Population

What IQ scores are so extreme that only 1 percent of the population would be expected to have them? This is a little different from the earlier questions about the "lower third" and the "upper 2 percent" because this question does not specify which end of the curve is of interest. To answer this type of question, divide the proportion in half, placing each part at one extreme end of the curve. For the current question, the problem is to find the scores that leave one-half of 1 percent (.0050) in each tail of the curve, as seen in **Figure 5.16**. This 1 percent problem turns up in other contexts in statistics.

Because you are starting with a proportion and looking for a score, look in column C of Table C for .0050. By convention, the z score to use is 2.58. You can again use $X = \mu + (z)(\sigma)$ to determine the raw score (IQ). Thus,

$$X = 100 + (2.58)(15) \quad \text{and} \quad X = 100 + (-2.58)(15)$$
$$= 100 + 38.7 \qquad\qquad\qquad = 100 - 38.7$$
$$= 138.7 \text{ or } 139 \qquad\qquad\quad = 61.3 \text{ or } 61$$

Thus, IQs of 139 or higher and 61 or lower are so extreme that they occur in only 1 percent of the population. Ninety-nine percent of the population, of course, has scores in between.

CLUE TO THE FUTURE

The idea of finding scores and proportions that are extreme in either direction will come up again after Chapter 6. In particular, the extreme 5 percent and the extreme 1 percent will be important.

23. What IQ scores are so extreme that they are achieved by only 5 percent of the population? Set this problem up using the "extreme 1 percent" example as a model.

24. What is the probability that a randomly selected person has an IQ higher than 129 *or* lower than 71?

25. Look at Figure 5.9 and suppose that the union leadership decided to ask for $0.85 per hour as a minimum wage. For those 185,822 workers, the mean was $0.99 with a standard deviation of $0.17. If $0.85 per hour was established as a minimum, how many workers would be affected?

26. Look at Figure 5.8 and suppose that a timber company decided to harvest all trees 8 inches DBH (diameter breast height) or larger from a 100-acre tract. On a 1-acre tract there were 199 trees with $\mu = 13.68$ and $\sigma = 4.83$. How many trees would be expected to be harvested from 100 acres? ■

COMPARISON OF THEORETICAL AND EMPIRICAL ANSWERS

You have been using the theoretical normal distribution to find probabilities and to calculate scores and proportions of IQs, diameters, wages, and other measures. Earlier in this chapter, the claim was made that *if* the empirical observations are distributed like a normal curve, accurate predictions can be made. A reasonable question is, How accurate are all these predictions I've just made? A reasonable answer can be fashioned from a comparison of the predicted proportions (from the theoretical curve) to the actual proportions (computed from empirical data). Figure 5.7 is based on 261 IQ scores of fifth-grade public school students. You worked through examples that produced proportions of people with IQs higher than 120, lower than 90, and between 90 and 110. These actual proportions can be compared with those predicted from the normal distribution. **Table 5.2** shows these comparisons.

As you can see by examining the Difference column of Table 5.2, the accuracy of the predictions ranges from excellent to not so good. Some of this variation can be explained by the fact that the mean IQ of the fifth-grade students was 101 and the standard deviation 13.4. Both the higher mean (101, compared with 100 for the normal curve) and the lower standard deviation (13.4, compared with 15) are due to the systematic exclusion of very-low-IQ children from regular public schools. Thus, the actual proportion of students with IQs lower than 90 is less than predicted, which is the result of the fact that our school sample is not representative of all 10- to 11-year-old children. This problem of sampling is the major topic of the next chapter.

Although IQ scores *are* distributed approximately normally, many other scores are not. Karl Pearson recognized this, as have others. In 1989 Theodore Micceri made this point again in an article titled, "The Unicorn, the Normal Curve, and Other Improbable Creatures." Caution is always in order when you are using theoretical distributions to

TABLE 5.2 Comparison of predicted and actual proportions

IQs	Predicted from normal curve	Calculated from actual data	Difference
Higher than 120	.0918	.0920	.0002
Lower than 90	.2514	.2069	.0445
Between 90 and 110	.4972	.5249	.0277

make predictions about empirical events. However, don't let undue caution prevent you from getting the additional understanding that statistics offers.

OTHER THEORETICAL DISTRIBUTIONS

In this chapter you learned a little about rectangular distributions and binomial distributions and quite a bit about normal distributions. Later in this book you will encounter other distributions such as the t distribution, the F distribution, and the chi square distribution. In addition to the distributions in this book, mathematical statisticians have identified yet others, all of which are useful in particular circumstances. Some have interesting names such as the Poisson distribution; others have complicated names (hypergeometric distribution is an example). In every case, however, a distribution is used because it provides reasonably accurate probabilities about particular events.

PROBLEMS

27. For human infants born weighing 5.5 pounds or more, the mean gestation period is 268 days, which is just less than 9 months. The standard deviation is 14 days (McKeown and Gibson, 1951). What proportion of the gestation periods would be expected to last 10 months or longer (300 days)?

28. The height of residential door openings in the United States is 6'8". Use the information in problem 17 to determine the number of males who have to duck to enter a room.

29. An imaginative anthropologist measured the stature of 100 hobbits (using the proper English measure of inches) and found the following:

$$\Sigma X = 3600 \qquad \Sigma X^2 = 130,000$$

Assume that the height of hobbits is normally distributed. Find μ and σ and answer the questions.

a. The Bilbo Baggins Award for Adventure is 32 inches tall. What proportion of the hobbit population is taller than the award?

    b. Three hundred hobbits entered a cave that had an entrance 39 inches high. The orcs chased them out. How many could exit without ducking?

    c. Gandalf is 46 inches tall. What is the probability that he is a hobbit?

30. Please review the objectives at the beginning of the chapter. Can you do what is asked?

6

Samples, Sampling Distributions, and Confidence Intervals

OBJECTIVES FOR CHAPTER 6

After studying the text and working the problems in this chapter, you should be able to:

1. Define a sampling distribution and a sampling distribution of the mean

2. Describe the Central Limit Theorem

3. Describe the effect of *N* (sample size) on the standard error of the mean

4. Use the *z*-score formula to find the probability that a sample mean or one more extreme was drawn from a population with a specified mean

5. Describe the concept of a confidence interval

6. Describe the *t* distribution

7. Know when to use the normal distribution and when to use the *t* distribution to calculate confidence intervals

8. Calculate confidence intervals using the normal distribution or the *t* distribution and write an interpretation

9. Define a random sample and obtain one if you are given a population

10. Define and identify biased sampling methods

This textbook is written with a particular audience in mind—students who want to understand statistical analyses and may also want to analyze data themselves. The book is not written for an audience that wants to understand the mathematics that statistical techniques are based on. The first part of *this* chapter, however, is about mathematical concepts created by mathematical statisticians. This material is included because I believe that for you to truly understand what is going on when data are analyzed with an inferential statistical test, you must understand these concepts.

Let me remind you about populations and samples. A population is all the numbers or measures of a specified group. A sample is a subset of that population. As for the relative importance of populations and samples, you already know that the important thing is the population. The *particular* aspect of the population that is of interest is often the mean. Thus, the parameter, μ, is what the researcher wants to know about.

However, for researchers, it is seldom possible to obtain data on an entire population. A common solution, which you are probably quite familiar with, is to use a sample as a convenient way to find out about an unmeasurable population. A sample can be obtained and from it a mean, $\bar{X}$, can be calculated. The sample mean, $\bar{X}$, gives you information regarding μ, which is what you are really interested in.

One question remains. How accurately does $\bar{X}$ estimate μ? Almost no one would expect $\bar{X}$ to be exactly the same as μ, but how close is it? A statistic, developed by mathematical statisticians, provides an answer to the question of the relationship of $\bar{X}$ to μ.

The statistic that you will learn to calculate and interpret is a **confidence interval**. Calculating a confidence interval requires a sample and a sampling distribution. Both samples and sampling distributions are important for *all* inferential statistics. The next section explains the mathematical idea of a sampling distribution.

*Be especially attentive to the concept of a sampling distribution. In inferential statistics, the idea of a **sampling distribution** is clearly the central concept. This book on "Tales of Distributions" is a book of tales of sampling distributions. Every chapter after this one is about some kind of sampling distribution; thus, you might consider this paragraph to be a superclue to the future.*

Please note the difference between the general idea of a sampling distribution and the specific idea of a sampling distribution of the mean. A sampling distribution is always for a particular statistic. Some important sampling distributions are: the sampling distribution of the correlation coefficient, the sampling distribution of the variance, and the sampling distribution of the mean. In this chapter, the topic is the sampling distribution of the mean.

SAMPLING DISTRIBUTION OF THE MEAN

To begin your education on sampling distributions, I will describe how I constructed an *empirical* sampling distribution of the mean.[1] A sampling distribution is always for a particular sample size. For my example, I used a sample size of eight.

I started with a clearly defined population for which μ can be found. (With μ known, the accuracy of the confidence interval technique can later be checked.) As a population, I used the numbers in **Table 6.1**.

T A B L E   6.1   20 numbers used as a population

9	7	11	13	10	8	10	6	8	8
10	12	9	10	5	8	9	6	10	11

[1] The actual sampling distributions you will use for problems in this text will be theoretical sampling distributions, which are based on mathematical formulas. For most people, however, an empirical sampling distribution, based on selecting actual numbers, provides a more comprehensible introduction.

PROBLEM

1. What is the mean, μ, and the standard deviation, σ, of the population of numbers in Table 6.1?

Once a population was obtained, I proceeded with the second step in constructing a sampling distribution of the mean. Using a technique called random sampling (to be explained in a section that follows), I drew samples (with $N = 8$) from the population. For my empirical sampling distribution, I obtained 200 samples.[2] As you might expect, this sampling was done by a computer program.

Besides drawing samples, the program also calculated the mean of each sample. As a last step, I constructed a frequency polygon from the 200 sample means. The result is **Figure 6.1**.

Notice in Figure 6.1 how nicely centered the sampling distribution is about the population mean of 9.00. Most of the sample means ($\bar{X}$'s) are fairly close to the population parameter, μ.

Thus, to characterize a sampling distribution of the mean:

1. Every sample is drawn randomly from a specified population.
2. The sample size (N) is the same for all samples.
3. The number of samples is very large.
4. The mean ($\bar{X}$) is calculated for each sample.
5. The sample means are arranged into a frequency distribution.

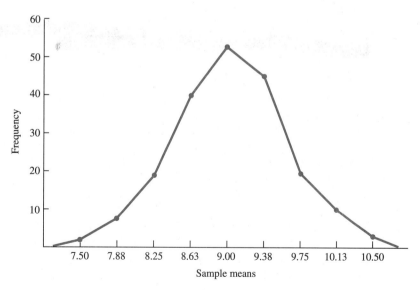

FIGURE 6.1 Empirical sampling distribution of the mean from population in Table 6.1. For each sample mean, $N = 8$.

[2] A theoretical sampling distribution includes *all* possible samples.

For sampling distributions of statistics other than the mean, you would follow these same steps except for the fourth step. There, you would calculate the appropriate sample statistic for the sampling distribution you are creating.

As further evidence of the importance of sampling distributions, statisticians use special names for their mean and standard deviation. The mean is called the **expected value** and the standard deviation is called the **standard error**.[3] This terminology is used for all sampling distributions and not just for the sampling distribution of the mean. I will have only a little more to say about expected value, but if you continue your study of statistics, you will encounter it again. The standard error, however, is used many times in this text.

I hope that when you looked at Figure 6.1, you were at least suspicious that it might be the ubiquitous normal curve. It is. Now you are in a position that educated people often find themselves: What you learned in the past (how to use the normal curve) can be used for a different problem—describing the relationship between $\bar{X}$ and μ.

Of course, the normal curve is a theoretical curve, and I presented you with an empirical curve that only appears normal. I would like to let you prove for yourself that the form of a sampling distribution of the mean is a normal curve, but, unfortunately, that requires mathematical sophistication beyond that assumed for this course. So I will resort to a time-honored teaching technique—an appeal to authority.

THE CENTRAL LIMIT THEOREM

The authority I appeal to is mathematical statistics, which has proved a theorem called the **Central Limit Theorem**. This important theorem says:

> For any population of scores, regardless of form, the sampling distribution of the mean will approach a normal distribution as N (sample size) gets larger. Furthermore, the sampling distribution of the mean will have a mean (the expected value) equal to μ and a standard deviation (the standard error) equal to $\sigma/\sqrt{N}$.

This appeal to authority resulted in a lot of information. Now you know not only that sampling distributions of the mean approach normal curves but also that, if you know the population parameters σ and μ, you can determine the parameters of the sampling distribution. The standard error of the sampling distribution is the standard deviation of the population (σ) divided by the square root of the sample size. The expected value of the sampling distribution will be the same as the population mean, μ. One final point: The standard error of the mean is symbolized $\sigma_{\bar{X}}$, and the expected value of the mean is symbolized $E(\bar{X})$.

The most remarkable thing about the Central Limit Theorem is that it works *regardless* of the form of the original population. The left part of **Figure 6.2** shows two populations you are familiar with. One is the rectangular distribution of playing cards (Figure 5.1), and the other is the bimodal distribution of choices of numbers between 1

---

[3] In this and in other statistical contexts, the term *error* means deviations or random variation. The word *error* is left over from the 19th century, when random variation was referred to as the "normal law of error." Of course, *error* sometimes also means mistake, so you will have to be alert to the context when this word appears (or you may make an error).

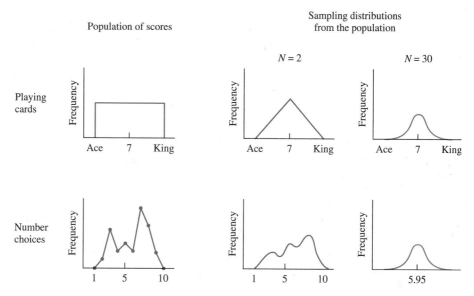

FIGURE 6.2 Populations of playing cards and number choices. Sampling distributions from each population with sample sizes of $N = 2$ and $N = 30$.

and 10 (Figure 5.5). The next two panels are sampling distributions for $N = 2$ and $N = 30$ that show the form of the distribution becoming more normal as N increases. Figure 6.2 illustrates that, regardless of the form of the population, the form of the sampling distribution of the mean approaches normal if N is large enough.

How large must N be for the sampling distribution of the mean to be normal? The traditional answer is 30 or more. However, if the population itself is symmetrical, sampling distributions of the mean will be normal with N's much smaller than 30. On the other hand, if the population is severely skewed, N's of more than 30 will be required.

Finally, the Central Limit Theorem does not apply to all sample statistics. Sampling distributions of the median, standard deviation, variance, and correlation coefficient are not normal distributions. The Central Limit Theorem does apply to the mean, though, which is a most important and popular statistic. (In a frequency count of all statistics, the mean is the mode.)

PROBLEMS

2. The standard deviation of a sampling distribution is called the _____ and the mean is called the _____.

3. a. Write the steps needed to construct an empirical sampling distribution of the range.

   b. What is the name of the standard deviation of this sampling distribution?

   c. Think about it, and then write something about the expected value of the range compared to the population range.
4. Describe the Central Limit Theorem in your own words.
5. In Chapter 4 you learned how to use a regression equation to predict a Y score if you are given an X score. That Y score is a statistic, so it has a sampling distribution with its own standard error. What could you conclude if the standard error were very small? Very large? ■

■ Calculating the Standard Error of the Mean

Calculating the standard error of the mean is fairly simple. I will illustrate with an example that we will use again. Recall that for the population in Table 6.1, you found σ equal to 2.0. For a sample size of eight, the standard error of the mean is

$$\sigma_{\bar{X}} = \frac{\sigma}{\sqrt{N}} = \frac{2.0}{\sqrt{8}} = \frac{2.0}{2.828} = 0.707$$

■ The Effect of Sample Size on the Standard Error of the Mean

As you can see by looking at the formula for the standard error of the mean, $\sigma_{\bar{X}}$ becomes smaller as N gets larger. **Figure 6.3** shows four sampling distributions of the mean, all based on the population of numbers in Table 6.1. The sample sizes are 2, 4, 8, and 16. A sample mean of 10 is included in all four figures as a reference point. Notice that, as N increases, a sample mean of 10 becomes less and less likely. The importance of sample size will become more apparent as this book progresses.

■ Determining Probabilities about Sample Means

To summarize where we are at this point: Mathematical statisticians have produced a mathematical invention, the sampling distribution. The sampling distribution of the mean, they tell us, is a normal curve. Fortunately for you, having worked problems in Chapter 5 about normally distributed *scores*, you are in a position to check this claim about normally distributed *means*.

    The check is fairly straightforward, given that we already have the 200 sample means from the population in Table 6.1. To make this check, we determine the proportion of sample means that are above a specified point, *using the theoretical normal curve* (Table C). We can then compare this theoretical proportion to the proportion of the 200 sample means that are *actually* above the specified point. If the two figures are similar, we have evidence that the normal curve can be used to answer questions about sample means.

    The z score for a sample mean drawn from a sampling distribution with a mean μ and a standard error $\sigma/\sqrt{N}$ is

$$z = \frac{\bar{X} - \mu}{\sigma_{\bar{X}}}$$

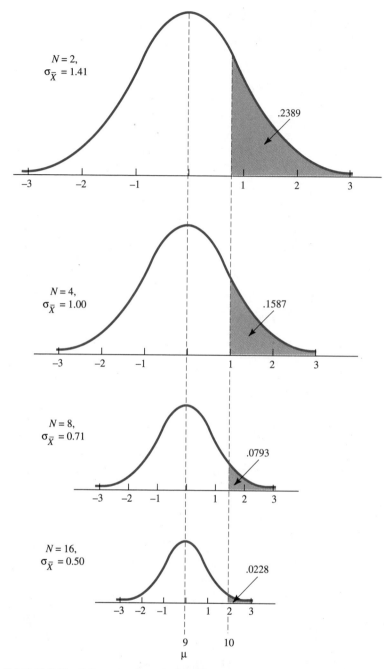

FIGURE 6.3 Sampling distributions of the mean for four different sample sizes. All samples are drawn from the population in Table 6.1. Note how a sample mean of 10 becomes rarer and rarer as $\sigma_{\bar{X}}$ becomes smaller.

Any sample mean will do for this comparison; I will use 10.0. The mean of the population is 9.0 and the standard error (for $N = 8$) is 0.707. Thus,

$$z = \frac{\bar{X} - \mu}{\sigma_{\bar{X}}} = \frac{10.0 - 9.0}{0.707} = 1.41$$

The proportion of the normal curve that is above a z value of 1.41 is .0793. When I looked at the sample means from which I constructed Figure 6.1, I found that 13 of 200 had means of 10 or more, a proportion of 0.650. Missing by less than $1\frac{1}{2}$ percent isn't bad. The normal curve model passes this test.[4]

PROBLEMS

*6. When the population parameters are known, the standard error of the mean is $\sigma_{\bar{X}} = \sigma/\sqrt{N}$. The following table gives four σ values and four N values. For each combination, calculate $\sigma_{\bar{X}}$ and enter it in the appropriate cell of the table.

		σ		
	1	2	4	8
N 1	1	2	4	8
4	0.5	1	2	4
16	0.25	0.5	1	2
64	0.125	0.25	0.5	1

7. On the basis of the table you constructed in problem 6, write a precise verbal statement about the relationship between $\sigma_{\bar{X}}$ and N.

8. In order to reduce $\sigma_{\bar{X}}$ to one-fourth its size, you must increase N by how much?

9. For the population in Table 6.1, and for samples with $N = 8$, what proportion of the sample means would be 8.5 or less?

10. For the population in Table 6.1, and for samples with $N = 16$, what proportion of the sample means would be 8.5 or less? 10 or greater?

11. As you know from the previous chapter, for IQs: $\mu = 100$ and $\sigma = 15$. What is the probability that a first-grade classroom of 25 students who were chosen randomly from the population would have a mean IQ of 105 or greater? 90 or less? ■

AN APPLICATION OF THE CENTRAL LIMIT THEOREM

Now, let's take our mathematical model into the world of data analysis to see how it is used to answer researchers' questions. Here's a summary of what the Central Limit

[4] As you have probably suspected, the validity of the normal curve model for the sampling distribution of the mean has been established with a mathematical proof rather than an empirical check.

Theorem requires if you are to use sample data and the normal curve to reach conclusions about a population.

1. To calculate a z score so you can enter the normal curve table, you must know σ, the standard deviation of the population.
2. To be certain that the normal curve will produce accurate probabilities about the population, the sample must be
   a. large
   b. drawn randomly from the population

Perhaps you have some reservations about how researchers can meet all three of these requirements. As the chapter progresses, I'll explain what to do if a particular requirement cannot be met. I'll begin with a problem for which all requirements can be satisfied.

Imagine a late-afternoon conversation between a sophomore and a junior at State University. The sophomore is musing, "I wonder what the average family income is for students on our campus? I'll bet it is less than the rest of the country."

"Well, the average family income for college students in the U.S. is $49,000," said the junior, after examining Astin (1995).[5] "Surely the campus average is that much."

"I don't think so," the sophomore replied. "I know lots of students who have only their own resources or come from pretty poor families. I'll bet you a dollar the mean for students here at State U. is below the national average."

"You're on," grinned the junior.

Together the two went out with a pencil and pad and asked 10 students how much their family income was. The mean of these 10 answers was $43,000.

Now the sophomore grinned. "I told you so; the mean here is $6000 less than the national average."

The disappointed junior immediately began to review their procedures and then responded, "Actually, this mean of $43,000 is meaningless—here's why. Those 10 students aren't representative of the entire student body. They're late-afternoon students, and lots of them support themselves with temporary jobs while they go to school. Most students are supported from home by parents who have permanent and better-paying jobs. This sample is no good. We need results for the whole campus or at least from a representative sample."

To get the results for the whole student body, the two went the next day to the director of the financial aid office, who, unfortunately, told them that family incomes for the student body are not public information.

The two were back to sampling. To find out how to obtain a representative sample, they got advice from a friendly statistics professor. Forty students, selected randomly, were identified. After three days of phone calls, visits, and call backs, the 40 replies produced a mean of $46,500.

"Pay," demanded the sophomore.

"OK OK, . . . here!"

At this point let's leave our two co-investigators, both of whom were thinking what you and other thoughtful people would think. How accurate *is* a random sample? Is this sample mean of $46,500 close to the State U. population mean, or could it be off by a

---

[5] The most recent figures on family income of college students are usually published in the *Chronicle of Higher Education* about the second week in January.

lot? Obviously, another random sample would produce a different mean. Could it be quite different?

Fortunately, calculating a *confidence interval* will reduce the uncertainty for the two students and others who understand statistics. However, short of measuring the entire population, you will be left with *some* uncertainty. **If you agree to use a sample, you agree to accept some uncertainty about the results**. One of the beauties of statistics is that it allows you to measure this uncertainty. If a great deal of uncertainty exists, the sensible thing to do is suspend judgment until you get more information. However, if there is very little uncertainty, the sensible thing to do is to reach a conclusion, even though there is a small risk of being wrong. *Reread this paragraph. It is important.*

CATEGORIES OF INFERENTIAL STATISTICS

Two of the major categories of inferential statistics are *hypothesis testing* and *estimation*. **Hypothesis testing**, which allows you to make a yes/no decision about a population, is the principal topic of the next two chapters.

The other kind of statistical inference, estimation, takes two forms: point estimation and confidence intervals. With **point estimation** one number is calculated that is the best estimator of the parameter of the population. A confidence interval is a range of values defined by a lower and an upper limit. The interval is expected, with a specified degree of confidence, to contain the parameter of the population.

CONFIDENCE INTERVALS

Suppose you draw a random sample of 30 scores from a population whose mean is unknown. The scores range from 60 to 90, and the sample mean is 76. Because $\bar{X}$ is the best estimator of the parameter, μ, you might just say, "I estimate that $\mu = 76$." (This is an example of a point estimate.) How confident would you be that μ is exactly 76? Not very, I hope. Although 76 is the best estimate, no one would be surprised to get a sample mean of 76 from a population with $\mu = 75$ (or 75.5, 75.8, and so on). Thus, you would not be very confident if you estimated μ with a specific point.

At the other extreme, how confident would you be if you said, "I estimate that the interval 60 to 90 contains μ"? Very, I imagine. For most purposes, however, such a wide interval would not be specific enough. The problem, as you may have surmised, is to trade off some of the specificity that goes with a point estimate for some of the confidence that comes with a wider interval (or, conversely, trade some of the confidence that goes with a wide interval for more specificity).

Fortunately, a sampling distribution of the mean can be used to establish both the interval and the confidence. You start by choosing the degree of confidence you want for your interval. The most common choice is 95 percent confidence. After calculating the lower and upper limits of the interval, you can state that you are 95 percent confident that *the interval contains the parameter.*

Here is the rationale for such confidence. Suppose you define a population of scores. A random sample is drawn and the mean ($\bar{X}$) is calculated. Using this sample mean and the formulas in the next section, you can calculate a confidence interval. (I will use a 95 percent confidence interval in this explanation.)

Next, suppose that from this population many more random samples are drawn, and for each sample mean a 95 percent confidence interval is calculated. For most of the samples, $\bar{X}$ will be close to μ and the lower and upper limits of the confidence interval will capture μ.

Occasionally, of course, a sample produces a $\bar{X}$ far from μ, and the confidence interval about $\bar{X}$ does not contain μ. The method of calculating confidence limits, however, allows you to hold these faulty intervals to some acceptable minimum, such as 5 percent. The result of all this is a method that produces confidence intervals, 95 percent of which contain μ.

With actual data, you have *one* sample and calculate *one* interval. You do not know whether or not the lower and upper limits capture μ, but the method you use makes you 95 percent confident that it does.

■ Calculating the Limits of a Confidence Interval Using the Normal Distribution

When the normal curve is appropriate, use these formulas for the lower and upper limits of a confidence interval about a population mean:

$$\text{LL} = \bar{X} - z(\sigma_{\bar{X}})$$

$$\text{UL} = \bar{X} + z(\sigma_{\bar{X}})$$

where $\bar{X}$ is the mean of the sample from the population

z is a z value from the normal curve

$\sigma_{\bar{X}}$ is the standard error of the mean, calculated by the formula:

$$\frac{\sigma}{\sqrt{N}}$$

where σ is the standard deviation of the population

In **Figure 6.4,** vertical lines separate the middle 95 percent of the sample means from the upper $2\frac{1}{2}$ percent and the lower $2\frac{1}{2}$ percent. To find the z scores that correspond to the 95 percent confidence interval, look in Table C for a z score that separates an upper $2\frac{1}{2}$ percent from the rest of the curve. That z score is 1.96. Of course, a z score of -1.96 separates the lower $2\frac{1}{2}$ percent of the curve. Thus, the z score in the formula of confidence interval limits is determined by the degree of confidence you want to have in the interval.

Although a 95 percent confidence interval is common, different situations call for different amounts of confidence. Sometimes 99 percent is needed, sometimes only 50 percent. The z values of ± 2.58 leave one-half of 1 percent in each tail and give you a 99 percent confidence interval; z values of ± 0.67 leave 25 percent in each tail and give you a 50 percent confidence interval. For other confidence intervals, find corresponding z values in Table C.

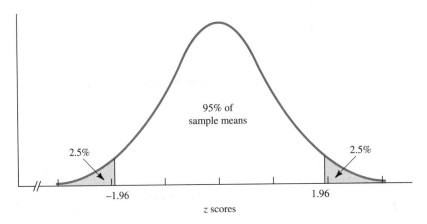

FIGURE 6.4 Sampling distributions of the mean when σ is known and sample size is adequate

Now we are ready to return to the two students who wondered what the mean family income was for their school and whether it was larger or smaller than the U.S. mean of $49,000. They had gone to the trouble of obtaining a random sample ($N = 40$) of the family incomes of State U. students, finding that their sample mean was $46,500.

If we calculate a 95 percent confidence interval about their sample mean, we will have 95 percent confidence that the interval has "captured" the parameter. In this case, the parameter is the mean family income for State U. students.

First, let's see if we have met the requirements for using the normal curve, as outlined on page 141. Although we don't know σ (the standard deviation of the population of State U. students), we find in Astin (1995), our source for the U.S. mean of $49,000, that σ is $12,000 for all students in the United States. We will accept $12,000 as the σ for State U. students. This will allow us to calculate $\sigma_{\bar{X}}$.

$$\sigma_{\bar{X}} = \frac{\sigma}{\sqrt{N}} = \frac{12,000}{\sqrt{40}} = 1897$$

The other requirement, a large random sample, has been met.

The lower and upper limits about the sample mean of $46,500 will be

$$LL = \bar{X} - z(\sigma_{\bar{X}}) = 46,500 - 1.96(1897) = \$42,782$$

$$UL = \bar{X} + z(\sigma_{\bar{X}}) = 46,500 + 1.96(1897) = \$50,218$$

Here's the interpretation. You are 95 percent confident that the interval $42,782 to $50,218 contains the mean family income for all students at State U. You can imagine that if the two students in our story had this confidence interval and understood what it meant, there might have been another conversation, with the junior speaking:

> "I'm sorry I paid you before I knew about confidence intervals. The confidence interval about our sample mean, which was calculated so it would be likely to contain our school's μ, *also* contains the U.S. μ. This seems to indicate that the usual fluctuation in samples *might* account for our sample mean being less than $49,000. If the State U. mean *is* actually $49,000, the confidence interval captured it."

"Yeah," said the sophomore, "but that *if* is a big one. What if the campus mean is actually $46,500. Then the sample we got was right on the nose. And after all, a sample mean is an *unbiased estimator* of the population mean."

"I see your point. And I see mine, too. It seems like either of us could be correct. That leaves me uncertain about whether or not the campus mean is below the national mean."

"Me, too. Well, come on. I'll buy you a cup of coffee with my winnings."

Not all statistical stories end with so much uncertainty. However, I said that one of the advantages of sampling is that you can measure uncertainty. You measured it, and in this case there was a lot. Remember, if you agree to use a sample, you agree to accept some uncertainty about the results.

Often, there is a practical way to reduce uncertainty: Increase sample size. You have already worked some problems that revealed that a fourfold increase in N reduces the standard error by half. Applying this principle to our story, suppose the two students made the effort to get a random sample of 400 students, and again the sample produced a mean of $46,500. Now $\sigma_{\bar{X}}$ becomes

$$\sigma_{\bar{X}} = \frac{\sigma}{\sqrt{N}} = \frac{12,000}{\sqrt{400}} = 600$$

and the confidence interval becomes

$$LL = \bar{X} - z(\sigma_{\bar{X}}) = 46,500 - 1.96(600) = \$45,324$$

$$UL = \bar{X} + z(\sigma_{\bar{X}}) = 46,500 + 1.96(600) = \$47,676$$

The first part of the interpretation should have a familiar ring. The degree of your confidence is 95 percent that the mean family income for students at State U. is between $45,324 and $47,676. With these data, the mean family income for students in the U.S. ($49,000) falls well *outside* the confidence interval. This means that we have more than 95 percent confidence that the mean family income for students at State U. is *below* the national mean.

ERROR DETECTION

The *sample* mean is always exactly in the middle of the confidence interval. The interpretation is based on whether or not the interval captures a particular *population* mean.

One other caution about sample size is in order. Increasing sample size does not automatically guarantee a reduction in uncertainty. You know that any new sample will probably produce a new mean. For example, the sample with $N = 400$ might have produced a mean of $48,300.

PROBLEMS

12. Calculate a 95 percent confidence interval about a State U. sample mean of $48,300 with $N = 400$ and $\sigma = \$12,000$. Write a conclusion about the relationship between the campus μ and the U.S. mean of $49,000.

13. Two categories of inferential statistics are _____ and _____.

14. Will a 99 percent confidence interval be wider or narrower than a 95 percent confidence interval?

15. Calculate a 99 percent confidence interval about a State U. sample mean of $46,500 with $N = 400$ and $\sigma = \$12,000$. Write a conclusion about the relationship between the campus μ and the U.S. mean of $49,000.

16. What z values are needed for confidence intervals of 90 percent? 98 percent?

RANDOM SAMPLES

When it comes to finding out about a population, the best sample is a random sample. In statistics, random refers to the method used to obtain the sample; it does not mean haphazard or unplanned. Any method that allows every possible sample of size N an equal chance to be selected produces a **random sample**. To obtain a random sample, you must do the following:

1. Define the population of numbers (scores).
2. Identify every member of the population.
3. Select numbers in such a way that every sample has an equal probability of being selected.

To illustrate, I will use the population of numbers in Table 6.1. This allows us to satisfy requirements 1 and 2. Here are two methods of selecting numbers so that all the possible samples have an equal probability of being selected. For this illustration, $N = 7$.

One method of getting a random sample is to write each number in the population on a slip of paper, put the 20 slips in a box, jumble them around, and draw out seven slips. The numbers on the slips are a random sample. This method works fine if the slips are all the same size, they are jumbled thoroughly, and there are only a few members of the population. If the population is large, this method is tedious.

A second (usually easier) method of selecting a random sample is to use a table of random numbers, such as Table B in Appendix B. To use the table, you must first assign an identifying number to each of the population numbers, as I did in **Table 6.2**. The population numbers do not have to be arranged in any order.

Each population number is identified by a two-digit number. Now turn to **Table B** and pick an intersection of a row and a column. Any haphazard method will work; close your eyes and stab a place with your finger. Suppose you found yourself at row 80, columns 15–19 (page 367). Find that place. Reading horizontally, you find the digits 82279. You need only two digits to identify any member of our population, so you might as well use the first two, 8 and 2, which are in columns 15 and 16. Unfortunately,

TABLE 6.2 **Assignment of identifying numbers to a population of numbers**

ID number	Population number	ID number	Population number
01	9	11	10
02	7	12	12
03	11	13	9
04	13	14	10
05	10	15	5
06	8	16	8
07	10	17	9
08	6	18	6
09	8	19	10
10	8	20	11

82 is larger than any of the identifying numbers, so it won't produce a number for the sample, but at least you are started. From this point you can read two-digit numbers in any direction—up, down, or sideways—but the decision should be made before you look at the numbers. If you decide to read down, you find 04, which corresponds to population number 13, so 13 becomes the first number in the sample. The next identifying number is 34, which again does not correspond to a population number. Indeed, the next ten numbers are too large. The next usable ID number is 16, which places an 8 in the sample. Continuing the search, we reach the bottom of the table. At this point you can go in any direction; I moved to the right and started back up the two outside columns (18 and 19). The first number, 83, was too large, but the next identifying number 06 corresponded to an 8, which went into the sample. Next, a 15 and a 20 identified population numbers of 5 and 11 for the sample. The next number that is between 01 and 20 is 15, but it has already been used, so it should be ignored. Identifying numbers of 18 and 02 produce population numbers of 6 and 7, which completes the sampling task. Thus, the random sample of seven consists of the following numbers: 13, 8, 8, 5, 11, 6, and 7.

PROBLEMS

*17. A random sample is supposed to yield a statistic similar to the population parameter. Find the mean of the random sample of seven selected by your text.

18. Calculate a 95 percent confidence interval for the sample mean you found in problem 17 and write an interpretation. ■

What is this table of random numbers? In Table B (and in any set of random numbers), the probability of occurrence of any digit from 0 to 9 at any place in the table is the same: .10. Thus, you are just as likely to find 000 as 123 or 397. Incidentally, you cannot generate random numbers out of your head. Certain sequences begin to recur, and (unless warned) you will not include enough repetitions like 000 and 555. If warned, you will produce too many.

Here are some suggestions for using a table of random numbers efficiently.

1. Make a check beside the identifying number of a score when it is chosen for the sample. This helps to prevent duplications.
2. If the population is large (more than 100), it is more efficient to get all the identifying numbers from the table first. As you select them, put them in some rough order to help prevent duplications. After you have all the identifying numbers, go to the population to select the sample.
3. If the population has exactly 100 members, let 00 be the identifying number for 100. By doing this, you can use two-digit identifying numbers, each one of which matches a population score. This same technique works for populations of 10 or 1000 members.

PROBLEMS

19. Draw a random sample of 10 from the population in Table 6.1.
20. Draw a random sample with $N = 12$ from the following scores:

76	47	81	70	67	80	64	57	76	81
68	76	79	50	89	42	67	77	80	71
91	72	64	59	76	83	72	63	69	
78	90	46	61	74	74	74	69	83	

BIASED SAMPLES

A **biased sample** is one obtained by a method that systematically underselects or overselects from certain groups within the population. Thus, in a biased sampling technique, every sample of a given size does *not* have an equal opportunity of being selected. With biased sampling techniques, you are much more likely to get a nonrepresentative sample than you are with random sampling.

For example, it is reasonable to conclude that some results based on mailed questionnaires are not valid because the samples are biased. Usually an investigator defines the population, identifies each member, and mails the questionnaire to a randomly selected sample. Suppose that 70 percent of the recipients respond. Can valid results for the population be based on the questionnaires returned? Probably not. There is good reason to suspect that the 70 percent who responded are different from the 30 percent who did not. Thus, although the population is made up of both kinds of people, the sample reflects only one kind. Therefore, the sample is biased. The probability of bias is particularly high if the questionnaire elicits feelings of pride or despair or disgust or apathy in *some* of the recipients.

PROBLEMS

21. Suppose a large number of questionnaires about educational accomplishments are mailed out. Do you think that some recipients will be more likely to return the questionnaire than others? Which ones? If the sample is biased, will it overestimate or underestimate the educational accomplishments of the population?

22. Sometimes newspapers sample opinions of the local population by printing a "ballot" and asking readers to mark it and mail it in. Evaluate such a sampling technique. ■

A very famous case of a biased sample occurred in a poll that was to predict the results of the 1936 election for president of the United States. The *Literary Digest* (a popular magazine) mailed 10 million "ballots" to those on their master mailing list, a list of more than 10 million people compiled from "all telephone books in the U.S., rosters of clubs, lists of registered voters," and other sources. More than 2 million "ballots" were returned and the prediction was clear: Alf Landon by a landslide over Franklin Roosevelt. As you may have learned, the actual results were just the opposite; Roosevelt got 61 percent of the vote.

From the 10 million who had a chance to express a preference, 2 million very interested persons had selected themselves. This 2 million had more than its proportional share of those who were disgruntled with Roosevelt's depression-era programs. The 2 million ballots were a biased sample; the results were not representative of the population. In fairness, it should be noted that the *Literary Digest* used a similar master list in 1932 and predicted the popular vote within 1 percentage point.[6]

Obviously, researchers want to avoid biased samples; random sampling seems like the appropriate solution. The problem, however, is step 2 on page 141. For almost all research problems, it is impossible to identify every member of the population. Thus, researchers are confronted with somewhat of a dilemma.

RESEARCH SAMPLES

Researchers want to avoid a biased sample, but they know that, for most research problems, it is impossible to obtain a random sample. Populations that researchers are interested in, such as college students, newborn infants, and people with passive-aggressive personality disorders, just cannot be sampled from randomly.[7] This means that a nonrandom sample will have to do if research is to be conducted.

Researchers have a number of ways of maintaining respect for their work even though they cannot use random samples. The best of these ways is to use *random assignment* of participants to groups. With random assignment, the available participants are partitioned into groups using a table of random numbers, which

[6] See the *Literary Digest*, August 22, 1936, and November 14, 1936.

[7] Technically, the population consists of the measurements of the members and not the members themselves.

creates groups that are equivalent at the start of an experiment. Samples that have been randomly assigned have many of the same characteristics as random samples.

In addition, researchers who are aware of the problems of sampling may develop and then describe their efforts to avoid a biased sample. Also, one experiment does not stand alone. Other researchers are conducting similar investigations. Usually, the results based on research samples are true for other samples from the same population.

The fact that a conclusion is not based on a random sample should not alarm you. Nonrandom (though carefully controlled) samples are used by a very wide variety of people and organizations. The public opinion polls reported by Gallup, Lou Harris, and the Roper organization are based on carefully selected samples. ASCAP, the musicians' union, samples the broadcasts of more than 9000 U.S. radio stations and then collects royalties for members based on sample results. Inventory procedures in large retail organizations involve sampling. What you know about your blood is based on the analysis of a small sample. (Thank goodness!)

The reason for this widespread use of sampling is that it works; you can find out about the whole thing by examining just a little of it. Thus, although the sampling distributions you will be using are all based on random samples, they are quite useful for research samples, which are usually not random samples.

A SAMPLING DISTRIBUTION WHEN σ IS UNKNOWN

Having admitted that researchers cannot meet the "requirement" that their samples be randomly drawn from the populations they are interested in, let's address another problem—knowing σ. Except in fields that have been heavily researched, you seldom know σ. Without σ, you cannot calculate $\sigma/\sqrt{N}$, which you need for the formulas for confidence limits. What can you do if you don't have σ? Do you have a suggestion?

One fairly common suggestion is to use $\hat{s}$ in place of σ. (Was this your solution?) Substituting $\hat{s}$ into the formula for the standard error is a solution that was used in the early part of the 20th century. Unfortunately, $\hat{s}$ is a statistic, so its value varies from sample to sample. Thus, a sampling distribution based on $\hat{s}$ is just one of many possible sampling distributions. How close is it to the sampling distribution based on σ? You see the problem, I'm sure.

Early researchers (around 1900) dealt with the question of the accuracy of $\hat{s}$ by using huge samples, which assured them (and us) that any errors would be negligible. Remember Karl Pearson and Alice Lee's data on 1376 pairs of father–daughter heights? That's an example of a huge sample that will produce an $\hat{s}$ identical to σ, for all practical purposes.

Other researchers, however, could not gather that much data. One of those was W. S. Gosset (1876–1937), who worked for Arthur Guinness, Son & Co., a brewery in Dublin, Ireland. Gosset had studied chemistry and mathematics at Oxford, and his job at the Guinness company was to make scientifically based recommendations about brewing. Many of these recommendations were based on sample data.

Gosset was familiar with the normal curve and the strategy of using large samples to accurately estimate σ. Unfortunately, though, the samples he had were small, and

Gosset recognized that small sample $\hat{s}$ values were not accurate estimators of σ, and thus the normal curve could not be relied on for accurate probabilities.

Gosset is remembered today because he worked out the mathematics of curves based on $\hat{s}$. He confirmed that the form of this sampling distribution was not normal and that it depended on sample size, with a different distribution for each N. These distributions make up a family of curves that have come to be called the *t* **distribution**.[8]

THE *t* DISTRIBUTION

The solution to the problem of what you do when you don't know σ is to use the *t* distribution instead of the normal distribution. That is, the sampling distribution of the mean, using $\hat{s}$ as an estimator of σ, is a *t* distribution. As it turns out, other statistics are distributed as *t*. In this text, the *t* distribution is used for confidence intervals and four other applications.[9]

The different *t* distributions are identified by their **degrees of freedom**. There is a different *t* distribution for each degree of freedom from 1 to ∞. Determining the proper degrees of freedom (abbreviated *df*) for a particular problem can become fairly complex. For the problems in this chapter, however, the formula is simple: $df = N - 1$. Thus, if the sample consists of 12 members, $df = 11$. In later chapters I will give you a more thorough explanation of degrees of freedom (and additional formulas). **Figure 6.5** is a picture of three *t* distributions, each based on a different number of degrees of freedom.

■ The *t* Distribution and the Table of *t*

Figure 6.4 (page 144) illustrated a concept that is common in inferential statistics— dividing a sampling distribution into a 95 percent portion and a 5 percent portion. This

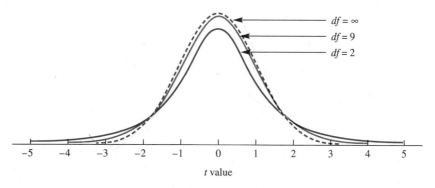

F I G U R E 6.5 Three different *t* distributions

[8] Gosset spent several months in 1906–07 studying with Karl Pearson in London. During this period, the *t* distribution was developed.

[9] Traditionally, the *t* distribution is written with a lowercase *t*. A capital *T* is used for another distribution (see Chapter 13) and as a standardized test score. However, because some computer programs do not print lowercase letters, *t* becomes *T* on some printouts (and often in text based on that printout). Be alert.

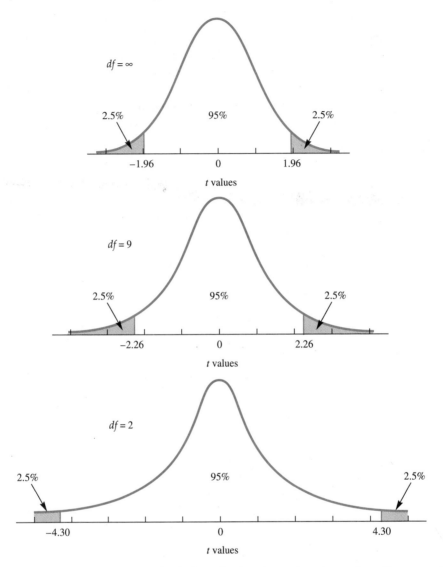

FIGURE 6.6 Three t distributions showing the t values that enclose 95 percent of the cases

same idea is shown in Figure 6.6, which separates the three t distributions of Figure 6.5. Each curve shows the points on the X axis that divide the middle 95 percent from the extreme 5 percent, half of which (2.5 percent) is in each tail. Examine **Figure 6.6** closely.

Note in Figure 6.6 that the greater the degrees of freedom, the more compact the curve. Perhaps when you studied the curves it occurred to you that the t distribution for $df = \infty$ might be a normal curve. It is. As df approaches ∞, the t distribution approaches the normal distribution.

Perhaps you also wondered where those values of ± 4.30, ± 2.26, and ± 1.96 came from. The answer is in Table D in Appendix B. Table D is really a partial version of 34 t distributions. Turn to **Table D** and note that there are 34 different degrees of freedom in the left-hand column. Across the top under "Confidence interval percents," you see six percent values that are commonly used when confidence intervals are calculated. Follow the 95 percent column down to $df = 2, 9$, and ∞, and you will find the values of 4.30, 2.26 and 1.96—the same values seen in Figure 6.6.

Table D differs in several ways from the normal curve table you have been using. In the normal curve table, the z scores are in the margin and the probability figures are in the body of the table. In Table D, the t values are in the *body* of the table and the probability figures are in the *headings* at the top. In the normal curve table there are hundreds of probability figures, and in Table D there are only six. These six are the ones commonly used by researchers who are calculating confidence intervals or testing hypotheses.[10]

Finally, it sometimes happens that you have a problem with degrees of freedom intermediate between two dfs given in Table D (for example, 37 df). In intermediate cases, be conservative and use the t distribution with *fewer* degrees of freedom (i.e., 30 df).

◼ Calculating the Limits of a Confidence Interval Using the t Distribution

To find the lower and upper limits of a confidence interval about a population mean when using the t distribution, use the formulas

$$LL = \bar{X} - t(s_{\bar{X}})$$

$$UL = \bar{X} + t(s_{\bar{X}})$$

where $\bar{X}$ is the mean of the sample from the population
t is a value from the t distribution table
$s_{\bar{X}}$ is the standard error of the mean

The formula for $s_{\bar{X}}$ will be quite familiar to you:

$$s_{\bar{X}} = \frac{\hat{s}}{\sqrt{N}}$$

where $\hat{s}$ is the standard deviation of the sample (using $N - 1$)

A confidence interval is useful for evaluating quantitative claims such as those made by manufacturers. For example, some light bulb manufacturers claim that their 60-watt bulbs will last for 1000 hours. On the average, is this true? One way to answer the question is to gather data on the number of hours that 60-watt bulbs burn. The following are fictitious data:

$$N = 16$$

$$\Sigma X = 15,600$$

$$\Sigma X^2 = 15,247,500$$

---

[10] As you examined Table D, you may have wondered about the other two headings across the top of the table. Those will be used beginning in the next chapter.

PROBLEM

*23. For the data on the lives of 60-watt light bulbs, what is the mean ($\bar{X}$), the standard deviation ($\hat{s}$), and the standard error of the mean ($s_{\bar{X}}$)? ■

I will calculate a 95 percent confidence interval about this sample mean and use it to evaluate the claim that bulbs last 1000 hours. In situations such as advertisement claims, there is no hope of knowing σ, the population standard deviation, so $\hat{s}$ and the t distribution are appropriate. With $N = 16$, $df = N - 1 = 15$. The t value in Table D for a 95 percent confidence interval with $df = 15$ is 2.131. Using the answers you found in problem 23,

$$\text{LL} = \bar{X} - t(s_{\bar{X}}) = 975 - 2.131(12.5) = 923.36 \text{ hours}$$

$$\text{UL} = \bar{X} + t(s_{\bar{X}}) = 975 + 2.131(12.5) = 1001.64 \text{ hours}$$

Thus, you are 95 percent confident that the interval of 923.36 to 1001.64 hours contains the true mean lighting time of 60-watt bulbs. The advertised claim of 1000 hours *is* in the interval, but just barely. A reasonable interpretation is to recognize that samples are changeable, variable things and that you do not have good evidence showing that the lives of bulbs are shorter than claimed. But because the interval *almost* does not capture the advertised claim, you might have an incentive to gather more data.

To obtain a 99 percent or a 90 percent confidence interval, use the t values under those headings in Table D.

As another example, normal body temperature is commonly given as 98.6°F. In Chapters 2 and 3 you were given some data based on actual measurements (Mackowiak, Wasserman, and Levine, 1992) that showed a mean temperature of *less* than 98.6°F. Have we been wrong all these years?

Perhaps the study's sample mean is within the expected range of samples taken from a population with a mean of 98.6°F. A 99 percent confidence interval will help in choosing between these two alternatives. The summary statistics were:

$$N = 41$$
$$\Sigma X = 4026.2$$
$$\Sigma X^2 = 395,393.36$$

These summary statistics produce the following:

$$\bar{X} = 98.2$$
$$\hat{s} = 0.716$$
$$s_{\bar{X}} = 0.112$$
$$df = 40$$

The next task is to find the appropriate t value in Table D. Look under the heading for a 99 percent confidence interval for 40 df and find 2.704. Calculating 99 percent confidence limits about the sample mean:

$$LL = \bar{X} - t(s_{\bar{X}}) = 98.2 - 2.704(0.112) = 97.9°F$$

$$UL = \bar{X} + t(s_{\bar{X}}) = 98.2 + 2.704(0.112) = 98.5°F$$

What conclusion do you reach about the 98.6°F standard? We have been wrong all these years. On the basis of actual measurements, we are 99 percent confident that normal body temperature is between 97.9°F and 98.5°F.[11]

In summary, confidence intervals for population means produce an interval statistic (lower and upper limits) that is destined to contain μ 95 percent of the time (or 99 or 90). Whether or not a particular confidence interval contains the unknowable μ is uncertain, but you have control over the degree of that uncertainty.

The concept of confidence intervals has been known to be troublesome. Here is another explanation of confidence intervals, this time with a picture.

Look at **Figure 6.7**, which shows

1. A population of scores (top curve)[12]
2. A sampling distribution of the mean when $N = 25$ (small curve)
3. Twenty 95 percent confidence intervals based on random samples ($N = 25$) from the population

On each of the 20 horizontal lines, the endpoints represent the lower and upper limits, and the filled circle shows the mean. As you can see, nearly all the confidence intervals have "captured" μ. One has not. Find it. Does it make sense to you that 1 out of 20 of the 95 percent confidence intervals would not contain μ?

WHEN TO USE THE NORMAL CURVE
AND WHEN TO USE THE t DISTRIBUTION

Both the normal distribution and the t distribution are sampling distributions that are used to give you probabilities about sample statistics. In this chapter, the statistics were the sample mean and the confidence interval, but both distributions are also used for other statistics, too. When the statistics in question are the sample mean or the confidence interval, use the normal curve if you know σ and if the sample size is adequate. The traditional definition of adequate, remember, is 30 or more, although if the population is symmetrical, sample sizes much smaller than 30 produce approximately normal sampling distributions.

In most research situations, however, σ is not known, and often sample sizes are small. As a consequence, the t distribution is the one usually found in research articles.

---

[11] C. R. A. Wunderlich, a German investigator in the mid-19th century, was influential in establishing 98.6°F as the normal body temperature.

[12] It is not necessary that the population be normal.

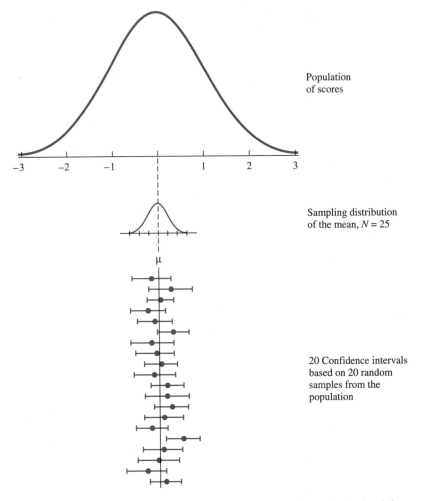

Population of scores

Sampling distribution of the mean, $N = 25$

μ

20 Confidence intervals based on 20 random samples from the population

FIGURE 6.7 Twenty 95 percent confidence intervals, each calculated from a random sample of 25 from a normal population. Each sample mean is represented by a dot.

PROBLEMS

24. When is the normal curve the appropriate sampling distribution for constructing confidence intervals and when is the t distribution?

25. A social worker conducted an 8-week assertiveness training workshop. Afterward, the 14 clients took the Door Manifest Assertiveness Test, the national mean of which is 24.0. Use the data that follow to construct a 95 percent confidence interval

about the sample mean. Write an interpretation about the effectiveness of the workshop.

$$\Sigma X = 378 \qquad \Sigma X^2 = 10,400$$

26. This problem gives you practice in constructing confidence intervals and in discovering how to reduce their size (without sacrificing confidence). Smaller confidence intervals tell you the value of μ more precisely.
    a. How wide is the 95 percent confidence interval about a sample mean when $\hat{s} = 4$ and $N = 4$?
    b. How much smaller is the 95 percent confidence interval about a sample mean when N is increased fourfold to 16? ($\hat{s} = 4$)?
    c. Compared to your answer in part a, how much smaller is the 95 percent confidence interval about a sample mean when N remains 4 and $\hat{s}$ is reduced to 2?

27. In Figure 6.7, the 20 lines that represent confidence intervals vary in length. What aspect of the sample causes this variation?

28. Using Figure 6.7 as a model of 95 percent confidence intervals, imagine a graphic with confidence intervals based on $N = 100$.
    a. How many of the 20 intervals would be expected to capture μ?
    b. Would the confidence intervals be wider or narrower? By how much?

29. Using Figure 6.7 as a model, imagine that the lines are 90 percent confidence intervals rather than 95 percent confidence intervals.
    a. How many of the 20 intervals would be expected to capture μ?
    b. Would the confidence intervals be wider or narrower?

30. In the following figure, look at the curve of performance over trials, which is generally upward except for the dip at trial 2. Is the dip just a chance variation, or is it a reliable, nonchance drop? By indicating the lower and upper confidence limits for each mean, a graph can convey at a glance the reliability that goes with each point. Calculate a 95 percent confidence interval for each mean. Pencil in the lower

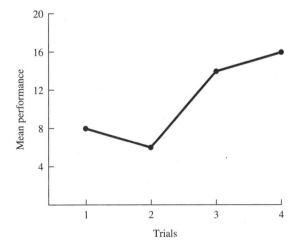

and upper limits for each point on the graph. Write a sentence explaining the dip at trial 2.

	Trials			
	1	2	3	4
$\bar{X}$	8	6	14	16
$\hat{s}$	5	5	5	5
N	25	25	25	25

31. Here's a problem you probably were expecting. Review the objectives at the beginning of the chapter. You *are* using this valuable, memory-consolidating technique, aren't you?   ■

You are in the midst of material on inferential statistics, which, as you probably recall, is a technique that allows you to use sample data to help decide about populations. In the last chapter you studied and worked problems on *confidence intervals*. A confidence interval allows you to infer something about the size of a parameter (the mean) of an unmeasured population.

The next chapter, Chapter 7, covers the basics of *hypothesis testing*, a more widely used inferential statistics technique. Hypothesis testing results in a yes/no decision about a population parameter. The decision about the population parameter is based on a probability that is obtained from a sampling distribution. So far, you have worked with two sampling distributions, the normal distribution and the *t* distribution.

In Chapter 7 you will test hypotheses about population means and population correlation coefficients. In Chapter 8 you will test hypotheses about the difference between the means of two populations. In both of these two chapters you will determine probabilities with a *t* distribution.

In addition to hypothesis testing, the next two chapters will introduce the *effect size index*, a measure of the amount of difference a variable causes.

Hypothesis Testing and Effect Size: One-Sample Tests

Let's start with a researcher who has a belief about how things are out there in the state of nature. This belief is often expressed as a comparison, such as: Group A is better than the national average, or The effect of this treatment is to reduce time in therapy.

Hypothesis testing is used by researchers to gather support for beliefs that involve a comparison. It is an objective method that allows a researcher to reach conclusions such as "better," "less time," or "more than," *if the data come out according to the researcher's expectations*. For this example, let's say our researcher thinks that the relationship is "more than."

Here's an introduction to the logic of hypothesis testing. Please attend closely. To begin, the researcher says, "Suppose I'm wrong, and the relationship is really 'equal' rather than 'more than.' If the relationship is actually 'equal,' then the data I gather will, of course, be in the realm of what is expected if the relationship is 'equal'; I wouldn't be surprised to get data showing 'a little more than' or 'a little less than.' Of course, if the relationship is actually 'equal,' I would *not* expect to get data showing 'much more than.'

"On the other hand, if I'm right and the relationship is 'more than,' then the data will certainly show 'more than' rather than 'equal' or 'less than.' And, if the data show 'much more than,' then I think I'd be justified in saying that the 'equal' hypothesis is false. In this case, the data support the idea I had from the beginning: The relationship is 'more than.'"

In hypothesis testing, the hypothesis being tested is a hypothesis about equality. The researcher thinks that the equality hypothesis is not true, and by showing that the data do not fit it, the equality hypothesis can be rejected.

The end point of hypothesis testing is a decision. Decision makers in fields that range from anthropology and business to zoology and zymurgy use hypothesis testing because it gives them an objective way to determine if a relationship is "more than" or "less than." Of course, hypothesis-testing decisions are based on data from samples, so some uncertainty always goes with the decision.

Every chapter from this point on assumes that you understand the concepts in hypothesis testing, so I will flesh out these ideas with care as the chapter progresses. In all, this book covers 14 different statistical tests that use hypothesis-testing logic. All of these statistical tests are used to reach conclusions about populations by using data from samples. The first of these tests allows you to test a hypothesis about a population mean, μ_0, by finding the probability that sample data with a mean $\bar{X}$ came from the population with the mean μ_0.

HYPOTHESIS TESTING—AN EXAMPLE PROBLEM

Determining whether a sample with a mean $\bar{X}$ came from a population with a mean μ_0 is useful for evaluating quantitative claims, such as that of the Frito-Lay® corporation, which claims that their bags of Doritos® tortilla chips contain 269.3 grams. In this case, the equality hypothesis is that $\mu_0 = 269.3$ grams. What do you think of such a claim? When you buy such a bag, how much do you get?

Of course, you could weigh one bag and obtain an exact answer for that one bag. Suppose the bag weighed less than 269.3 grams. The conclusion that this observation supports, "*This bag* has less than claimed," is not very far-reaching. A conclusion about the *population* of actual weights would be much more interesting.

The population of weights of Doritos bags does have a mean, which I will symbolize as μ_1. Of course, it would be impossible to know what this mean, μ_1, is, so sampling is in order. By using a hypothesis-testing technique called the *t* test, you can reach a conclusion about μ_1, the population of actual weights.

The result of hypothesis testing is not an exact value for μ_1 or even a confidence interval that captures μ_1. Rather, the result of hypothesis testing is a description of the *relationship* of μ_1 to μ_0. Here are the three possible relationships between μ_1 and μ_0, followed by what the relationships would mean in the Doritos example.

$\mu_1 = \mu_0$. The contents of Doritos bags are about what would be expected if the company is trying to put 269.3 grams into each package.

$\mu_1 < \mu_0$. The contents of Doritos bags weigh less than the company claims.

$\mu_1 > \mu_0$. The contents of Doritos bags weigh more than the company claims.

To reach a decision about μ_1 (and to have an example for this chapter), I weighed the contents of eight packages of Doritos tortillas (Nacho cheese flavor) on a scale in the lab. Some summary statistics were:

$N = 8$

$\Sigma X = 2165.4$ grams

$\Sigma X^2 = 586,121.56$ grams

The first thing to do with summary statistics is to calculate descriptive statistics to see what they tell you. Stop and decide which descriptive statistics would be helpful. What would you expect to find out from them?

As for the descriptive statistics:

$\bar{X} = 270.675$ grams

$\hat{s} = 0.523$ gram

Most important, the sample mean is *more than* the company claim of 269.3 grams. Also, note that the standard deviation about the sample mean is quite small.

The next step for researchers would be to use these sample statistics to reach a conclusion about the population of actual weights. For you to be in a position to take this step, you need a more thorough understanding of hypothesis testing, including some conventions that researchers use.

THE LOGIC OF HYPOTHESIS TESTING

This section presents logic and terms that will be important throughout the remainder of this book. Note taking, including questions for which you want answers, is probably a good idea here.

You use hypothesis testing when you want to know a parameter (such as the mean) of a population that cannot be measured. You can, however, gather data with a sample from that population and calculate a statistic (such as the mean).

The basic logic of hypothesis testing is to set up two hypotheses that cover all possibilities about the parameter. The first is a hypothesis of equality that states that the population parameter is exactly a specified value. The second hypothesis is that the population parameter is *not* the specified value. You can see that logically one of these hypotheses *must* be true. *Tentatively* assume that the equality hypothesis is correct. Using a sampling distribution that will give accurate probabilities if the equality hypothesis is correct, find the probability of obtaining the statistic that you actually found in your representative sample.

If the probability is very small, give up your tentative assumption that the equality hypothesis is correct. Giving up the equality hypothesis leaves you with the other hypothesis, which you then conclude is correct. To restate this: If you can show that the data that were actually obtained are unlikely when the equality hypothesis is true, you should reject the equality hypothesis.

If the probability is large, then the sample data are consistent with the equality hypothesis. Unfortunately, such data are *also* consistent with the other hypothesis. Given this state of affairs, the data are not helpful in eliminating one of the hypotheses.

Now I want to expand this explanation by giving you the statistical terms that are used in hypothesis testing, working with the Doritos data as a concrete example.

1. Begin with a population mean you want to know about and data from a sample from the population. For the Doritos example, the population mean is that of *all* bags; the sample mean ($N = 8$) was 270.675 grams.

2. Recognize two hypotheses: A hypothesis of equality, called the **null hypothesis** and symbolized H_0 ("H sub oh" or "H sub zero"), and an **alternative hypothesis**, symbolized H_1 ("H sub one"):

    H_0: The mean weight of Doritos bags is 269.3 grams, the weight claimed by the Frito-Lay company.

    H_1: The mean weight of Doritos bags is not 269.3 grams.

3. Tentatively assume that H_0, the null hypothesis, is true. When the null hypothesis is true, whatever difference there is between the population mean and the sample mean is the result of chance variation. For the Doritos example, a true null hypothesis would mean that the difference between 269.3 and 270.675 is just a chance fluctuation.

4. Choose a sampling distribution that shows the probabilities of differences. These probabilities are correct when the null hypothesis is true and only chance is at work.

5. By subtraction, find the difference between the sample statistic and the parameter specified in H_0. For the Doritos data, $270.675 - 269.3 = 1.375$ grams.

6. Compare the difference obtained from the data to the differences shown by the sampling distribution, and conclude that the difference obtained was either expected or unexpected.

    a. Expected. Differences of this size are likely when the null hypothesis is true. Unfortunately, such differences are also likely when alternative hypotheses are true. In this case, the data have not allowed you to choose between the null hypothesis and the alternative hypothesis.

    b. Unexpected. Differences this large are highly improbable when the null hypothesis is true. The most reasonable conclusion is that the null hypothesis is false and should be rejected. This leaves you with the alternative hypothesis, H_1, which is that the population parameter is not that specified by the null hypothesis.

7. Write a conclusion about the population that was measured. For the Doritos data, tell what the sample and hypothesis testing have revealed about the weight of tortillas in Doritos bags.

Central to the logic of hypothesis testing is establishing a null hypothesis (step 2). It is usually a good idea to state the null hypothesis explicitly by using the formal language of mathematics. For the Doritos data:

$$H_0: \mu_0 = 269.3 \text{ grams}$$

In words, this null hypothesis says that the mean of the population of Doritos weights is equal to the published claim, which is 269.3 grams. The step 3 tactic of tentatively assuming that the null hypothesis is true is common to all hypothesis testing.

Of course, the researcher hopes to reject H_0. If the data allow rejection of H_0, then the alternative hypothesis, H_1, is accepted. Because the researcher has a hoped-for conclusion in mind from the start, H_1 is selected so that it will encompass the researcher's belief. In practice, a researcher chooses one of the three H_1's, *before the data are collected*. The three alternative hypotheses for the Doritos data are:

1. $H_1: \mu_1 \neq 269.3$ grams. This alternative says simply that the mean of the population of Doritos weights is not 269.3 grams; it doesn't specify whether the actual mean is larger or smaller than 269.3 grams. If you reject H_0 and accept this H_1, you must examine the sample mean to determine whether its population mean is larger or smaller than the one specified by H_0.
2. $H_1: \mu_1 < 269.3$ grams. This alternative hypothesis says that the sample is from a population with a mean less than 269.3 grams.
3. $H_1: \mu_1 > 269.3$ grams. This alternative hypothesis says that the sample is from a population with a mean larger than 269.3 grams.

For the Doritos example, the proper alternative hypothesis is $H_1: \mu_1 \neq 269.3$ grams because this alternative allows us to conclude that the Doritos bags contain more *or* less than the company claims. The section on one-tailed and two-tailed tests, which comes later in this chapter, will give you more information on choosing an alternative hypothesis.

In summary, the null hypothesis meets with one of two fates at the hands of the data. It may be *rejected*, which allows you to accept an alternative hypothesis, or it may be *retained*. If it is retained, *it is not proved as true*; it is simply retained as one among many possibilities. Hypothesis testing is a procedure in which a rejected null hypothesis leads to a strong conclusion involving "greater than" (or "less than"). A retained[1] null hypothesis leaves you where you began—with both the null hypothesis and an alternative hypothesis.

CLUE TO THE FUTURE

You have just completed the section that explains the reasoning that is at the heart of inferential statistics. This reasoning is a part of every test in every chapter that follows in this book (and will also turn up in some of your future reading).

---

[1] Some textbooks use the term *accepted* instead of *retained*.

1. Is the null hypothesis a statement about a statistic or a parameter?
2. Distinguish between the null hypothesis and the alternative hypothesis, giving the symbol for each.
3. In your own words, outline the logic of hypothesis testing.
4. Agree or disagree: The text's sample of Doritos weights is a random sample. ■

USING THE t DISTRIBUTION FOR HYPOTHESIS TESTING

You will need some way to determine probabilities if you are going to test the hypothesis that a sample with a mean, $\bar{X}$, came from a population with a mean, μ_0. Fortunately, you already know about sampling distributions, a statistical device that was invented to reveal probabilities. Even more fortunately, the sampling distribution you need for the Doritos problem is a t distribution, which you learned about in the previous chapter. For the task at hand (testing the null hypothesis that $\mu_0 = 269.3$ grams using a sample of eight), a t distribution with 7 df will give you correct probabilities.

The t distributions you will use will be the theoretical ones in Table D, which you used for Chapter 6 problems. The explanation that comes next, however, is a description of a procedure for generating an *empirical t* distribution with 7 df.

Start with a set of scores that you designate as your population. Calculate the mean, which is a parameter. Draw a random sample ($N = 8$) and calculate a sample mean, which is a statistic. Subtract the parameter from the statistic, producing a difference. Draw another random sample ($N = 8$) and repeat the procedure. Do this many, many times and then arrange all the differences into a frequency distribution.

Think carefully for a moment about the mean (the expected value) of this distribution of differences. Stop reading and decide what its numerical value will be.

The mean of the sampling distribution of differences is zero because positive and negative differences occur equally often, giving you an algebraic sum close to zero. If you were successful on this problem, congratulations! Few students get it right when they first read the chapter.

Here are three reminders. (1) Large differences between the sample mean and the population mean will be found in the tails of the distribution. (2) The area covered by any set of differences is the same as the probability of obtaining these differences. (3) The standard deviation of this distribution is called the standard error of the mean.

Figure 7.1 shows a theoretical t distribution with 7 df. It is based on the assumption that the sample of Doritos weights is from a population with a mean of 269.3 grams. That is, *the figures in Figure 7.1 are based on the assumption that the null hypothesis is true*.

Examine Figure 7.1, paying careful attention to the three sets of values on the abscissa. The t values and the probability figures are true for any problem that uses a t

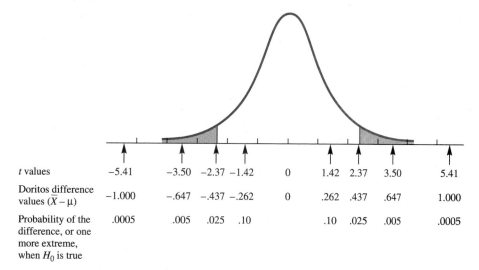

t values	−5.41		−3.50	−2.37	−1.42	0	1.42	2.37	3.50		5.41
Doritos difference values ($\overline{X} - \mu$)	−1.000		−.647	−.437	−.262	0	.262	.437	.647		1.000
Probability of the difference, or one more extreme, when H_0 is true	.0005		.005	.025	.10		.10	.025	.005		.0005

F I G U R E 7.1   A *t* distribution. The *t* values and probability figures are those that go with 7 *df*. The difference values are true only for the Doritos data.

distribution with 7 *df*. Sandwiched between are differences between sample means and the hypothesized population mean—differences that are specific to the Doritos data.

In your examination of Figure 7.1, note the following characteristics:

1. The mean is zero.
2. As the size of the difference between the sample mean and the population mean increases, *t* values increase.
3. As differences and *t* values become larger, probabilities become smaller.
4. As always, small probabilities go with small areas at each end of the curve.

Using Figure 7.1, we can make a decision about the null hypothesis for the Doritos data. (Perhaps you have already anticipated the conclusion?) The difference between the mean weight of my sample and the advertised weight was 1.375 grams. This difference is not even shown on the distribution, so it must be extremely rare—*if the null hypothesis is true*. How rare is it? According to Figure 7.1, differences the size of +1.000 gram or more are expected only 5 times in 10,000 (.0005). A difference of −1.000 or greater *also* has a probability of .0005. Adding these two probabilities will give .001, a step that I will explain in the section on one- and two-tailed tests. A difference of ±1.375 or more is even less likely than ±1.000. Thus, *if the population mean is 269.3 grams and only chance is at work*, the probability of getting the difference I did is less than 1 in 1000 ($p < .001$).

The reasonable thing to do is abandon (reject) the null hypothesis. With the null hypothesis gone, only the alternative hypothesis remains: H_1: $\mu_1 \neq 269.3$ grams. Knowing that the actual weight is *not* 269.3 grams, the next step is to ask whether it is more than or less than 269.3 grams. The sample mean was greater, so a conclusion can be written: "The contents of Doritos bags weigh more than the company claims, $p < .001$."

A PROBLEM AND THE ACCEPTED SOLUTION

The hypothesis that the sample of Doritos weights came from a population with a mean of 269.3 grams was so unlikely ($p < .001$) that it was easy to reject the hypothesis. But what if the probability had been .01, or .05, or .25, or .50? The problem that this section addresses is where on this continuum should you change your decision from "reject the null hypothesis" to "retain the null hypothesis."

It may already be clear to you that any solution will be an arbitrary one. Breaking a continuum into two parts often leaves you uncomfortable near the break. Nevertheless, those who use statistics have adopted a solution.

■ Setting α

The generally accepted solution to the question of where to break the continuum is to use the .05 level of probability. The choice of where to make this break is called *setting* α (alpha). If $\alpha = .05$ and $p \leqslant .05$, then reject H_0. If $\alpha = .05$ and $p > .05$, then retain H_0. Thus, if $p = .03$, or .01, or .001, reject H_0. If the probability is .051 or greater, retain H_0. When H_0 is rejected, the difference is described as **statistically significant**. Sometimes, when the context is clear, this is shortened to simply "significant." When H_0 is retained (the difference is not significant) the abbreviation **NS** is often used.

■ Significance Level

The actual probability figure that you obtain from the data is referred to as the **significance level**. Thus, $p < .001$ is an expression of the level of significance of the difference. In some statistical reports, an α level is not specified; only significance levels are given. Thus, in the same report, some differences may be reported as significant at the .001 level, some at the .01 level, and some at the .05 level. Regardless of how the results are reported, however, researchers view .05 as an important cutoff (Nelson, Rosenthal, and Rosnow, 1986). When .10 or .20 is used as an α level, a justification should be given.

■ Rejection Region

The area of the sampling distribution that includes all the differences that have a probability *equal to or less than* α is called the **rejection region**.[2] Thus, if a difference between the sample mean and the null hypothesis mean is in the rejection region, H_0 is rejected. In **Figure 7.1**, the rejection region is shaded (for an α level of .05).

Although widely adopted, an α level of .05 is not used by everyone. Some investigators use the .01 level in their research. For an α level of .01, the rejection region consists of the extreme one-half of 1 percent (.005) in each tail of the sampling distribution. In Figure 7.1, differences greater than ± 0.647 gram are required to reject H_0 at the .01 level.

---

[2] Some texts use the term **critical region** instead of rejection region.

■ Critical Values

Most statistical tables do not provide you with probability figures for every outcome of a statistical test. What they provide is statistical test values that correspond to commonly chosen α values. These specific test values are called **critical values**.

I will illustrate critical values using the t distribution table (**Table D** in Appendix B). The t values in that table are called critical values when they are used in hypothesis testing. The α levels that researchers use are listed across the top in rows 2 and 3. Degrees of freedom are in column 1. Remember that if you have a df that is not in the table, it is conventional to use the next *smaller df* or to interpolate a t value that corresponds to the exact df.

The t values in the table separate the rejection region from the rest of the sampling distribution (sometimes referred to as the *acceptance region*). Thus, data-produced t values that are larger than the critical value fall in the rejection region, and the null hypothesis is rejected. The critical value itself is in the rejection region.

I will indicate critical values with an expression such as $t_{.05} (14\ df) = 2.145$. This expression indicates the sampling distribution that was used (t), α level (.05), degrees of freedom (14), and critical value from the table (2.145, for a two-tailed test).

PROBLEMS

5. True or false? Sampling distributions used for hypothesis testing are based on the assumption that H_0 is false. F

6. Distinguish between α level and significance level. What is the largest α level you can use without giving a justification? Largest w/out justification is .05

7. Circle all the phrases that go with an event that is in the rejection region of a sampling distribution.

    p is small
    Retain the null hypothesis
    Accept H_1
    Middle section of the sampling distribution

8. Use Table D to find critical values for two-tailed tests for
    a. an α level of .01; $df = 17$
    b. an α level of .001; $df = 46$
    c. an α level of .05; $df = \infty$

THE ONE-SAMPLE t TEST

Some students suspect that it is not necessary to construct an entire sampling distribution such as the one in Figure 7.1 each time you have a hypothesis you want to test. You may be such a student and, if so, you are correct. This section shows an analysis of the Doritos data using the **one-sample t test**. The formula for the one-sample t test is

$$t = \frac{\bar{X} - \mu_0}{s_{\bar{X}}} \qquad df = N - 1$$

where $\bar{X}$ is the mean of the sample

   μ_0 is the hypothesized mean of the population

   $s_{\bar{X}}$ is the standard error of the mean

The question to be answered is whether the Frito-Lay corporation is justified in claiming that their bags of Doritos tortilla chips weigh 269.3 grams. The null hypothesis is that the mean weight of the chips is 269.3 grams. The alternative hypothesis is that the mean weight is *not* 269.3 grams. A sample of eight bags produced a mean of 270.675 grams with a standard deviation (using $N - 1$) of 0.523 gram. To test the hypothesis that the sample weights are from a population with a mean of 269.3 grams, calculate a t value using the one-sample t test. Thus,

$$t = \frac{\bar{X} - \mu_0}{s_{\bar{X}}} = \frac{\bar{X} - \mu_0}{\frac{\hat{s}}{\sqrt{N}}} = \frac{270.675 - 269.3}{\frac{0.523}{\sqrt{8}}} = 7.44 \qquad 7 \ df$$

To interpret a t value of 7.44 with 7 df, turn to Table D and find the appropriate critical value. For α levels for two-tailed tests, go down the .05 column until you reach 7 df. The critical value there, 2.365, is smaller than 7.44, so the difference of 1.375 grams is significant at the .05 level. However, the critical value at the .001 level, 5.408, is also smaller than 7.44, so the difference is also significant at the .001 level. The generic conclusion is, "Reject the null hypothesis at the .001 level." A more informative conclusion is one that tells the story of the variable that was measured. Thus, after noting that the sample mean is larger than H_0, you can write, "The contents of Doritos bags weigh more than the company claims, $p < .001$." Obviously, the second conclusion, which names the variable that was measured, is a better conclusion.

The algebraic sign of the t in the t test is ignored when you are finding the critical value for a two-tailed test. Of course, the sign tells you whether the sample mean is larger or smaller than the null hypothesis mean, which is an important point in any statement of the results.

I want to address two additional issues about the Doritos conclusion; both have to do with generalizability. The first issue is the nature of the sample: It most certainly was not a random sample of the population in question. Because of this, my conclusion is not as secure as it would be for a random sample. How much less secure is it? I don't really know, and inferential statistics provides me with no guidelines on this issue. So, I state both my conclusion (contents weigh more than company claims) and my methods (nonrandom sample, $N = 8$) and then see whether this conclusion will be challenged or supported by others with more or better data. (I'll just have to wait to see who responds to my claim.)

The other generalizability issue is whether the conclusion applies to other-sized bags of Doritos, to other Frito-Lay products, or to other manufacturers of tortilla chips.

These are good questions but, again, inferential statistics does not give any guidelines. You'll have to use aids such as your experience, library research, correspondence with the company, or more data gathering.

Finally, I want to address the question of the usefulness of the one-sample t test. The t test is useful for checking manufacturers' claims (where an advertised claim provides the null hypothesis), comparing a sample to a known norm (such as an IQ score of 100), and comparing behavior against a "no error" standard (as might be established in studying lying or visual illusions). In addition to its use in analyzing data, the one-sample t test is probably the simplest example of hypothesis testing, which is an important consideration when you are first being introduced to the topic.

AN ANALYSIS OF POTENTIAL MISTAKES

To some forward-thinking students, the idea of adopting an α level of 5 percent seems preposterous.

"You gather the data," they say, "and if the difference is large enough, you reject H_0 at the .05 level. You then state that the sample came from a population with a mean other than the hypothesized one. But in your heart you are uncertain. Perhaps a rare event happened."

This line of reasoning is fairly common. Many thoughtful students take the next step.

"I don't have to use the .05 level. I'm going to use an α level of 1 in 1 million. That way I can reduce the uncertainty."

It is true that adopting a .05 α level leaves some room for mistaking a chance difference for a real difference. It is probably clear to you that lowering the α level (to a probability such as .01 or .001) *reduces* the probability of this kind of mistake. Unfortunately, lowering the α level *increases* the probability of a different kind of mistake.

Look at **Table 7.1**, which shows the two ways to make a mistake.

In Table 7.1, cell 1 shows the situation when the null hypothesis is true and you retain it—your sample data led you to a correct decision. However, if H_0 was true and your sample data led you to reject it (cell 2), you would have made a mistake called a **Type I error**. The probability of a Type I error is symbolized by α.

T A B L E 7.1 **Type I and Type II errors***

		The true situation in the population	
		H_0 true	H_0 false
The decision made on the basis of sample data	Retain H_0	1. Correct decision	3. Type II error β
	Reject H_0	2. Type I error α	4. Correct decision

* In this case, error means mistake.

On the other hand, suppose H_0 is really false (the second column). If, on the basis of your sample data, you retain H_0 (cell 3), you have made a mistake—a **Type II error**, the probability of which is symbolized by β (beta). If your sample data led you to reject H_0 (cell 4), you made a correct decision.

You already have some familiarity with α from using it as a decision rule for rejecting or retaining H_0. When an α level of .05 is adopted, and the data produce a $p < .05$, the researcher rejects H_0 and concludes that the difference is significant. The researcher could be wrong; if so, a Type I error has been made. The probability of a Type I error is controlled by the α level that was adopted.

The proper way to think of α and a Type I error is in terms of "the long run." **Figure 7.1** is a theoretical sampling distribution of differences between sample means and a parameter. It is a picture of repeated sampling when the null hypothesis is true. All those differences were obtained with samples drawn from the null hypothesis population, but some of the large differences were so rare that they occurred less than 5 percent of the time (the rejection region). When you gather your data, you do not know if your sample is from the null hypothesis population or not. If your difference is so large that it falls in the rejection region, you conclude that the sample was not from the null hypothesis population. Although this may be a Type I error, the probability of such errors is not more than .05.

Calculating β is a more complicated matter. For one thing, a Type II error can be committed only when the sample is from a population that is different from the hypothesized one. Naturally, the more different the population is, the more likely you are to detect it, and, thus, the lower β is. Other factors that affect β will be covered in Chapter 8 in a section called "How to Reject the Null Hypothesis—The Topic of Power."

The general relationship between α and β is an inverse one. As α goes down, β goes up; that is, if you insist on a larger difference between means before you call the difference nonchance, you are less likely to detect a real—but small—nonchance difference. The following description will illustrate this relationship.

Figure 7.2 shows two populations. The population on the left has a mean $\mu_0 = 10$;

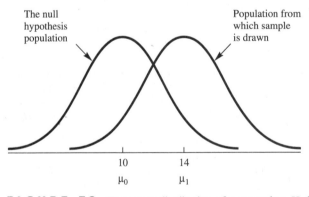

FIGURE 7.2 Frequency distribution of scores when H_0 is false

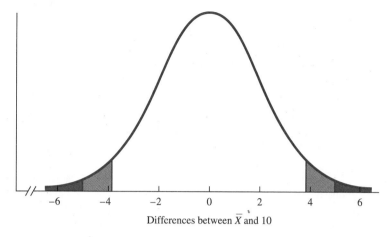

Differences between $\bar{X}$ and 10

FIGURE 7.3 A sampling distribution from the population on the left in Figure 7.2

the one on the right has a mean $\mu_1 = 14$. An investigator draws a large sample from the population on the right and uses it to test the null hypothesis H_0: $\mu_0 = 10$. In the real world of data, every decision carries some uncertainty, but in this textbook example the correct decision is clear: Reject the null hypothesis.

The actual decision, however, will be made by evaluating the difference between the sample mean $\bar{X}$ and the mean specified by the null hypothesis, $\mu_0 = 10$.

Now, let's suppose that the sample produced a mean of 14. That is, the sample tells the exact truth about its population. Under these circumstances, will the hypothesis-testing procedure produce a correct decision?

Figure 7.3 shows the sampling distribution for this problem. As you can see, if the 5 percent α level is used (*all* the green areas), a difference of 4 falls in the rejection region. Thus, if $\alpha = .05$, you will correctly reject the null hypothesis. However, if a 1 percent α level is used (the *dark* green areas only), a difference of 4 *does not* fall in the rejection region. Thus, if $\alpha = .01$, you would not reject H_0. This failure to reject the false null hypothesis is a Type II error.

At this point, I can return to the discussion of setting the α level. The suggestion was, Why not reduce the α level to 1 in 1 million? From the analysis of the potential mistakes, you can answer that when you lower the α level so you can reduce the chance of a Type I error, you increase β, the probability of a Type II error. More protection from one kind of error increases the liability to another kind of error.

Most who use statistics adopt an α level (usually at .05) and let β fall where it may. The actual calculation of β will not be discussed in this book.

Finally, there *are* ways to reduce uncertainty when you are designing and conducting studies that use statistical analyses. The section in Chapter 8, "How to Reject the Null Hypothesis—The Topic of Power," gives advice on reducing uncertainty. However, to quote a phrase in a famous statistics textbook, "If you agree to use a sample, you agree to accept *some* uncertainty about the results."

THE MEANING OF p IN $p < .05$

The p in $p < .05$ is the probability of getting the sample statistic if H_0 is true. This is a simple definition that is easy to memorize. Nevertheless, the meaning of p is commonly misinterpreted. Everitt and Hay (1992) report that among 70 academic psychologists, only 3 scored 100 percent in a six-item test on the meaning of p. Here is what p is not:

p is not the probability that H_0 is true
p is not the probability of a Type I error
p is not the probability of making a wrong decision
p is not the probability that the sample statistic is due to chance

Sampling distributions that are used to determine probabilities are always ones that assume the null hypothesis is true. Thus, the probabilities these sampling distributions show for a particular statistic are accurate when H_0 is true. Thus, p is the probability of getting the sample statistic if H_0 is true.

ONE- AND TWO-TAILED TESTS

Earlier, you read that researchers choose one of three possible alternative hypotheses before the data are gathered. The choice of an alternative hypothesis depends on the conclusions you want to have available after the statistical analysis is finished. The general expression of these alternative hypotheses and the conclusions they allow you are:

Two-tailed test
$H_1: \mu_1 \neq \mu_0$. The population means differs, but there is no indication about the direction of the difference. Conclusions that are possible: The sample is from a population with a mean *less* than that of the null hypothesis *or* the sample is from a population with a mean *greater* than that of the null hypothesis.

One-tailed tests
$H_1: \mu_1 < \mu_0$. Conclusion that is possible: The sample is from a population with a mean less than that of the null hypothesis.
$H_1: \mu_1 > \mu_0$. Conclusion that is possible: The sample is from a population with a mean greater than that of the null hypothesis.

If you want to have two different conclusions available to you—that the sample is from a population with a mean *less* than that of the null hypothesis *or* that the sample is from a population with a mean *greater* than that of the null hypothesis—you must have a rejection region in each tail of the sampling distribution. Such a test is called a **two-tailed test of significance** for reasons that should be obvious from **Figure 7.3** and **Figure 7.1**.

If your only interest is in showing (if the data will permit you) that the sample comes from a population with a mean greater than that of the null hypothesis or from a population with a mean less than that of the null hypothesis, you should run a **one-**

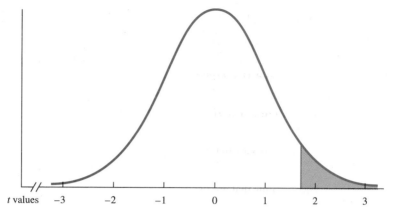

t values   −3   −2   −1   0   1   2   3

F I G U R E 7.4 A one-tailed test of significance, with $\alpha = .05$, $df = 30$

tailed test of significance, which puts all of the rejection region into one tail of the sampling distribution. **Figure 7.4** illustrates this for a t distribution with 30 df.

You will find probability figures for two-tailed tests in the second row of **Table D**; probability figures for one-tailed tests are in row 3.

For most research situations, a two-tailed test is the appropriate one. The usual goal of a researcher is to discover the way things actually are, and a one-tailed test does not allow you to discover, for example, that the populations are exactly reversed from the way you expect them to be. Researchers always have expectations (but they sometimes get surprised).

In many applied situations, a one-tailed test is appropriate. Some new procedure, product, or person, for example, is being compared to an already existing standard. The only interest is in whether the new is better than the old; that is, only a difference in one direction will produce a decision to change. In such a situation, a one-tailed test seems appropriate.

I chose a two-tailed test for the Doritos data because I wanted to be able to conclude that the Frito-Lay corporation was giving more than they claimed *or* that they were giving less than they claimed. Of course, the data would have their say about my conclusion, but either outcome was interesting to me.

I'll end this section on one- and two-tailed tests by telling you that statistics instructors sometimes smile or even laugh at the subtitle of this book. The phrase "tales of distributions" brings to mind "tails of distributions."

PROBLEMS

9. Distinguish among α, Type I error, significance level, and p.

10. What is a Type II error?

11. Suppose the actual situation is that the sample comes from a population that is not the one hypothesized by the null hypothesis. If the significance level is .05, what is the probability of making a Type I error? (Careful on this one.)

12. Suppose a researcher chose to use $\alpha = .01$ rather than .05. What would be the effect on the probability of Type I and Type II errors?

*13. Project Head Start is a program for preschool children from educationally disadvantaged environments. Head Start began over 25 years ago, and the data that follow are representative of early research on the effects of the program.

   A group of children participated in the program for a year and then enrolled in the first grade. They took an IQ test, the national norm of which was 100. Use a t test to test the hypothesis that the Head Start program had no effect on the children's IQ. Begin your work by stating the null hypothesis and end by writing a conclusion about the effects of the program.

$\Sigma X = 5250$
$\Sigma X^2 = 560,854$
$N = 50$

14. In addition to weighing Doritos chips, I also weighed six Snickers candy bars. Test the company's advertised claim that the net weight of a bar is 58.7 grams by performing a t test on the following data, which are in grams. Write a conclusion about the data.

   59.1   60.0   58.6   60.2   60.8   62.1

*15. The NEO-PI is a personality test. The E stands for extroversion, one of the personality characteristics the test measures. The population mean for the extroversion test is 50. Suppose 13 people who were successful in used car sales took the test, producing a mean of 56.10 and a standard deviation of 10.00. Perform a t test and write a conclusion. ■

STOP FOR TEST

EFFECT SIZE

Hypothesis testing answers the question of whether the population a sample is drawn from is different from the population specified by the null hypothesis. If the answer is yes, the next question a reasonable person would ask is, How *big* is the difference? The issue of this difference between the mean of the population the sample is from and the mean of the null hypothesis population is the issue of **effect size**.

   Of course, researchers have no control over effect size; it is determined by the way things are out there in nature. I will give you a way to measure effect size (an effect size index) and a convention used by experienced researchers to decide if the effect size is small, medium, or large.

   Effect size is an important topic usually covered in intermediate books under a discussion of power. In this text you will learn to calculate an effect size *index* for several statistical tests, but you will not do a complete power analysis (although factors that affect power will be discussed in the next chapter).

The effect size index for a one-sample t test is called d and is given by the formula

$$d = \frac{\bar{X} - \mu_0}{\sigma}$$

The value of d is evaluated using conventions proposed by Cohen (1969) and widely adopted by researchers.

Small effect	$d = .20$
Medium effect	$d = .50$
Large effect	$d = .80$

There is some empirical validation for these conventions. Cohen (1992) notes that the average effect size index in published studies approximates a medium effect.

In problem 13, you concluded that the Head Start program has a statistically significant effect on the IQ of preschool children. You found that the mean IQ of children in the program (105) was significantly higher than 100. But how big is this effect? The statistic d provides an answer. Knowing that $\sigma = 15$,

$$d = \frac{\bar{X} - \mu_0}{\sigma} = \frac{105 - 100}{15} = .33$$

An effect size index of .33 indicates that the effect of Head Start is midway between a small effect and a medium effect.

Many research situations, unlike those using IQ scores, do not provide σ. In these cases, researchers use $\hat{s}$, calculated from sample data, as an acceptable approximation of σ.

Using the effect size index as more than a descriptive index requires an understanding of power analysis. Simply as a descriptive index, however, it alerts you to the importance of the question, How *big* was the significant difference?

OTHER SAMPLING DISTRIBUTIONS

You have been learning about the sampling distribution of the mean. There are times, though, when the statistic necessary to answer a researcher's question is not the mean. For example, to find the degree of relationship between two variables, you need a correlation coefficient. To determine whether a treatment causes more variable responses, you need a standard deviation. And, as you know, proportions are commonly used statistics. In each of these cases (and indeed, for any statistic), researchers often use the basic hypothesis-testing procedure you have just learned. Hypothesis testing always involves a sampling distribution of the statistic.

Where do you find sampling distributions for statistics other than the mean? In the rest of this book you will encounter new statistics and new sampling distributions. Tables E–L in Appendix B represent sampling distributions from which probability figures can be obtained. In addition, some statistics have sampling distributions that are t distributions or normal curves. Still other statistics and their sampling distributions are covered in other books.

Along with every sampling distribution comes a standard error. Just as every statistic has its sampling distribution, every statistic has its standard error. For example, the standard error of the median is the standard deviation of the sampling distribution of the median. The standard error of the variance is the standard deviation of the sampling distribution of the variance. Worst of all, the standard error of the standard deviation is the standard deviation of the sampling distribution of the standard deviation. If you follow that sentence, you probably understand the concept of standard error quite well.

The next step in your exploration of sampling distributions will be brief, although useful and (probably) interesting. The t distribution will be used to answer a question about correlation coefficients.

PROBLEMS

16. The table below has blanks for you to fill in. Your task is to look up critical values in Table D, decide if H_0 should be rejected or retained, and give the probability figure that characterizes your confidence in that decision. I designed these problems so that some common misconceptions will be revealed—just in case you have picked up one or two.

N	α level	Two-tailed or one-tailed test	Critical value	t-test value	Reject H_0 or retain H_0	p value
10	.05	two	————	2.25	————	———
20	.01	one	————	2.57	————	———
35	.05	two	————	2.03	————	———
6	.001	two	————	6.72	————	———
24	.05	one	————	1.72	————	———
56	.02	two	————	2.41	————	———

17. Calculate the effect size index for Doritos data on page 163 and interpret it using the conventions used by researchers.

18. In problem 15 you found that people who are successful in used car sales have significantly higher extroversion scores than those in the general population. What is the effect size index for those data and is the effect small, medium, or large?

19. In Chapter 2 and Chapter 6 you worked with data on normal body temperature. You found that the mean body temperature was 98.2°F rather than the conventional 98.6°F. The standard deviation for those data was 0.7°F. Calculate an effect size index and interpret it. ■

USING THE t DISTRIBUTION TO TEST THE SIGNIFICANCE OF A CORRELATION COEFFICIENT

In Chapter 4 you learned to calculate a Pearson product-moment correlation coefficient, a descriptive statistic that indicates the degree of relationship between two variables in a bivariate distribution. This section is on testing the statistical significance of correlation coefficients. The techniques of inferential statistics will be used to test the hypothesis that a sample-based r came from a population with a parameter coefficient of .00.

The present-day symbol for the Pearson product-moment correlation coefficient of a population is ρ (rho). The null hypothesis that is being tested is:

$$H_0: \rho = .00$$

The alternative hypothesis is:

$$H_1: \rho \neq .00 \text{ (a two-tailed test)}$$

For $\rho = .00$, the sampling distribution of r is a t distribution with a mean of .00 and a standard error of $s_r = \sqrt{(1 - r^2)/(N - 2)}$. The test statistic for a sample r is a t value calculated from the formula

$$t = \frac{r - \rho}{s_r}$$

Note that this formula has the same form as that of the t test you used to test the significance of a sample mean. This form, which you will see again, shows a difference between a statistic and a parameter divided by the standard error of the statistic.

An algebraic manipulation produces a formula that is easier to use:

$$t = (r)\sqrt{\frac{N - 2}{1 - r^2}} \qquad df = N - 2, \text{ where } N = \text{number of } pairs$$

This test is the first of several instances in which you will find a t value whose df is *not* one less than the number of observations. For testing the null hypothesis, $H_0: \rho = .00$, $df = N - 2$. (Having fulfilled my promise to explain one- and two-tailed tests, I hope you will accept my promise to explain degrees of freedom more fully in the next chapter.)

Do you remember your work with correlation coefficients in Chapter 4?[3] Think about a sample $r = -.40$. How likely is a sample $r = -.40$ if the population $\rho = .00$? Does it help to know that the coefficient is based on a sample of 22 pairs? Let's answer these questions by testing the $r = -.40$ against the $\rho = .00$, using a t test. Applying the formula for t,

$$t = (r)\sqrt{\frac{N - 2}{1 - r^2}} = (-.40)\sqrt{\frac{22 - 2}{1 - (-.40)^2}} = -1.95$$

$$df = N - 2 = 22 - 2 = 20$$

Table D shows that, for 20 df, a t value of ± 2.09 (two-tailed test) is required to reject the null hypothesis. The obtained t for $r = -.40$, where $N = 22$, is less than the tabled t, so the null hypothesis is retained. That is, a coefficient of $-.40$ would be expected by chance alone more than 5 times in 100 from a population in which the true correlation is zero.

---

[3] If not, take a few minutes to review.

In fact, for $N = 22$, $r = \pm.43$ is required for significance at the .05 level and $r = \pm.54$ for the .01 level. As you can see, even medium-sized correlations can be expected by chance alone for samples as small as 22 when the null hypothesis is true. Most researchers strive for N's of 30 or more for correlation problems.

The critical values under "α level for two-tailed test" in Table D are ones that allow you to reject the null hypothesis if the sample r is a large *positive* coefficient or a large *negative* one.

PROBLEM

20. Determine whether the following Pearson product-moment correlation coefficients are significantly different from .00:

   a. $r = .62$; $N = 10$ **b.** $r = -.19$; $N = 122$
   c. $r = .50$; $N = 15$ **d.** $r = -.34$; $N = 64$ ∎

Here is a think problem. Suppose you have a summer job in an industrial research laboratory testing the statistical significance of hundreds of correlation coefficients based on varying sample sizes. An α level of .05 has been adopted by the management, and your task is to determine whether each coefficient is "significant" or "not significant." How can you construct a table of your own, using Table D and the t formula for testing the significance of a correlation coefficient, that will allow you to label each coefficient without working out a t value for each? Imagine or sketch out your answer.

The table you mentally designed already exists. One version is reproduced in Table A in Appendix B. The α values of .10, .05, .02, .01, and .001 are included there. Use **Table A** in the future to determine whether a Pearson product-moment correlation coefficient is significantly different from .00, but remember that this table is based on the t distribution.

A word of caution is in order about testing ρ's other than .00. The sampling distributions for $\rho \neq .00$ are *not* t distributions. Also, if you want to know whether the *difference* between two sample-based coefficients is statistically significant, you should not use a t distribution because it will not give you correct probabilities. Fortunately, if you have questions that can be answered by testing these kinds of hypotheses, you won't find it difficult to understand the instructions given in intermediate-level texts. (See Ferguson and Takane, 1989, p. 205, or Howell, 1992, p. 251.)

This chapter has introduced you to two uses of the t distribution:

1. Testing a sample mean against a hypothesized population mean
2. Testing a sample correlation coefficient against a hypothesized population correlation coefficient of .00

The t distribution has other uses, too, and it has been important in the history of statistics.

Here is a little more background on W. S. Gosset, who invented the t distribution so he could assess probabilities for experiments conducted by the Guinness brewery. Gosset was one of a small group of scientifically trained employees who developed ways for the Guinness Company to improve its products. (They were a very early research and development team.) Ideas were judged by experimenting with changes in recipes and the brewing process. In addition, the Guinness Company maintained 10 farms in the principal barley-growing regions of Ireland. Experiments on new strains of barley and hops from these farms were analyzed with the newly derived *t* distribution.

Gosset wanted to publish his work on the *t* distribution in *Biometrika*, a journal founded in 1901 by Francis Galton, Karl Pearson, and W. R. F. Weldon. However, at the Guinness Company there was a rule that employees could not publish (the rule there, apparently, was publish *and* perish). Because the rule was designed to keep brewing secrets from escaping, there was no particular ferment within the company when Gosset, in 1908, published his new mathematical distribution under the pseudonym "Student." The distribution later came to be known as "Student's *t*." (No one seems to know why the letter *t* was chosen. E. S. Pearson surmises that *t* was simply a "free letter"; that is, no one had yet used *t* to designate a statistic.) Gosset worked for the Guinness Company all his life, so he continued to use the pseudonym "Student" for his publications in mathematical statistics. Gosset was devoted to his company, working hard and rising through the ranks. He was appointed head brewer a few months before his death in 1937.[4]

WHY .05?

The story of how science and other disciplines came to adopt .05 as the arbitrary point separating differences attributed to chance from those not attributed to chance is not usually covered in textbooks. The story takes place in England at the turn of the century.

It appears that the earliest explicit rule about α was in a 1910 article by Wood and Stratton, "The Interpretation of Experimental Results," which was published in *The Journal of Agricultural Science*. Thomas B. Wood, the principal editor of this journal, advised researchers to take "30 : 1 as the lowest odds which can be accepted as giving practical certainty that a difference . . . is significant" (Wood and Stratton, 1910). "Practical certainty" meant "enough certainty for a practical farmer." A consideration of the circumstances of Wood's journal provides a plausible explanation of his advice to researchers.

At the turn of the century, farmers in England were doing quite well. For example, Biffen (1905) reported that wheat production averaged 30 bushels per acre compared to 14 in the United States, 10 in Russia, and 7 in Argentina. Politically, of course, England was near the peak of its influence.

In addition, a science of agriculture was beginning to be established. Practical scientists, with an eye on increased production and reduced costs, conducted studies of weight gains in cattle, pigs, and sheep; amounts of dry matter in milk and mangels (a kind of beet); and yields of many kinds of grain. In 1905, *The Journal of Agricultural Science* was founded so these results could be shared around the world. Conclusions were reached after calculating probabilities, but the researchers avoided or ignored the problem created by probabilities between .01 and .25.

[4] For more information, see "Gosset, W. S.," in *Dictionary of National Biography. 1931–1940* (London: Oxford University Press, 1949), or see McMullen and Pearson (1939, pp. 205–253).

Wood's recommendation in his "how to interpret experiments" article was to adopt an α level that would provide a great deal of protection against a Type I error. (Odds of $30:1$ against the null hypothesis convert to $p = .0323$.) A Type I error in this context would be a recommendation that, when implemented, did not produce improvement. I think that Wood and his colleagues wanted to be *sure* that when they made a recommendation, it would be correct.

A Type II error (failing to recommend an improvement) would not cause much damage; farmers would just continue to do well (though not as well as they might). A Type I error, however, would, at best, result in no agricultural improvement *and* in a loss of credibility and support for the fledgling institution of agriculture science.

To summarize my argument, it made sense for the agricultural scientists to protect themselves and their institution against Type I errors. They did this by not making a recommendation unless the odds were $30:1$ against a Type I error.

When tables were published as an aid to researchers, probabilities such as .10, .05, .02, and .01 replaced odds. So $30:1$ ($p = .0323$) gave way to .05. Those tables were used by researchers in many areas besides agriculture, who appear to have adopted .05 as the most important significance level. (See Nelson, Rosenthal, and Rosnow, 1986, for a survey of psychologists.)

PROBLEMS

21. In Chapter 4 you encountered data for 11 countries on per capita cigarette consumption and the male death rate from lung cancer 20 years later. The correlation coefficient was .74. After consulting Table A, write a response to the statement, "The correlation is just a fluke; chance is the likely explanation for this correlation."

22. One of my colleagues gathered data on the time students spent on a midterm examination and their grade. The X variable is time and the Y variable is examination grade. Calculate the r. Use Table A to determine if it is statistically significant.

$$\Sigma X = 903 \qquad \Sigma X^2 = 25{,}585 \qquad \Sigma XY = 46{,}885$$
$$\Sigma Y = 2079 \qquad \Sigma Y^2 = 107{,}707 \qquad N = 42$$

23. Please review the objectives at the beginning of the chapter. Can you do them? ■

Hypothesis Testing and Effect Size: Two-Sample Tests

OBJECTIVES FOR CHAPTER 8

After studying the text and working the problems in this chapter, you should be able to:

1. Describe the logic of a simple experiment

2. Explain the testing of a null hypothesis when there are two samples

3. Explain some of the reasoning in finding degrees of freedom

4. Distinguish between independent-samples designs and correlated-samples designs

5. Calculate *t*-test values for both independent-samples designs and correlated-samples designs and write interpretations

6. List and explain the assumptions for using the *t* distribution

7. Calculate and interpret an effect size index

8. Describe the factors that affect the rejection of the null hypothesis

9. Distinguish between statistically significant results and important results

In Chapter 7 you learned to use statistical hypothesis testing to answer questions about a population when you have one sample available. In this chapter, the same hypothesis-testing reasoning will be used, but you will have data from two samples. Two-sample hypothesis testing is heavily used by researchers who analyze data from experiments.

One of the greatest benefits of studying statistics is that it helps you understand experiments and the experimental method. The experimental method is probably the most powerful method we have of finding out about natural phenomena. Besides being powerful, experiments can be interesting. They can answer such questions as:

1. Would you rather have a pigeon or a person searching for you if you were lost at sea?
2. Does the removal of 20 percent of the cortex of the brain have an effect on the memory of tasks learned before the operation?
3. Can you *reduce* people's ability to solve a problem by educating them?

The plan in this chapter is to discuss the simplest kind of experiment and then show you how the logic of hypothesis testing, which you studied in the last chapter, can be expanded to answer questions like those above.

A SHORT LESSON ON HOW TO DESIGN AN EXPERIMENT

The basic ideas of a simple two-group experiment are not very complicated.

The logic of an experiment: Start with two equivalent groups. Treat them exactly alike except for one thing. Measure both groups and attribute any statistically significant difference between the two to the one way in which they were treated differently.

This summary of an experiment is described more fully in **Table 8.1**. The question that the experiment in the table sets out to answer is, What is the effect of Treatment A on a person's ability to perform task Q? In formal, statistical terms, the question is, For task Q scores, is the mean of the population of those who have had Treatment A different from the mean of the population of those who have not had Treatment A?

To conduct this experiment, a group of participants is identified and two samples are assembled. These samples should be (approximately) equivalent. Treatment A is then administered to one group (generically called the **experimental group**) but not to the other group (generically called the **control group**). Except for Treatment A, both groups are treated in exactly the same way; that is, extraneous variables are held constant or balanced out for the two groups. Both groups perform task Q, and the mean score for each group is calculated.

The two sample means almost surely will differ. The question is whether the difference is due to Treatment A or is just the usual difference that would be expected of two samples from the same population.[1] This question can be answered by applying hypothesis-testing logic.

The first step in this logic is to assume there is no difference between two population means. Next, let the data tell you whether the assumption is reasonable. If the assumption is not reasonable, you are left with only one alternative: The populations have different means.

This generalized example has an independent variable with two levels (Treatment A and no Treatment A) and one dependent variable (scores on task Q). The word **treatment** is recognized by all experimentalists; it refers to different levels of the independent variable. The experiment in Table 8.1 has two treatments.

The experimental procedure is a versatile one. Experiments have been used to decide a wide variety of issues such as how much sugar to use in a cake recipe, what kind of tea tastes the best, whether a drug is useful in treating cancer, and the effect of alcoholic parents on the personality of their children.

In many experiments, it is obvious that there are *two* populations of participants to begin with—for example, a population of men and a population of women. The question, however, is whether they are equal on the dependent variable.

---

[1] The idea of one population with two samples is the same statistically as two identical populations with one sample from each.

TABLE 8.1 **Summary of a Simple Experiment**

Key words	Tasks for the researcher	
Population	A population of participants	
	↙	↘
Assignment	Randomly assign one-half of *available* participants to the experimental group.	Randomly assign one-half of *available* participants to the control group.
	↓	↓
Independent variable	Give Treatment A to participants in experimental group.	Withhold Treatment A from participants in control group.
	↓	↓
Dependent variable	Measure participants' behavior on task Q.	Measure participants' behavior on task Q.
	↓	↓
Descriptive statistics	Calculate mean score on task Q, $\bar{X}_e$.	Calculate mean score on task Q, $\bar{X}_c$.
	↘	↙
Inferential statistics	Compare $\bar{X}_e$ and $\bar{X}_c$ using inferential statistics. Reject or retain the null hypothesis.	
	↓	
Interpretation	Write a conclusion about the effect of Treatment A on task Q scores.	

In some experimental designs, participants are randomly assigned to treatments by the researcher; in others, the researcher uses a group of participants who have already been "treated" (for example, being males or being children of alcoholic parents). In either of these designs, the methods of inferential statistics are the same, although the interpretation of the first kind of experiment is usually less open to attack.[2]

The *raison d'être*[3] of experiments is to be able to tell a story about the universal effect of one variable on a second variable. All the work with samples is just a way to get evidence to support the story, but samples, of course, have chance built into them. Hypothesis testing takes chance factors into account when the story is told.

[2] I am bringing up an issue that is beyond the scope of this statistics book. Courses with titles such as "Experimental Design" and "Research Methods" cover the intricacies of interpreting experiments.
[3] *Raison d'être* means "reason for existence." (Ask someone who can pronounce French phrases to teach you this one. If you know how to pronounce the phrase, use it in conversation to identify yourself as someone to ask.)

HYPOTHESIS TESTING: THE TWO-SAMPLE EXAMPLE

The researcher obtains data by following the procedures dictated by the design of the experiment. Once the data are gathered, some sort of hypothesis testing usually follows.

Hypothesis testing with two samples is similar to hypothesis testing with one sample; certainly, the logic is very similar. If you stayed with the section in Chapter 7 called "The Logic of Hypothesis Testing" until you understood it, you'll find it easy to add the few new points here. If you aren't confident of your understanding, read, take notes, ask yourself questions, reread, ask questions of others, and so on until you do feel confident.

Here is the logic of hypothesis testing for a two-sample experiment. In a well-designed, well-executed experiment, all imaginable results are included in the statement: Either Treatment A has an effect or it does not have an effect. Begin by making a tentative assumption that Treatment A does *not* have an effect. Gather data. Using a sampling distribution based on the assumption that Treatment A has no effect, find the probability of the data obtained. If the probability is low, abandon your tentative assumption and conclude that Treatment A has an effect. If the probability is not low, you are back where you began: The data have not given you evidence against either of the two logical possibilities.

Combining the logic of hypothesis testing with the language of an experiment:

1. Begin with two logical possibilities, the null hypothesis, H_0, and an alternative hypothesis, H_1.

   H_0: Treatment A *does not* have an effect; that is, the mean of the population of scores of those who receive Treatment A is equal to the mean of the population of scores of those who do not receive Treatment A. The difference between population means is zero. In statistical language:

   $$H_0: \mu_A - \mu_{no\,A} = 0 \qquad \text{or} \qquad H_0: \mu_A = \mu_{no\,A}$$

   H_1: Treatment A *does* have an effect; that is, the mean of the population of scores of those who receive Treatment A is *not* equal to the mean of the population of scores of those who do not receive Treatment A. The alternative hypothesis, which should be chosen before the data are gathered, can be two-tailed or one-tailed. An alternative hypothesis consists of *one* of the three H_1's that follow.

   Two-tailed alternative: $H_1: \mu_A \neq \mu_{no\,A}$

   In the example of the simple experiment, this hypothesis says that Treatment A has an effect but does not indicate whether the treatment improves or disrupts performance on task Q. A two-tailed test allows either conclusion.

   One-tailed alternatives: $H_1: \mu_A > \mu_{no\,A}$
   $H_1: \mu_A < \mu_{no\,A}$

   If the first one-tailed alternative is chosen, then no outcome of the experiment can lead to the conclusion that Treatment A produces lower scores than no treatment. If the second is chosen, the conclusion that Treatment A improves scores is not available to you.

2. Tentatively assume that Treatment A has no effect (that is, assume H_0). If H_0 is true, the two samples will be alike except for the usual variations in samples. Thus, the difference in sample means is tentatively assumed to be due to chance.

3. Decide on an α level. (Usually, $\alpha = .05$.)

4. Choose an appropriate inferential statistical test. This test will have (a) a test statistic that can be calculated from the data, (b) a sampling distribution of the test statistic that shows its distribution when H_0 is true, and (c) a critical value for the α level. (For my two-group example, a t test and the t distribution with its critical values are appropriate.)

5. Calculate a test statistic using the sample data.

6. Compare the test statistic to the critical value from the sampling distribution. For the t distribution, if the data-based t value is larger than the critical value in the table, reject H_0. If the data-based t value is smaller than the critical value, retain H_0. In more general terms, if the test statistic's probability is less than α, reject H_0. If larger, retain H_0.

7. Write a conclusion that uses the terms of the experiment. Your conclusion should describe the relationship of the dependent variable to the independent variable for the population of participants.

Researchers sometimes are interested in statistics other than the mean, and they often have more than two samples in an experiment. These situations call for sampling distributions covered later in this book—sampling distributions such as F, chi square, U, and the normal distribution. I described the steps in this seven-item list somewhat broadly so that it will make sense not only as an explanation for two-sample t tests, but also for analyses you will be doing in the chapters that follow. I hope you will turn down the corner of this page and come back and reread this section several times.

The sampling distribution that you will use to analyze the two-group experiments in this chapter is the t distribution. Not only are sample means distributed as t, but differences between sample means are also distributed as t.[4]

To understand *which* t distribution you need, you will need to know more about degrees of freedom. But first, some problems.

PROBLEMS

1. In your own words, outline the logic of an experiment.
2. In the experiment described in Table 8.1, what procedure was used to ensure that the two samples were equivalent before treatment?
3. For each of the two experiments that follow, identify the independent and dependent variables and write a statement of the null hypothesis:

---

[4] This statement is true *if* the assumptions discussed at the end of this chapter are true.

> **a.** A psychologist compared people with Type A personalities (high drive, time-oriented, and angry) to people with Type B personalities (who do not exhibit these characteristics) on a test that measured satisfaction in life.
>
> **b.** A sociologist compared the number of years served in prison by those convicted of robbery and those convicted of embezzlement.

4. In your own words, outline the logic of hypothesis testing for a two-group experiment.

5. What is the purpose of an experiment, according to your textbook?   ■

DEGREES OF FREEDOM

In your use of the t distribution, you have been determining degrees of freedom by rule-of-thumb techniques: $N - 1$ when you are determining whether a sample mean $(\bar{X})$ comes from a population with a mean, μ_0, and $N - 2$ when you are determining whether a correlation coefficient, r, is significantly different from .00. Now it is time to explain the concept more thoroughly.

The "freedom" in *degrees of freedom* refers to the freedom of a number to have any possible value. If you were asked to pick two numbers and there were no restrictions, both numbers would be free to vary (take any value) and you would have two degrees of freedom. If a restriction is imposed—say, that $\Sigma X = 0$—then one degree of freedom is lost because of that restriction; that is, when you now pick the two numbers, only one of them is free to vary. As an example, if you choose 3 for the first number, the second number *must be* -3. Because of the restriction that $\Sigma X = 0$, the second number is not free to vary. In a similar way, if you were to pick five numbers with a restriction that $\Sigma X = 0$, you would have four degrees of freedom. Once four numbers are chosen (say, -5, 3, 16, and 8), the last number (-22) is determined.

The restriction that $\Sigma X = 0$ may seem to you to be an "out-of-the-blue" example and unrelated to your earlier work in statistics, but some of the statistics you have calculated have had such a restriction built in. For example, when you found $s_{\bar{X}}$, as required in the formula for the one-sample t test, you used some algebraic version of

$$s_{\bar{X}} = \frac{\hat{s}}{\sqrt{N}} = \frac{\sqrt{\dfrac{\Sigma(X - \bar{X})^2}{N - 1}}}{\sqrt{N}}$$

The restriction that is built in is that $\Sigma(X - \bar{X})$ is always zero and, in order to meet that requirement, one of the X's is determined. All X's are free to vary except one, and the degrees of freedom for $s_{\bar{X}}$ is $N - 1$. Thus, for the problem of using the t distribution to determine whether a sample came from a population with a mean μ_0, $df = N - 1$. Walker (1940) summarizes this reasoning by stating: "A universal rule holds: The number of degrees of freedom is always equal to the number of observations minus the number of necessary relations obtaining among these observations." A necessary relationship for $s_{\bar{X}}$ is that $\Sigma(X - \bar{X}) = 0$.

Another approach to explaining degrees of freedom is to emphasize the parameters that are being estimated by statistics. The rule with this approach is that df is equal to the number of observations minus the number of parameters that are estimated with a sample statistic. In the case of $s_{\bar{X}}$, 1 df is subtracted because $\bar{X}$ is used as an estimate of μ.

Now, let's turn to your use of the t distribution to determine whether a correlation coefficient, r, is significantly different from .00. The test statistic was calculated using the formula

$$t = (r)\sqrt{\frac{N-2}{1-r^2}}$$

The r in the formula is a *linear* correlation coefficient, which is based on a linear regression line. The formula for the regression line is

$$Y' = a + bX$$

There are two parameters in this regression formula, a and b. Each of these parameters costs one degree of freedom, so the df for testing the significance of correlation coefficients is $N-2$.

These explanations of df for the one-sample t test and for testing the significance of r will prepare you for the reasoning I will give for determining df when you are analyzing two-group experiments.

INDEPENDENT-SAMPLES AND CORRELATED-SAMPLES DESIGNS

An experiment with two groups can be either an **independent-samples design**[5] or a **correlated-samples design**.[6] These different designs require different t-test formulas, so you must decide what kind of design you have before you analyze the data.

You cannot tell the difference between the two designs by knowing the independent variable or the dependent variable. And, after the data are analyzed, you cannot tell the difference if you know the t-test value or if you have the interpretation of the experiment. To tell the difference, you must attend to the *procedures* of the experiment.

The procedure to attend to is how participants are assigned to treatments. If participants are assigned in a way that allows a score in one treatment to be *paired* with a score in the other treatment, you have a correlated-samples design. A correlated-samples design always consists of *pairs* of scores. If pairing is not part of the procedure, you have an independent-samples design.

CLUE TO THE FUTURE

Most of the rest of this chapter is organized around independent-samples and correlated-samples designs. In Chapters 9 and 10, the procedures you will learn are appropriate only for independent samples. Chapter 11 is a correlated-samples chapter. Three-fourths of Chapter 13 is also organized around these two designs.

---

[5] An independent-samples design is also referred to as a between-subjects design.
[6] A correlated-samples design is also referred to as a within-subjects design.

■ Correlated-Samples Design

The widely used correlated-samples design[7] may come about in a number of ways. Fortunately, the actual arithmetic of calculating a *t*-test value is the same for any of three correlated-samples designs. The three types of designs are natural pairs, matched pairs, and repeated measures.

Natural Pairs.   In a **natural-pairs** investigation, the researcher does not assign the participants to one group or the other; the pairing occurs naturally, prior to the investigation. **Table 8.2** identifies one way in which natural pairs may occur in family relationships—in this case, fathers and sons. In such an investigation, you might ask whether fathers are shorter than their sons (or more religious, or more racially prejudiced, or whatever). Notice, though, that it is easy to decide that these are correlated-samples data: There is a logical pairing of the scores. That is, the two 74s belong together because they both came from the Smiths. Thus, pairing is based on coming from the same family.

Did you notice that Table 8.2 is the same as Table 4.2, which illustrated a bivariate distribution? A bivariate distribution is a characteristic of a correlated-samples design.

Matched Pairs.   In some situations, the researcher has control over the ways pairs are formed and a match can be arranged. One method is for two participants to be paired on the basis of similar scores on a pretest that is related to the dependent variable. This ensures that the two groups are fairly equivalent at the beginning of the experiment.

For example, in an experiment on the effect of hypnosis on problem solving, participants might be paired on the basis of IQ scores. IQ scores are known to be related to problem-solving ability. One member of each pair is randomly assigned to the hypnosis group (they solve problems while hypnotized) or the control group (they solve problems without hypnosis). If the two groups differ in problem-solving ability, you have some assurance that it wasn't because they were different to start with because the groups were equivalent in IQ, which is related to problem solving.

Another variation of **matched pairs** is the split-litter technique used with animal subjects. A pair from a litter is split, and one is put into each group. In this way, the

T A B L E 8.2   Illustration of a correlated-samples design

Father	Height (in.) X	Son	Height (in.) Y
Michael Smith	74	Mike, Jr.	74
Christopher Johnson	72	Chris, Jr.	72
Matthew Williams	70	Matt, Jr.	70
Joshua Brown	68	Josh, Jr.	68
Andrew Jones	66	Andy, Jr.	66
James Miller	64	Jim, Jr.	64

[7] Some texts use the terms *paired samples* or *dependent samples* instead of *correlated samples*.

genetic makeup of one group is matched with that of the other. The same technique has been used in human experiments with twins or siblings. Gosset's barley experiments used this design; Gosset started with two similar subjects (adjacent plots of ground) and assigned them at random to one of two treatments.

A third example of the matched-pairs technique occurs when a pair is formed *during* the experiment. Two subjects are "yoked" so that what happens to one, happens to the other (except for the independent and dependent variable). For example, if the procedure allows an animal in the experimental group to get a varying amount of food (or exercise or punishment), that same amount of food (or exercise or punishment) is administered to the "yoked" control subject.

The difference between the matched-pairs design and a natural-pairs design is that, with the matched pairs, the investigator can randomly assign one member of the pair to a treatment. In the natural-pairs design, the investigator has no control over assignment. Although the statistics are the same, the natural-pairs design is usually open to more interpretations than the matched-pairs design.

Repeated Measures. A third kind of correlated-samples design is called a **repeated-measures** design because more than one measure is taken on each participant. This design may take the form of a before-and-after experiment. A pretest is given, some treatment is administered, and a post-test is given. The mean of the scores on the post-test is compared with the mean of the scores on the pretest to determine the effectiveness of the treatment. Clearly, two scores should be paired: the pretest and the post-test scores of each participant. In such an experiment, each person is said to serve as his or her own control.

All three of these procedures for forming groups have one thing in common: The first score in one treatment is paired with the first score in the second treatment, the second score in the first treatment is paired with the second score in the second treatment, and so forth. In correlated-samples designs, one level of the independent variable is labeled X and the other level is labeled Y.

■ Independent-Samples Design

In an independent-samples design,[8] there is no reason to pair a score in one group with a particular score in the second group. Often in this design, the whole pool of participants is assigned in a random fashion to the groups. The design that is outlined in Table 8.1 is an independent-samples design. In independent-samples designs, one level of the independent variable is labeled X_1 and the other level is labeled X_2.

One caution is in order. Random assignment may be used with both of these designs. With an independent-samples design, the *pool* of participants is assigned randomly to groups. With correlated-samples designs, one member of a *pair* is assigned randomly.

Finally, as for similarities, both these designs have one independent variable that has two levels. Both have one dependent variable on which every participant has a score. Both designs lend themselves to either a one-tailed or a two-tailed test.

---

[8] Some texts use the terms *unpaired* or *uncorrelated* for this design.

The basic difference between the designs is that with a correlated-samples design there is a logical reason to pair up scores from the two groups; with the independent-samples design there is not.

PROBLEMS

6. Describe the two ways to state the rule governing the number of degrees of freedom.

7. Identify each of the following as an independent-samples design or a correlated-samples design. For each, identify the independent and dependent variables. Work all six problems before checking your answers.

   a. An investigator gathered many case histories of situations in which identical twins were raised apart—one in a "good" environment and one in a "bad" environment. The group raised in the "good" environment was compared with that raised in the "bad" environment on attitude toward education.

   *b. A researcher counted the number of aggressive encounters among children who were playing with worn and broken toys. Next, the children watched other children playing with new, shiny toys. Finally, the first group of children resumed playing with the worn and broken toys and the researcher again counted the number of aggressive encounters. This procedure was repeated for five additional groups of children. (For a similar experiment, see Barker, Dembo, and Lewin, 1941.)

   c. Alaskans with seasonal affective disorder (depression) spent two hours a day under bright artificial light. Before treatment the mean depression score was 20.0, and at the end of one week of treatment the mean depression score was 6.4. (See Hellekson, Kline, and Rosenthal, 1986.)

   d. To determine the effect of REM (rapid eye movement) sleep on mood, a researcher formed two groups: One was deprived of REM sleep and the other was not. Upon reporting to the sleep lab, participants were paired and randomly assigned to one of the two groups. When those in the deprivation group began to show REM, they were awakened, thus depriving them of REM sleep. The other member of the pair was then awakened for an equal length of time during non-REM sleep. The next day all participants filled out a mood questionnaire.

   e. To determine the effect of REM (rapid eye movement) sleep on mood, a researcher formed two groups: One was deprived of REM sleep and the other was not. Upon reporting to the sleep lab, participants were randomly assigned to one of the two groups. When those in the deprivation group began to show REM, they were awakened, thus depriving them of REM sleep. The other participants were awakened during non-REM sleep. The next day all participants filled out a mood questionnaire.

   f. A college dean faced a problem that suggested an experiment. Thirty-two freshmen had applied for the sophomore honors course. Only 16 could be accepted, so the dean flipped a coin for each applicant, with the result that 16 were selected. At graduation, she compared the mean grade-point average of those who had taken the course with those who had not to see if the sophomore honors course had any effect on GPA.   ■

THE t TEST FOR INDEPENDENT-SAMPLES DESIGNS

When a t test is used to decide whether two populations have the same mean, the null hypothesis is

$$H_0: \mu_1 = \mu_2$$

where the subscripts 1 and 2 are assigned arbitrarily to the two populations. If this null hypothesis is true, any difference between the two sample means will be due to chance. The task is to establish an α level, calculate a t-test value, and compare that value with a critical value of t in Table D. If the t value calculated from the data is larger than the critical value (that is, less probable than α), reject H_0 and conclude that the two samples came from populations with different means. If the data-based t value is not as large as the critical value, retain H_0. I expect that this sounds familiar to you. For an independent-samples design, the formula for the **t test** is

$$t = \frac{\bar{X}_1 - \bar{X}_2}{s_{\bar{X}_1 - \bar{X}_2}}$$

The term $s_{\bar{X}_1 - \bar{X}_2}$ is the **standard error of a difference**, and **Table 8.3** shows formulas for calculating it. Use the formula at the top of the table when the two samples have an unequal number of scores. In the situation where $N_1 = N_2$, the formula simplifies to those shown on the bottom of Table 8.3.

TABLE 8.3 Formulas for $s_{\bar{X}_1 - \bar{X}_2}$, the standard error of a difference, for independent-samples t tests

---

If $N_1 \neq N_2$:

$$s_{\bar{X}_1 - \bar{X}_2} = \sqrt{\left(\frac{\Sigma X_1^2 - \dfrac{(\Sigma X_1)^2}{N_1} + \Sigma X_2^2 - \dfrac{(\Sigma X_2)^2}{N_2}}{N_1 + N_2 - 2} \right) \left(\frac{1}{N_1} + \frac{1}{N_2} \right)}$$

If $N_1 = N_2$:

$$s_{\bar{X}_1 - \bar{X}_2} = \sqrt{s_{\bar{X}_1}^2 + s_{\bar{X}_2}^2}$$

$$= \sqrt{\left(\frac{\hat{s}_1}{\sqrt{N_1}} \right)^2 + \left(\frac{\hat{s}_2}{\sqrt{N_2}} \right)^2}$$

$$= \sqrt{\frac{\Sigma X_1^2 - \dfrac{(\Sigma X_1)^2}{N_1} + \Sigma X_2^2 - \dfrac{(\Sigma X_2)^2}{N_2}}{N_1(N_2 - 1)}}$$

This t-test formula is the "working formula" for the more general case in which the numerator is $(\bar{X}_1 - \bar{X}_2) - (\mu_1 - \mu_2)$. For the cases in this book, the hypothesized value of $\mu_1 - \mu_2$ is zero, which reduces the general case to the working formula shown above. Thus, the t test, like many other statistical tests, consists of a difference between a statistic and a parameter divided by the standard error of the statistic.

The formula for degrees of freedom for independent samples is $df = N_1 + N_2 - 2$. Here is the reasoning. For each sample, the number of degrees of freedom is $N - 1$ because, for each sample, a mean has been calculated with the restriction that $\Sigma(X - \bar{X}) = 0$. Thus, the total degrees of freedom is $(N_1 - 1) + (N_2 - 1) = N_1 + N_2 - 2$.

The following example shows the use of an independent-samples design to evaluate the effect of a drug on learning a complex problem-solving task. The drug

TABLE 8.4 Number of errors for each monkey during six days of training

Drug group, X_1	Placebo group, X_2
34	39
52	57
26	68
47	74
42	49
37	57
40	
ΣX 278	344
ΣX^2 11,478	20,520
N 7	6
$\bar{X}$ 39.71	57.33

Applying the t test,

$$t = \frac{\bar{X}_1 - \bar{X}_2}{s_{\bar{X}_1 - \bar{X}_2}} = \frac{\bar{X}_1 - \bar{X}_2}{\sqrt{\left(\dfrac{\Sigma X_1^2 - \dfrac{(\Sigma X_1)^2}{N_1} + \Sigma X_2^2 - \dfrac{(\Sigma X_2)^2}{N_2}}{N_1 + N_2 - 2}\right)\left(\dfrac{1}{N_1} + \dfrac{1}{N_2}\right)}}$$

$$= \frac{39.71 - 57.33}{\sqrt{\left(\dfrac{11,478 - \dfrac{(278)^2}{7} + 20,520 - \dfrac{(344)^2}{6}}{7 + 6 - 2}\right)\left(\dfrac{1}{7} + \dfrac{1}{6}\right)}}$$

$$= \frac{-17.62}{\sqrt{\left(\dfrac{437.43 + 797.33}{11}\right)(0.31)}} = \frac{-17.62}{5.90} = -2.99$$

$$df = N_1 + N_2 - 2 = 7 + 6 - 2 = 11$$

group (seven monkeys) received pills while the other group (six monkeys) was given an inert substance (a placebo). Pills and training on the task continued for six days. The total number of errors that each monkey made was the dependent variable.

The null hypothesis is that the drug has no effect on errors. If true, then any difference in sample means is the result of chance. A two-tailed test is called for because the investigator is interested in any effect the drug has, either to enhance or to inhibit performance.

The number of errors each animal made and the t test are shown in **Table 8.4**. Because the N's are unequal for the two samples, the longer formula for the standard error must be used. The t value for these data is -2.99, a negative value. For a two-tailed test, always use the absolute value of t. Thus, $|-2.99| = 2.99$.

To evaluate a data-based t value of 2.99 with 11 df, you need a critical value from **Table D**. Begin by finding the row with 11 df. The critical value in the column for a two-tailed test with $\alpha = .05$ is 2.201. That is, $t_{.05}(11) = 2.201$. Thus, the null hypothesis can be rejected. Because 2.99 is also larger than the critical value at the .02 level (2.718), the results would usually be reported as "significant at the .02 level."

The final step is to interpret the results. Stop for a moment and compose your interpretation of what this experiment shows.

Because the drug group averaged fewer errors than the placebo group (39.71 vs. 57.33), conclude that the drug *facilitated* learning ($p < .02$).

PROBLEMS

*8. Imagine being lost at sea, bobbing around in an orange life jacket. The Coast Guard sends out two kinds of observers to find you—fellow humans and pigeons. Each observer is successful; the search time in minutes for each observer is given in the table. Decide on a one- or two-tailed test, analyze the results, and write a conclusion.

Fellow humans	Pigeons
45	31
63	24
39	20

*9. Karl Lashley (1890–1958) did many studies to find out how the brain stores memories. In one experiment, rats learned a simple maze that had one blind alley. Afterward, they were anesthetized and operated on. For half the rats, 20 percent of the cortex of the brain was removed. For the other half, the same operation was performed, except that no brain tissue was removed (a sham operation). After the rats recovered, they were given 20 trials in the simple maze, and the number of errors was recorded. Identify the independent variable and the dependent variable, and then analyze the data in the table. Write a conclusion about the effect of a 20 percent loss of cortex on retention by rats of a simple maze task.

	Percent of cortex removed	
	0	20
ΣX	208	252
ΣX^2	1706	2212
N	40	40

10. The general manager of a large office force is considering a new word processing package. To compare it to the one in use, she picked 20 "new hires" and randomly assigned them to training on the old package or the new package. After two weeks of training and work, each employee completed the same "test procedure." The number of minutes required is shown in the table. Because two of the workers were absent several times and one quit, the N was 17. Analyze the data and draw a conclusion.

New package	Old package
4.3	6.3
4.7	5.8
6.4	7.2
5.2	8.1
3.9	6.3
5.8	7.5
5.2	6.0
5.5	5.6
4.9	

*11. Two sisters who attended different universities began a friendly discussion of the relative intellectual capacities of their fellow students. Each was sure that she had to compete with brighter students than her sister did. Though both were majoring in literature, each had an appreciation for quantitative thinking, so they decided to settle their discussion with ACT admission scores of freshmen at the two schools. Suppose they bring you the following data and ask for an analysis. The N's represent all freshmen for one year.

	The U.	State U.
$\overline{X}$	21.4	21.5
$\hat{s}$	3	3
N	8000	8000

The sisters agreed to set $\alpha = .05$. Begin by deciding whether to run a one-tailed or a two-tailed test. Run the test and interpret your results. Be sure to carry several decimal places in your work on this problem. ■

THE t TEST FOR CORRELATED-SAMPLES DESIGNS

The formula for a t test for data from a correlated-samples design has a familiar theme: a difference between means divided by the standard error of a difference. The standard error of a difference between means of correlated samples is symbolized $s_{\bar{D}}$.

One formula for a t test between correlated samples is

$$t = \frac{\bar{X} - \bar{Y}}{s_{\bar{D}}}$$

where $s_{\bar{D}} = \sqrt{s_{\bar{X}}^2 + s_{\bar{Y}}^2 - 2r_{XY}(s_{\bar{X}})(s_{\bar{Y}})}$

$df = N - 1$, where $N = $ number of pairs

The number of degrees of freedom in the correlated-samples case is the number of *pairs* minus one. Although each pair has two values, once one value is determined, the other is restricted to a similar value. (After all, they are called *correlated* samples.) In addition, another degree of freedom is subtracted when $s_{\bar{D}}$ is calculated. This loss is similar to the loss of 1 df when $s_{\bar{X}}$ is calculated.

As you can see by comparing the denominator of the correlated-samples t test with the denominator of the independent samples t test (Table 8.3, for $N_1 = N_2$), the difference in the two procedures is the term $2r_{XY}(s_{\bar{X}})(s_{\bar{Y}})$. Of course, when $r_{XY} = 0$, this term drops out of the formula, and the standard error is the same as for independent samples.

Also notice what happens to the standard error term in the correlated-samples case where $r > 0$: The standard error is *reduced*. Such a reduction will increase the size of t. Whether this reduction will increase the likelihood of rejecting the null hypothesis depends on how much t is increased because there are fewer degrees of freedom in a correlated-samples design than in an independent-samples design.

The formula $s_{\bar{D}} = \sqrt{s_{\bar{X}}^2 + s_{\bar{Y}}^2 - 2r_{XY}(s_{\bar{X}})(s_{\bar{Y}})}$ is used only for illustration purposes. The algebraically equivalent but arithmetically easier *direct-difference method* does not require you to calculate r. To find $s_{\bar{D}}$ by the direct-difference method, find the difference between each pair of scores, calculate the standard deviation of these difference scores, and divide the standard deviation by the square root of the number of pairs.

Thus,

$$t = \frac{\bar{X} - \bar{Y}}{s_{\bar{D}}} = \frac{\bar{X} - \bar{Y}}{s_D/\sqrt{N}}$$

where $s_D = \sqrt{\dfrac{\Sigma D^2 - \dfrac{(\Sigma D)^2}{N}}{N - 1}}$

$D = X - Y$

$N = $ number of *pairs* of scores

Here is an example of a correlated-samples design and a t-test analysis. Suppose you were interested in the effects of interracial contact on racial attitudes. You have a fairly reliable test of racial attitudes in which high scores indicate more positive attitudes. You administer the test one Monday morning to a biracial group of fourteen 12-year-old girls who do not know each other but who have signed up for a week-long community day camp. The campers then spend the next week taking nature walks, playing ball, eating lunch, swimming, making things, and doing the kinds of things that camp directors dream up to keep 12-year-old girls busy. On Saturday morning, the girls are again given the racial attitude test. Thus, the data consist of 14 pairs of before-and-after scores. The null hypothesis is that the mean of the population of "after" scores is equal to the mean of the population of "before" scores or, in terms of the specific experiment, that a week of interracial contact has no effect on racial attitudes.

Suppose the data in **Table 8.5** were obtained. Using the sum of the D and D^2 columns in Table 8.5, you can find s_D:

$$s_D = \sqrt{\frac{\Sigma D^2 - \frac{(\Sigma D)^2}{N}}{N-1}} = \sqrt{\frac{897 - \frac{(-81)^2}{14}}{13}} = \sqrt{32.95} = 5.74$$

$$s_{\bar{D}} = \frac{s_D}{\sqrt{N}} = \frac{5.74}{\sqrt{14}} = 1.53$$

T A B L E 8.5 Hypothetical data from a racial attitudes study

| Name* | Racial attitude scores | | D | D^2 |
	Before day camp X	After day camp Y		
Brittany	34	38	−4	16
Ashley	22	19	3	9
Jessica	25	36	−11	121
Amanda	31	40	−9	81
Sarah	27	36	−9	81
Megan	32	31	1	1
Caitlin	38	43	−5	25
Samantha	37	36	1	1
Stephanie	30	30	0	0
Katherine	26	31	−5	25
Emily	16	34	−18	324
Lauren	24	31	−7	49
Kayla	26	36	−10	100
Rachel	29	37	−8	64
Sum	397	478	−81	897
Mean	28.36	34.14		

*The girls' names are the 14 most common in the United States, in order (Dunkling, 1993).

Thus,

$$t = \frac{\bar{X} - \bar{Y}}{s_{\bar{D}}} = \frac{28.36 - 34.14}{1.53} = \frac{-5.78}{1.53} = -3.78$$

$$df = N - 1 = 14 - 1 = 13$$

Because $t_{.01}$ (13 df) $= 3.012$, a t value of 3.78 is significant beyond the .01 level; that is, $p < .01$. The "after" mean was larger than the "before" mean; therefore, the conclusion is that racial attitudes were significantly more positive after a week of camp than before.

You might note that $\bar{X} - \bar{Y} = \bar{D}$, the mean of the difference scores. In this problem, $\Sigma D = -81$ and $N = 14$, so $\bar{D} = \Sigma D/N = -81/14 = -5.78$.

Gosset, who discovered the t distribution, preferred the correlated-samples design. In his agricultural experiments, there were significant correlations between the yields of the old barley and the new barley grown on adjacent plots. This correlation reduced the standard-error term in the denominator of the t test, making the correlated-samples design more sensitive than the independent-samples design for detecting a difference between means. In behavioral science experiments today, the correlated-sample design is widely used.

PROBLEMS

12. Give a formula and definition for each of the following symbols.

a. s_D **b.** D

c. $s_{\bar{D}}$ **d.** t (verbal definition)

e. $\bar{Y}$

13. The table shows data based on the experiment described in problem 7, part b. Reread problem 7, part b, and analyze the scores in this table. Write a conclusion.

Before	After
16	18
10	11
17	19
4	6
9	10
12	14

*14. Here are some data from a consulting job that I worked on some time ago. A lawyer asked me to analyze salary data for employees at a large rehabilitation center to determine whether there was any evidence of sex discrimination. For all employees, men made about 25 percent more than women did. To rule out explanations like "more educated" and "more experienced," a subsample of employees with bachelor's degrees was separated out. From this group, men and women with equal experience were paired, resulting in an N of 16 pairs. For the data in the table, compare the mean annual salaries of women and men. In your conclusion (which you can direct to the judge), explain whether education, experience, or chance could account for the observed difference. *Note:* You may have to do some thinking to set up this problem. Start by looking at the data you have.

	Women, X	Men, Y
ΣX or ΣY	\$204,516	\$251,732
ΣX^2 or ΣY^2	2,697,647,000	4,194,750,000
ΣXY	3,314,659,000	

*15. Decide whether the following study is an independent- or correlated-samples experiment and analyze the data. Participants were randomly divided into two groups. Six participants found their reaction time (RT) to an auditory "go" signal and then their RT to a visual "go" signal. Five participants found their RT first to the visual signal and then to the auditory signal. The scores for each person are arranged in rows for the 11 members of the class.

Auditory RT (seconds)	Visual RT (seconds)
0.16	0.17
0.19	0.23
0.23	0.20
0.14	0.19
0.19	0.19
0.18	0.22
0.21	0.23
0.18	0.16
0.17	0.21
0.16	0.19
0.17	0.18

16. Two groups were chosen randomly from a large sociology class to study the effects of primacy versus recency. Both groups were given a two-page description of a person at work, with a paragraph telling how the person had been particularly helpful to a new employee. For half of the descriptions, the "helping" paragraph was near the beginning (primacy), and for the other half the "helping" paragraph was near the end (recency). After reading, each participant wrote a page summary of how the worker would spend leisure time. The dependent variable was the number of positive adjectives (words like *good, exciting, cheerful*) in the page summary of leisure activities. Based on these data, write a conclusion about primacy and recency.

Recency	Primacy
13	10
14	17
4	9
8	11
11	12
6	5
6	10
10	16
14	14

17. Asthma occurs in as much as 5 percent of the population. In one experiment, a group whose attacks began before age 3 was compared with a group whose attacks began after age 6. The dependent variable was the number of months the asthma persisted. Analyze the difference between means and write a conclusion that tells what you have found.

	Onset of asthma	
	Before age 3	After age 6
$\bar{X}$	60	36
$\hat{s}$	12	6
N	45	45

ASSUMPTIONS WHEN USING THE t DISTRIBUTION

You can perform a t test on the difference between means on any two-group data you have or any that you can beg, borrow, buy, or steal. No doubt about it, you can easily come up with a t value using the formulas you have learned.

Once you have obtained a t value from your data, you can leap to the conclusion that the t distribution will give you an accurate probability figure. By consulting Table D you can *get* a probability figure and, using the .05 rule, retain or reject the null hypothesis.

Should you leap to the conclusion that the t distribution will be accurate? *Will* the t distribution reflect actual empirical probabilities?

The t distribution will give correct probabilities when the assumptions it is based on are true for the populations being analyzed. To give you an example, mathematical statisticians make three assumptions when they derive a t distribution for analysis of an independent-samples design:

For the two populations, the scores on the dependent variable

1. are normally distributed and
2. have variances that are equal.

In addition, the two samples

3. are randomly selected from their populations.[9]

Now let's return to the question, When will the *t* distribution produce accurate probabilities? The answer is, When random samples are obtained from populations that are normally distributed and have equal variances.

This may appear to be a tall order. It is, and in practice no one is able to demonstrate these characteristics exactly. The next question becomes, Suppose I am not sure that my data have these characteristics. Am I likely to reach the wrong conclusion if I use Table D?

There is no simple answer to your reasonable question. In the past, several studies have suggested that the answer is, No, you won't reach the wrong conclusion because the *t* test is a robust test. (*Robust* means that a test gives you fairly accurate probabilities even when the data do not meet the assumptions on which they are based.) However, Bradley (1978) points out that robustness has not been defined quantitatively and, furthermore, that several studies (largely ignored by textbooks) show sizable departures from accuracy in a variety of situations. [See, for example, Blair and Higgins (1985).] In actual practice, however, many researchers routinely use a *t* test unless one or more of these assumptions is clearly not justified.

EFFECT SIZE

An effect size index gives you important information that a *t* test doesn't provide. A *t* test can tell you that the independent variable had an *effect* by revealing that it is unlikely that the two samples came from the same population. An effect size index tells you *how much of an effect* the independent variable had.

The effect size index for a two-sample design is called *d*. The mathematical formula for *d* is

$$d = \frac{\mu_1 - \mu_2}{\sigma}$$

where μ_1 and μ_2 are the populations the samples come from
   σ is the population standard deviation (which is assumed to be the same for the two populations).

As you know, population parameters are unknowable, so calculating *d* requires estimates of the parameters. As you also know, $\bar{X}$ is used to estimate μ and $\hat{s}$ is used to estimate σ. The actual calculation of *d*, however, depends on whether you have an independent-samples design or a correlated-samples design.

■ Effect Size Index for Independent Samples

The formula for *d*, the effect size index in the independent-samples case, is:

$$d = \frac{\bar{X}_1 - \bar{X}_2}{\hat{s}}$$

---

[9] As mentioned before, *researchers* use random assignment where possible because random samples are usually impossible to obtain.

Calculating $\bar{X}_1$ and $\bar{X}_2$ requires no explanation. The calculation of $\hat{s}$ depends on whether or not the sample sizes are equal. When $N_1 = N_2$,

$$\hat{s} = \sqrt{N_1}\,(s_{\bar{X}_1 - \bar{X}_2})$$

where N_1 is the sample size for *one group*

By giving a little thought, you can see that this formula follows from your work in Chapter 7, where the standard error of the mean was found by dividing $\hat{s}$ by $\sqrt{N}$.

In the case where $N_1 \neq N_2$, you cannot find $\hat{s}$ directly from $s_{\bar{X}_1 - \bar{X}_2}$. To find $\hat{s}$ when $N_1 \neq N_2$, use the formula

$$\hat{s} = \sqrt{\frac{\hat{s}_1{}^2(df_1) + \hat{s}_2{}^2(df_2)}{df_1 + df_2}}$$

where df_1 and df_2 are the degrees of freedom ($N - 1$) for each of the two samples
$\hat{s}_1$ and $\hat{s}_2$ are the sample standard deviations

The formula for $s_{\bar{X}_1 - \bar{X}_2}$ for $N_1 \neq N_2$ (Table 8.3) does not include separate elements for $\hat{s}_1$ and $\hat{s}_2$. Thus, you will have to calculate $\hat{s}_1$ and $\hat{s}_2$ from the raw data for each sample.

■ Effect Size Index for Correlated Samples

The formula for d for correlated samples is

$$d = \frac{\bar{X} - \bar{Y}}{\hat{s}}$$

The formula for $\hat{s}$ for correlated samples is,

$$\hat{s} = \sqrt{N}\,(s_{\bar{D}})$$

■ Interpretation of d

The task of interpreting d has been simplified by conventions proposed by Cohen (1969) and widely adopted by other researchers.

Small effect $d = .20$
Medium effect $d = .50$
Large effect $d = .80$

To illustrate the interpretation of d, I will calculate an effect size index for two problems that you have already worked. For both of these, the t test revealed a significant difference.

When you compared the ACT scores at two state universities (problem 11), you found that the mean for State U. freshmen (21.5) was significantly higher than the mean for freshmen at The U. (21.4). The standard error of the difference for this $N_1 = N_2$ study was 0.0474. Thus,

$$d = \frac{\bar{X}_1 - \bar{X}_2}{\hat{s}} = \frac{21.5 - 21.4}{\sqrt{8000}(0.0474)} = \frac{0.1}{4.240} = 0.02$$

An effect size index of 0.02 is very, very small. Thus, although the difference between the two schools is statistically significant, the size of this difference seems to be of no consequence.

In problem 14, you found that there was a significant difference between the salaries of men and women. An effect size index will reveal whether this difference can be considered small, medium, or large.

$$d = \frac{\bar{X} - \bar{Y}}{\hat{s}} = \frac{12{,}782.25 - 15{,}733.25}{\sqrt{16(2872)}} = \frac{2951}{718} = 1.02$$

An effect size index of 1.02 qualifies as large. Thus, the statistical analysis of these data, which includes a t test and an effect size index, supports the conclusion that discrimination was taking place and that the effect was large.

HOW TO REJECT THE NULL HYPOTHESIS—THE TOPIC OF POWER

From your reading of this text and other material about experiments, you may have the impression that researchers do not really believe that the null hypothesis is true. They get excited when they reject H_0 and unhappy when they don't. That impression is correct. It is also reasonable. To reject H_0 is to be left with only one alternative, H_1, from which a conclusion can be drawn. To retain H_0 is to be left up in the air. You don't know whether the null hypothesis is really true or whether it is false and you just failed to detect it. So if you are going to design and run an experiment, you should maximize your chances of rejecting H_0.[10]

The question of whether or not an experiment will reject a false null hypothesis is a question of the **power** of the experiment. Power is defined by statisticians as $1 - \beta$, where β is the probability of a Type II error. Researchers use power analyses to help decide how large an N an experiment needs and to evaluate the worth of an experiment that retains the null hypothesis. Conducting a complete power analysis is covered in other textbooks.[11] I do want you to know, however, the three factors that determine the power of an experiment: effect size, standard error, and α.

1. *Effect size.* The larger the effect size, the more likely you are to reject H_0. For example, the larger the difference between the mean of the experimental group population and the mean of the control group population, the more likely it is that the samples will lead you to correctly conclude that the populations are different. Of course, determining effect size *before* the data are gathered is difficult, and experience helps.[12]

[handwritten margin note: Ability to find a significant difference when there is a difference]

---

[10] Some of you may object to this philosophy of deliberately setting out to reject H_0. You might argue that a scientist should try to discover the "true situation," which certainly may be that H_0 is true. Unfortunately, inferential statistics are poor tools for establishing the truth of H_0.

[11] See Howell (1995), Chapter 15; Aron and Aron (1994), Chapter 8; or Hurlburt (1994), Chapter 15.

[12] There is some danger in trusting your rational conclusions. Otto Loewi received a Nobel Prize in 1936 for demonstrating the chemical nature of nerve transmission in 1921. The idea for the experiment came to him during the night. He got out of bed, went to his laboratory, and performed an experiment that involved isolated frog hearts and some Ringer solution. Later he wrote, "If I had carefully considered it in the daytime, I would undoubtedly have rejected the kind of experiment I performed." Loewi goes on to explain that it seems improbable that the actual difference between the Ringer solution with the chemical and that without could be detected by the method he used. "It was good fortune that at the moment of the hunch I did not think but acted" (Loewi, 1963).

2. *The standard error of a difference.* Look at the formulas for *t* on pages 193 and 197. You can see that as $s_{\bar{X}_1 - \bar{X}_2}$ and s_D get smaller, *t* gets larger and you are more likely to reject H_0. Here are two ways you can reduce the size of the standard error:

Sample size. The larger the sample, the smaller the standard error of a difference. **Figure 6.3** shows that the larger the sample size, the smaller the standard error of the mean. The same relationship is true for the standard error of a difference. Cohen (1992) provides a helpful table of sample sizes required to reject H_0. This table, of course, makes assumptions about effect size. Many times, however, sample size is dictated by practical considerations—time, money, or availability of participants.

Sample variability. Reducing the variability in the sample data will produce a smaller standard error. You can reduce variability by using reliable measuring instruments, recording data correctly, being consistent, and, in short, reducing the "noise" or random error in your experiment.

3. α. The larger α is, the more likely you are to reject H_0. As you know, the conventional limit for α is .05. Values above .05 begin to raise doubts in readers' minds, although there has been a healthy recognition in recent years of the arbitrary nature of the .05 rule. As for researchers whose results come out with $p = .08$ (or near there), they are very likely to remain convinced that a real difference exists and proceed to gather more data, expecting that the additional data will establish a statistically significant difference.

I will close this section on power by asking you to imagine that you are a researcher directing a project that could make a Big Difference. (Because you are imagining this, the difference can be in anything you would like to imagine—the health of millions, the destiny of nations, your bank account, whatever.) Now suppose that the success or failure of the project hinges on one final statistical test. One of your assistants comes to you with the question, "How much power do you want for this last test?"

"All I can get," you answer.

If you examine the list of factors that influence power, you will find that there is only one item that you have some control over, and that is the standard error of a difference. Of the factors that affect the size of the standard error of a difference, the one that most researchers can best control is *N*. So, allocate plenty of power to important statistical tests—use large *N*'s.

SIGNIFICANT RESULTS AND IMPORTANT RESULTS

When I was a rookie instructor, I did some research with an undergraduate student, David Cervone, who was interested in hypnosis. We used two rooms in the library to conduct the experiment. When the results were analyzed, two of the groups were significantly different. One day, the librarian asked how the experiment had come out.

"Oh, we got some significant results," I said.

"I imagine David thought they were more significant than you did," was the reply.

At first I was confused. Why would David think they were more significant than I would? Point oh-five was point oh-five. Then I realized that the librarian and I were using the word *significant* in two quite different ways. He meant "important" and I meant "not due to chance."

By *significant*, I meant that the difference was a *reliable* one that would be expected to occur again if the study was run again. That is the meaning used so far in this chapter. However, let's pursue the librarian's meaning and ask about the *importance* of a difference.

Inferential statistics, including hypothesis testing and effect size indexes, are helpful in deciding about importance, but only in a negative way. If a difference is *not* statistically significant, then it probably isn't important (because it may not be a reliable difference that would occur again). If it *is* statistically significant, then you have to use considerations outside of statistics to decide about importance.

When you worked problem 11, you found that the difference in the college admissions scores between two schools was significant at the .05 level. However, the difference was only one-tenth of a point, which doesn't seem important at all. Problem 17, which showed that the number of months a child is subject to asthma attacks depends on whether the first attack occurred before age 3 or after age 6, also produced significant results. This difference, however, is important information for parents.

Arthur Irion (1976) captured the two meanings of the word *significant* when he reported that his study "reveals that there is significance among the differences although, of course, it doesn't reveal what significance the differences have."

The basic point is that a study that has statistically significant results may or may not have important results. You have to decide about the importance without the help of inferential statistics.

PROBLEMS

18. Your confidence that the probabilities are accurate when you run an independent-samples *t* test depends on certain assumptions. List them.

19. In problem 8 you found that pigeons were significantly better at spotting people in the ocean. How much better are pigeons? Calculate and interpret an effect size index for those data.

20. Problem 9 revealed that there was no significant loss in the memory of rats that had 20 percent of their cortex removed. What is the effect size index for those data? My interpretation will note that the nonsignificant *t* test was based on 80 rats, which is quite a large sample. Perhaps, with this hint, you will be able to write an interpretation based on both the effect size index and the *t* test.

21. In the example problem for the correlated-samples *t* test (page 198), you found that a week of camp activities significantly improved racial attitudes. Calculate and interpret an effect size index for those data.

22. Using words, define statistical power.

23. List the four factors that determine the power of an experiment. [If you can recall them without looking at the text, you know that you have been reading actively (and effectively).]

24. In your own words, distinguish between a significant difference and an important difference.

25. In an experiment in teaching German, the holistic method produced better scores than the lecture-discussion method. A report of the experiment included the phrase,

"$p < .01$." Write a description of what this phrase means. What event does the p refer to?

26. For an experiment on the effects of sleep on memory, eight volunteers were randomly divided into two groups. Everyone learned a list of ten nonsense syllables (later called consonant-vowel-consonants—CVCs). One group then slept for four hours, while the other group engaged in daytime activities. Then each subject recalled the CVCs. Two days later, everyone learned a new list. The group that had slept now engaged in daytime activities, and the group that had been active slept. After four hours, each person recalled CVCs. The scores in the table represent the number of CVCs recalled by each person under the two conditions. Perform a t test, calculate d, and write a sentence of interpretation. (This problem is modeled after a 1924 study by Jenkins and Dallenbach.)

Awake	Asleep
3	3
2	3
4	5
0	3
3	2
3	4
1	2
4	6

27. An educational psychologist was interested in the effect of set (previous experience) on a problem-solving task. [See Birch and Rabinowitz, (1951).] The task was to tie together two strings (A and B) that were suspended from the ceiling 12 feet apart. Unfortunately, a participant could not reach both strings at one time. The plight is shown in the illustration. The solution to the problem is to tie a weight to string A, swing it out, and then catch it on the return. The psychologist supplied the weight (an electrical light switch) under one of two conditions. In one condition, participants had previously wired the switch into an electrical circuit and used it to turn on a light. In the other condition, participants had no experience with the switch until they were confronted with the two-string problem. The question was whether having used the switch as an electrical device would have any effect on the time required to solve the problem. State the null hypothesis, analyze the data with a t test and an effect size index, and write a conclusion.

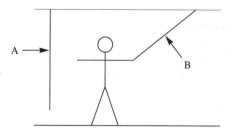

	Wired the switch into circuit	No previous experience with switch
$\bar{X}$	7.40 minutes	5.05 minutes
$\hat{s}$	2.13 minutes	2.31 minutes
N	20	20

28. A college French teacher wanted to test the claim that "concrete, immediate experience enhances vocabulary learning." As he drove from the college to the city one day (about 30 miles), he made an audiotape describing in French the terrain, signs, distances, and so forth. ("L'auto est sur le pont. Carrefour prochain est dangereux.") For the next part of his investigation, students in his French II class were ranked from 1 to 10. Numbers 1, 4, 5, 8, and 9 then listened to the tape while riding to the city (concrete, immediate experience) and numbers 2, 3, 6, 7, and 10 listened to the tape in the language laboratory. The next day a vocabulary test was given and the number of errors for each student was recorded. Think about how to set up these data, analyze them, and write a conclusion about immediate, concrete experience.

#1   7;   #2   7;   #3   15;   #4   7;   #5   12;
#6   22;   #7   24;   #8   12;   #9   21;   #10   32

29. Skim over the objectives at the beginning of the chapter as a way to consolidate what you have learned.   ∎

WHAT WOULD YOU RECOMMEND? Chapters 5–8

It is time again for you to do a set of *What would you recommend?* problems. Your task is to choose an appropriate statistical technique from among the several you have learned in the previous four chapters. No calculations are necessary. Reviewing is allowed. For each problem that follows, (1) recommend a statistic that will either answer the question or make the comparison, and (2) explain why you recommend that statistic.

A. The number of smiles by 10-month-old infants was recorded for 12 minutes. There was an audience during half the time and no audience during the other half. The researchers—Jones, Collins, and Hong (1991)—wanted to find out whether smiles are for communication (require an audience) or whether they express emotion (don't require an audience).

B. McCrae and Costa (1990, p. 105) describe an adult social worker who has a score of 73 on a personality inventory that measures gregariousness (the NEO Personality Inventory). The scores on this inventory are distributed normally with a population mean of 50 and a population standard deviation of 10. How could you find the percent of the population that is less gregarious than this social worker?

C. Suppose a psychologist conducted a workshop to help people improve their social skills. At the end of the workshop, all the participants filled out the NEO

Personality Inventory. Suppose the mean gregariousness score of the participants was 55. What statistical test could you run to determine whether the workshop produced a group whose mean gregariousness score was larger than the population mean?

D. A sampling of high school students takes the SAT test each year. The mean score for the students in a particular state is readily available. What statistical technique could you use that would give you a set of brackets that "captures" a population mean?

E. One interpretation of the statistical technique in problem D is that the brackets capture the *state's* population mean. This interpretation is not correct. Why not?

F. In a sample of 100 people, a correlation coefficient of .20 was found between the number of illnesses in the past year and scores on a measure of stress. How could you determine the likelihood of such a correlation coefficient if there were actually no relationship between the two variables?

G. Colleges and universities in the United States are sometimes divided into six categories: public universities, four-year public institutions, two-year public institutions, private universities, four-year private institutions, and two-year private institutions. In any given year, each category has a certain number of students. Which of the six categories does your institution belong to? How could you find the probability that a student, chosen at random, comes from that category?

H. Children (mean age = 6 years) who ate breakfast at school participated in an experiment to determine the effect of sugar on activity. On some days a particular child would receive sugar with breakfast and on other days the child would not. This was true for all the children in the study. Later in the day each child was observed and activity recorded. [See Rosen et al. (1988), who found that the effect of sugar was quite small.]

The *t* test, at which you are now skilled, is the traditional way to analyze the results from an experiment that is designed to compare two treatments. Your next step is to learn a technique for analyzing the results of an experiment that has more than two treatments.

This technique is called the **analysis of variance** (ANOVA for short, pronounced uh-ṅove-uh). ANOVA is based on the very versatile concept of assigning the variability in the data to various sources. (This often means that ANOVA identifies the variables that produce changes in the scores.) Social and behavioral scientists use ANOVA more frequently than any other inferential statistical technique. It was invented by Sir Ronald A. Fisher, an English biologist and statistician.

So the transition this time is from a *t* test and its sampling distribution, the *t* distribution, to ANOVA and its sampling distribution, the *F* distribution. In Chapter 9 you will use ANOVA to compare means from an independent-samples design that involves more than two treatments. Chapter 10 will show you that ANOVA can be used to analyze experiments in which there are two *independent variables*, each of which may have two or more levels of treatment. In Chapter 11 you will use ANOVA to compare means from a correlated-samples design that has more than two treatments (but just one independent variable).

The analysis of variance is a widely used statistical technique, and Chapters 9 through 11 are devoted to an introduction to its elementary forms. Intermediate and advanced books explain more sophisticated (and complicated) analysis-of-variance designs. [Howell (1992) is particularly accessible.]

9

Analysis of Variance: One-Way Classification

OBJECTIVES FOR CHAPTER 9

After studying the text and working the problems in this chapter, you should be able to:

1. Identify the independent and dependent variables in a one-way ANOVA

2. Explain the rationale of ANOVA

3. Define *F* and explain its relationship to *t*

4. Compute sums of squares, degrees of freedom, mean squares, and *F* for an ANOVA

5. Construct a summary table of ANOVA results

6. Interpret the *F* value for a particular experiment and explain what the experiment shows

7. Distinguish between *a priori* and *post hoc* tests

8. Use the Tukey Honestly Significant Difference (HSD) test to make all pairwise comparisons

9. List and explain the assumptions of ANOVA

10. Calculate and interpret *f*, an effect size index

In this chapter, you will work with the simplest analysis-of-variance design, a design called **one-way ANOVA**. The *one* in one-way refers to the one independent variable that is being analyzed. One-way ANOVA is used to find out if there are any statistically significant differences among three or more population means.[1] Experiments with more than two treatment levels are common in all disciplines that use statistics. Here are examples, some of which will be covered as you progress through this chapter:

    1. Samples of lower-, middle-, and upper-class persons were compared on attitudes toward religion.

---

[1] As will become apparent, ANOVA can be used also when there are just two groups.

2. Four groups learned a task. A different schedule of reinforcement was used for each group. Afterward, response persistence was measured.
3. Suicide rates were compared for countries that represented low, medium, and high degrees of modernization.
4. The effect of strychnine on memory was assessed. One group received strychnine; one control group received a saline injection and the other control group received no injection.

Except for the fact that these experiments have more than two groups, they are like those you analyzed with an *independent-samples t test* in Chapter 8. There is one independent and one dependent variable. The null hypothesis is that the population means are the same for all groups. The subjects in each group are independent of subjects in the other groups. The only difference is that the independent variable has more than two levels. To analyze such designs, use a one-way ANOVA.[2]

This chapter does not have "hypothesis testing" or "effect size" in the title (as did the two previous chapters on *t* tests). Nevertheless, ANOVA is a hypothesis-testing technique, and the question of effect size remains important when there are more than two treatments.

In example 1, the independent variable is social class, and it has three levels. The dependent variable is attitude toward religion. The null hypothesis is that the religious attitudes are the same in all three populations of social classes—that is, H_0:

$$\mu_{lower} = \mu_{middle} = \mu_{upper}$$

PROBLEM

*1. For examples 2, 3, and 4, identify the independent variable, the number of levels of the independent variable, the dependent variable, and the null hypothesis.   ■

A common first reaction to the task of determining if there is a difference among three or more population means is to run *t* tests on all possible pairs of sample means. For three populations, three *t* tests would be required ($\bar{X}_1$ vs. $\bar{X}_2$, $\bar{X}_1$ vs. $\bar{X}_3$, and $\bar{X}_2$ vs. $\bar{X}_3$). For four populations, six tests are needed.[3] This multiple *t*-test approach *will not work* (and not just because doing lots of *t* tests is tedious).

Here is the reasoning: Suppose you had 15 samples all drawn from the same population. (Because there is just one population, the null hypothesis is clearly true.) These 15 sample means will vary from one another as a result of chance factors. Now suppose you ran every possible *t* test (all 105 of them), retaining or rejecting each null hypothesis at the .05 level. How many times would you reject the null hypothesis? About five; that is, if a *t* test has a .05 probability of making a Type I error, and you run about 100 tests on samples all drawn from the same population, you'll have about five Type I errors when you finish.

Now, let's move from this theoretical analysis to the reporting of an experiment. Suppose you conducted a 15-group experiment, ran 105 *t* tests, and found five

---

[2] The ANOVA technique for analyzing a correlated-samples design is covered in Chapter 11.
[3] The formula for the number of combinations of *n* things taken two at a time is $[n(n-1)]/2$.

significant differences. If you then pulled out those five and said they were reliable differences (that is, differences that are not due to chance), people who understand the preceding paragraph would realize that you don't understand statistics (and would recommend that your manuscript not be published). You can protect yourself from such a disaster if you choose a technique that keeps the overall risk of a Type I error at an acceptable level (such as .05 or .01).

Sir Ronald A. Fisher (1890–1962) developed such a technique, called ANOVA. Fisher did brilliant work in genetics, but it has been overshadowed by his fundamental work in statistics. In genetics, it was Fisher who showed that Mendelian genetics was compatible with Darwinian evolution. (For a while after 1900, genetics and evolution were in opposing camps.) And, among other contributions, Fisher showed how a recessive gene could become established in a population.

In statistics, it was Fisher who invented ANOVA, the topic of this chapter and the next two. He also discovered the exact sampling distribution of r (1915), developed a general theory of estimation, and wrote *the* book on statistics. (*Statistical Methods for Research Workers* was first published in 1925 and went through 14 editions and several translations by 1973.)

Before getting into biology and statistics in such a big way, Fisher worked for an investment company for two years and taught in a public school for four years. [(See Greene (1966) or Yates (1981).]

RATIONALE OF ANOVA

The question that ANOVA answers is whether the populations that the samples come from have the same μ or whether at least one of the populations has a different μ. The reasoning (rationale) that leads up to an answer follows. I've grouped the reasoning into subsections on (1) hypothesis testing, and (2) the F distribution.

■ Hypothesis Testing

1. The populations you are comparing (the different treatments) may be exactly the same *or* one or more of them may have a different mean. These two descriptions cover all the logical possibilities and are, of course, the null hypothesis (H_0) and the alternative hypothesis (H_1).
2. Tentatively assume that the null hypothesis is correct. If the populations are all the same, any differences among sample means will be the result of chance.
3. Choose a sampling distribution that will show the probability of various differences among sample means when H_0 is true. The t distribution will work if you have two sample means but not if you have more than two. For more than two, the sampling distribution you need is the one that Fisher discovered, which is now called the F distribution.
4. Obtain data from the populations you are interested in. (This usually means you must conduct an experiment.) Perform calculations on the data using the procedures of ANOVA until you have an F value.

5. Compare your calculated F value to the critical value of F in the sampling distribution (F distribution). From this you can determine the probability of obtaining the data you did, *if the null hypothesis is true.*

6. Come to a conclusion about H_0. If the probability is .05 or less, reject H_0. With H_0 eliminated, you are left with H_1. If the probability is greater than .05, retain H_0, leaving you with both H_0 and H_1 as possibilities.

7. Tell the story of what the data show. If you reject H_0 when there are three or more treatments, a comparison between specific groups will require further data analysis. If you retain H_0, then your data leave you with both of the hypotheses. You don't have evidence against H_0.

I'm sure that the ideas in this list are more than somewhat vaguely familiar to you. In fact, I suspect that your only uncertainties are about step 3, the F distribution, and step 4, calculating an F value. The next several pages will explain the F distribution and then what to do with the data to get an F value.

■ The F Distribution

The **F distribution**, like the t distribution, is a sampling distribution. As such, it gives the probabilities of various outcomes when the null hypothesis is true (all samples come from identical populations). The F value that is calculated from the data is obtained by dividing one estimate of the population variance by a second estimate. Thus,

$$F = \frac{\text{estimate of } \sigma^2}{\text{estimate of } \sigma^2}$$

These two estimates of σ^2 are obtained by different methods. The numerator of the F ratio is obtained by a method that does a good job of estimating σ^2 *only* when H_0 is true. If H_0 is false, the estimate of σ^2 in the numerator will be too large.

The denominator of the F ratio is obtained by a method that does a good job of estimating σ^2 *regardless* of whether the null hypothesis is true or false.

Thus, when the null hypothesis is true, the expected value of F is about 1.00 because both numerator and denominator are estimators of σ^2, and $\sigma^2/\sigma^2 \cong 1.00$. If an F value is much larger than 1.00, there is cause to suspect that H_0 is false. Of course, when the null hypothesis is true, values somewhat larger or smaller than 1.00 are to be expected because of sampling fluctuation.

I have two additional explanations for you about the numerator and denominator of the F ratio. The first is an algebraic explanation and the second is a graphic explanation.

An algebraic explanation of the F ratio. Assume that the null hypothesis is true and that several treatments have produced a set of *treatment means.*

1. Using the treatment means as data, calculate a standard deviation. This standard deviation of treatment means is an acquaintance of yours, the standard error of the mean, $s_{\bar{X}}$.

2. Remember that $s_{\bar{X}} = \hat{s}/\sqrt{N}$. Squaring both sides gives $s_{\bar{X}}^2 = \hat{s}^2/N$. Multiplying both sides by N and rearranging gives you $\hat{s}^2 = N s_{\bar{X}}^2$.

3. $\hat{s}^2$ is, of course, an estimator of σ^2.

Thus, starting with a null hypothesis that is true and a set of treatment means, you can obtain an estimate of the variance of the populations that the samples come from. This estimate of σ^2 is called the *between-treatments estimate* (also called the between-means estimate and the between-groups estimate).

Notice that the validity of the between-treatments estimate of σ^2 depends on the assumption that the samples are all drawn from the same population. If one or more samples come from a population with a larger or smaller mean, the treatment means will vary more, and the between-treatments estimate will be larger.

The other estimate of σ^2 (the denominator) is obtained from the variability within each of the samples. Because each sample variance is an independent estimator of σ^2, averaging them produces a more reliable estimate. This estimate is called **error estimate** (also called the *within-groups estimate*). The error estimate is an unbiased estimate even if the null hypothesis is false. Finally, the two estimates can be compared. If the between-treatments estimate is much larger than the error estimate, the null hypothesis is rejected.

A graphic explanation of the *F* ratio.   Here is an illustrated explanation of the *F* ratio. **Figure 9.1** shows the situation when the null hypothesis is true ($\mu_A = \mu_B = \mu_C$). Three samples have been drawn from identical populations and a mean calculated for each sample. (See where the means are projected onto the dependent-variable line.) The means are fairly close together, so the variability among these three means (the between-treatments estimate) will be small. Look at **Figure 9.1**.

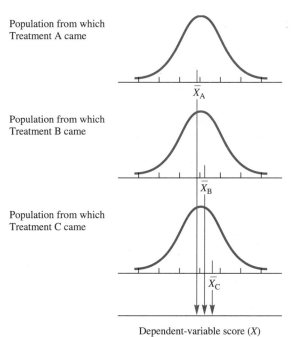

Population from which Treatment A came

$\bar{X}_A$

Population from which Treatment B came

$\bar{X}_B$

Population from which Treatment C came

$\bar{X}_C$

Dependent-variable score (X)

F I G U R E   9.1 H_0 is true. All three treatment groups are drawn from identical populations. Not surprisingly, samples from those populations produce means that are similar. (See arrow points on the X axis.)

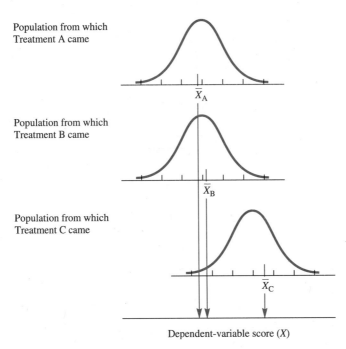

Population from which
Treatment A came

$\bar{X}_A$

Population from which
Treatment B came

$\bar{X}_B$

Population from which
Treatment C came

$\bar{X}_C$

Dependent-variable score (X)

FIGURE 9.2 H_0 is false. Two treatment groups come from identical populations. A third group comes from a population with a larger mean. The sample means show more variability than those in Figure 9.1.

In **Figure 9.2**, the null hypothesis is false (one group comes from a population with a larger μ). The projection of the means this time shows that population C produces a sample mean that increases the variability among the three sample means. Look at **Figure 9.2**.

By studying Figures 9.1 and 9.2, you can convince yourself that the between-treatments estimate is larger when the null hypothesis is false. So, if you have a small amount of variability between treatment means, retain H_0. If you have a large amount of variability between treatment means, reject H_0. Small and large, however, are relative terms and, in this case, they are relative to the population variance. A comparison of **Figures 9.3** and **9.1** illustrates that the amount of between-treatments variability *depends* on the population variance. In both Figures 9.1 and 9.3, the null hypothesis is true, but notice the projection of the sample means onto the dependent-variable line. There is more variability among the means that come from populations with greater variability. Figure 9.3, then, shows a large between-treatments estimate that is the result of large population variances and not the result of a false null hypothesis.

You can decide whether a large between-treatments estimate is due to a false null hypothesis rather than to a large population variance if you know the population variance. Fortunately, your data will produce such an estimate (the error estimate).

The F distribution is the sampling distribution that shows the probability of various ratios of variances when the null hypothesis is true. **Table F** in Appendix B gives you critical values of F for the .05 and the .01 levels of significance. The first version of this

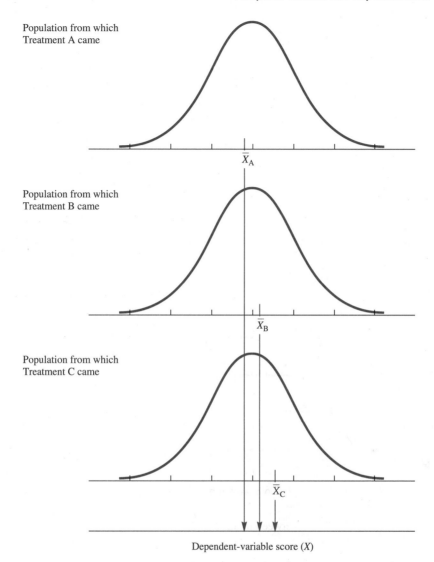

Population from which
Treatment A came

$\overline{X}_A$

Population from which
Treatment B came

$\overline{X}_B$

Population from which
Treatment C came

$\overline{X}_C$

Dependent-variable score (X)

FIGURE 9.3 H_0 is true. All three treatment groups are drawn from identical populations. Because the populations have larger variances than those in Figure 9.1, the sample means show more variability; that is, the arrow points are spaced farther apart.

table was compiled by George W. Snedecor of Iowa State University in 1934. At that time, Snedecor named the variance ratio F in honor of Fisher.

Table F allows you to compare an F value from an experiment with those listed in the table. *If the F value from the data is as large as or larger than the tabled value, reject the null hypothesis.* If the F value from the data is not as large as the tabled value, the null hypothesis must be retained.

Figure 9.4 is a picture of two F distributions. The curve that is based on 10 *df* in the numerator and 20 *df* in the denominator has the rejection region shaded (for an α

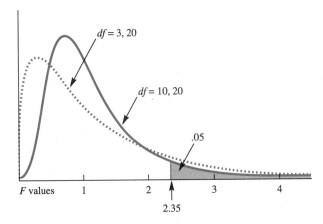

FIGURE 9.4 Two F distributions, showing their positively skewed nature. The distribution with $df = 10, 20$ shows a critical value of 2.35 for an α level of .05.

level of .05). If an ANOVA produced an F value (based on 10 and 20 df) that was equal to or greater than 2.35, the null hypothesis would be rejected.

There are a number of similarities between the F distribution and the t distribution. Like t, the F distribution is also a family of curves, with each member of the family characterized by different degrees of freedom. A particular F distribution depends on the degrees of freedom in both the numerator estimate of σ^2 and the denominator estimate of σ^2. As the number of degrees of freedom approaches infinity, the F curve approaches a normal distribution, which is true for t as well.

The mathematical relationship between t and F is $t^2 = F$ (for an F with 1 df in the numerator). For example, the square of a t value with 15 df is equal to F with 1, 15 df.

Finally, Fisher and Gosset were friends for years, each respectful of the other's work [Fisher's daughter reported their correspondence (Box, 1981).] Fisher recognized that the Student's t test was really a ratio of two measures of variability and that such a concept was applicable to experiments with more than two groups.[4]

PROBLEMS

2. Suppose three means came from a population with $\mu = 100$ and a fourth from a population with a smaller mean, $\mu = 50$. Will the between-treatments estimate of

[4] That is,

$$t = \frac{\bar{X}_1 - \bar{X}_2}{s_{\bar{X}_1 - \bar{X}_2}}$$

$$= \frac{\text{a range}}{\text{standard error of a difference between means}} = \frac{\text{a measure of variability between treatments}}{\text{a measure of variability within treatments}}$$

variability be larger or smaller than the between-treatments estimate of variability of four means drawn from a population with $\mu = 100$?

3. How can two estimates of a population variance be obtained?
4. Define F.
5. Interpret the meaning of the F values given here for the experiments described in examples 1 and 3 on the first two pages of this chapter.
   a. A very small F value for example 1
   b. A very large F value for example 3
6. In your own words, explain the rationale of ANOVA. Identify the two components that combine to produce an F value. Explain what each of these components measures, both when the null hypothesis is true and when it is false. ■

MORE NEW TERMS

Sum of squares   In the "Clue to the Future" on page 64 you read that certain parts of the computation of the standard deviation would be important in future chapters. That future is now. The term Σx^2 (the numerator of the basic formula for the standard deviation) is called the **sum of squares** (abbreviated SS). So, $SS = \Sigma x^2 = \Sigma(X - \bar{X})^2$. A more descriptive name for sum of squares is "sum of the squared deviations."

Mean square   **Mean square** (MS) is the ANOVA term for a variance $\hat{s}^2$. The mean square is a sum of squares divided by its degrees of freedom.

Grand mean   The **grand mean** is the mean of all scores; it is computed without regard for the fact that the scores come from different groups (samples).

tot   A subscript *tot* (total) after a symbol means that the symbol stands for all such numbers in the experiment; ΣX_{tot} is the sum of all X scores.

t   The subscript t after a symbol means that the symbol applies to a treatment group; for example, $\Sigma(\Sigma X_t)^2$ tells you to sum the scores in each group, square each sum, and then sum these squared values. N_t is the number of scores in one treatment group.

K K is the number of treatments in the experiment. This is the same as the number of levels of the independent variable.

CLUE TO THE FUTURE

In this chapter and the next two you will work problems with several sets of numbers. Unless you are using a computer package, you will need the sum, the sum of squares, and other values for each set. Most calculators will produce both the sum and the sum of squares with just one entry of each number. If you know how (or take the time to learn) to exploit your calculator's capabilities, you can spend less time on statistics problems and make fewer errors as well.

SUMS OF SQUARES

Analysis of variance is called that because the variability of all the scores in an experiment is divided between two or more sources. In the case of simple analysis of variance, just two sources contribute all the variability to the scores. One source is the variability between treatments, and the other source is the variability within the scores of each treatment. The sum of these two sources is equal to the total variability.

Now that you have spent some study time on the rationale of ANOVA, it is time to learn to do the calculations. **Table 9.1** shows fictional data from an experiment in which there were three treatments for mental patients. The independent variable was drug therapy. Three levels of the independent variable were used—Drug A, Drug B, and Drug C. The dependent variable was the number of psychotic episodes observed for each patient during the therapy period.

In this experiment (and in all one-way ANOVAs), the null hypothesis is that all the samples come from identical populations. Thus, in terms of treatment means in the drug experiment, H_0: $\mu_A = \mu_B = \mu_C$. If the null hypothesis is true, the differences among the psychotic episodes of the three treatment groups is due to sampling fluctuation and not to the drugs administered.

TABLE 9.1 Computation of sums of squares of fictional data from a drug study

	Drug A		Drug B		Drug C	
	X_A	$X_A{}^2$	X_B	$X_B{}^2$	X_C	$X_C{}^2$
	9	81	9	81	4	16
	8	64	7	49	3	9
	7	49	6	36	1	1
	5	25	5	25	1	1
Σ	29	219	27	191	9	27
$\bar{X}$	7.25		6.75		2.25	

$$\Sigma X_{tot} = 29 + 27 + 9 = 65$$

$$\Sigma X_{tot}^2 = 219 + 191 + 27 = 437$$

$$SS_{tot} = \Sigma X_{tot}^2 - \frac{(\Sigma X_{tot})^2}{N_{tot}} = 437 - \frac{65^2}{12} = 84.917$$

$$SS_{drugs} = \Sigma \left[\frac{(\Sigma X_t)^2}{N_t} \right] - \frac{(\Sigma X_{tot})^2}{N_{tot}}$$

$$= \frac{29^2}{4} + \frac{27^2}{4} + \frac{9^2}{4} - \frac{65^2}{12}$$

$$= 210.25 + 182.25 + 20.25 - 352.083 = 60.667$$

$$SS_{error} = \Sigma \left[\Sigma X_t^2 - \frac{(\Sigma X_t)^2}{N_t} \right] = \left(219 - \frac{29^2}{4} \right) + \left(191 - \frac{27^2}{4} \right) + \left(27 - \frac{9^2}{4} \right)$$

$$= 8.750 + 8.750 + 6.750 = 24.250$$

Check: $SS_{tot} = SS_{treat} + SS_{error} = 84.917 = 60.667 + 24.25$

In this particular experiment, the researcher believed that Drug C would be better than Drugs A or B, which were in common use. In this case, *better* means resulting in fewer psychotic episodes. The researcher's belief is an alternative hypothesis (sometimes called a private hypothesis or experimenter's hypothesis). This belief, however, is not ANOVA's alternative hypothesis, H_1. The alternative hypothesis in ANOVA is that one or more of the samples is from a population that is different from the rest. No *greater than* or *less than* is specified.

After you have read the description of an experiment and understand what it is about, the next step in an ANOVA is to calculate the *sum of squares*, which is one measure of variability. I will illustrate with calculations for the data in **Table 9.1**. First, please focus on the *total* variability as measured by the total sum of squares (SS_{tot}). As you will see, you are already familiar with SS_{tot}. To find SS_{tot}, subtract the grand mean from each score. Square these deviation scores and sum them:

$$SS_{tot} = \Sigma(X - \bar{X}_{tot})^2$$

Computationally, SS_{tot} is more readily obtained using the algebraically equivalent raw-score formula

$$SS_{tot} = \Sigma X_{tot}^2 - \frac{(\Sigma X_{tot})^2}{N_{tot}}$$

which you may recognize as the numerator of the raw-score formula for $\hat{s}$. SS_{tot} is equivalent to Σx^2. Its computation, as illustrated in **Table 9.1**, requires you to square each score and sum the squared values to obtain ΣX_{tot}^2. Next, the scores are summed and the sum squared. That squared value is divided by the total number of scores to obtain $(\Sigma X_{tot})^2/N_{tot}$. Subtraction of $(\Sigma X_{tot})^2/N_{tot}$ from ΣX_{tot}^2 yields the total sum of squares. For the data in Table 9.1,

$$SS_{tot} = \Sigma X_{tot}^2 - \frac{(\Sigma X_{tot})^2}{N_{tot}} = 437 - \frac{65^2}{12} = 84.917$$

Thus, the total variability of all the scores in Table 9.1 is 84.917 when measured by the sum of squares. This total comes from two sources: the between-treatments sum of squares and the error sum of squares. Each of these can be computed separately.

The *between-treatments sum of squares* (SS_{treat}) is the variability of the group means from the grand mean of the experiment, weighted by the size of the group:

$$SS_{treat} = \Sigma[N_t(\bar{X}_t - \bar{X}_{tot})^2]$$

SS_{treat} is more easily computed by the raw-score formula:

$$SS_{treat} = \Sigma\left[\frac{(\Sigma X_t)^2}{N_t}\right] - \frac{(\Sigma X_{tot})^2}{N_{tot}}$$

This formula tells you to sum the scores for each treatment group and then square the sum. Each squared sum is then divided by the number of scores in its group. These values (one for each group) are then summed, giving you $\Sigma[(\Sigma X_t)^2/N_t]$. From this sum is subtracted the value $(\Sigma X_{tot})^2/N_{tot}$, a value that was obtained previously in the computation of SS_{tot}.

When describing experiments in general, the term SS_{treat} is used. In a specific analysis, *between treatments* is changed to a summary word for the independent variable. For the experiment in Table 9.1, that word is *drugs*. Thus,

$$SS_{drugs} = \Sigma \left[\frac{(\Sigma X_t)^2}{N_t} \right] - \frac{(\Sigma X_{tot})^2}{N_{tot}}$$

$$= \frac{29^2}{4} + \frac{27^2}{4} + \frac{9^2}{4} - \frac{65^2}{12} = 60.667$$

The other source of variability is the *error sum of squares* (SS_{error}), which is the sum of the variability within each of the groups. SS_{error} is defined as

$$SS_{error} = \Sigma(X_1 - \bar{X}_1)^2 + \Sigma(X_2 - \bar{X}_2)^2 + \cdots + \Sigma(X_K - \bar{X}_K)^2$$

or the sum of the squared deviations of each score from the mean of its treatment group added to the sum of the squared deviations from all other groups for the experiment. As with the other SS values, one arrangement of the arithmetic is easiest. For SS_{error}, it is

$$SS_{error} = \Sigma \left[\Sigma X_t^2 - \frac{(\Sigma X_t)^2}{N_t} \right]$$

This formula tells you to square each score in a treatment group and sum them (ΣX_t^2). Subtract from this a value that you obtain by summing the scores, squaring the sum, and dividing by the number of scores in the group: $(\Sigma X_t)^2/N_t$. For each group, a value is calculated, and these values are summed to get SS_{error}. For the data in Table 9.1, you get

$$SS_{error} = \Sigma \left[\Sigma X_t^2 - \frac{(\Sigma X_t)^2}{N_t} \right]$$

$$= \left(219 - \frac{29^2}{4} \right) + \left(191 - \frac{27^2}{4} \right) + \left(27 - \frac{9^2}{4} \right) = 24.250$$

Please pause and look at what is happening in calculating the sums of squares, which is the first step in any ANOVA problem. The total variation (SS_{tot}) is found by squaring *every* number (score) and adding them up. From this total, you subtract $(\Sigma X_{tot})^2/N_{tot}$. In a similar way, the variation that is due to treatments is found by squaring the *sum of each treatment*, adding these up, and then subtracting the same component you subtracted before, $(\Sigma X_{tot})^2/N_{tot}$. This factor,

$$\frac{(\Sigma X_{tot})^2}{N_{tot}}$$

is sometimes referred to as the *correction factor* and is found in all ANOVA analyses. Finally, the error sum of squares is found by squaring each number *within* a treatment group and subtracting that group's correction factor. Summing the differences from all the groups gives you SS_{error}.

Another way of expressing these ideas is to say that SS_{error} is the variation *within* the groups and SS_{treat} is the variation *between* the groups. In a one-way ANOVA, the

FIGURE 9.5 $SS_{tot} = SS_{treat} + SS_{error}$

total variability (as measured by SS_{tot}), consists of these two components, SS_{treat} and SS_{error}. So,

$$SS_{tot} = SS_{treat} + SS_{error}$$

Thus, $84.917 = 60.667 + 24.250$ for the drug experiment. This relationship is shown graphically in **Figure 9.5**.

ERROR DETECTION

$SS_{tot} = SS_{treat} + SS_{error}$. All sums of squares are always zero or positive, never negative. This check will not catch errors made in summing the scores (ΣX) or in summing squared scores (ΣX^2).

PROBLEMS

*7. Here are three groups of numbers. This problem and the ones that use this data set are designed to illustrate some relationships and to give you practice calculating ANOVA values. To begin, show that $SS_{tot} = SS_{treat} + SS_{error}$.

X_1	X_2	X_3
5	6	12
4	4	8
3	2	4

*8. Emile Durkheim (1858–1917), a founder of sociology, believed that modernization produced social problems such as suicide. He compiled data from several European countries and published *Suicide* (1897), a book that helped establish quantification in social science disciplines. The data in this problem were created so that the conclusion you reach will mimic that of Durkheim (and present-day sociologists).

Using variables such as electricity consumption, newspaper circulation, and gross national product, 13 countries were classified as to their degree of modernization (low, medium, or high). The numbers in the table are suicide rates per 100,000 persons. For the data in the table, identify the independent and dependent variables and compute SS_{tot}, SS_{mod}, and SS_{error}.

Degree of modernization		
Low	Medium	High
4	17	20
8	10	22
7	9	19
5	12	9
		14

*9. The question of the best way to cure phobias has been answered. Bandura and colleagues' (1969) experiment provided individuals with one of four treatments for their intense fear of snakes. One group worked with a model—a person who handled a 4-foot king snake and encouraged others to imitate her. One group watched a film of adults and children who enjoyed progressively closer contact with a king snake. A third group received desensitization therapy, and a fourth group served as a control, receiving no treatment. The number of snake-approach responses after treatment is shown in the table. Name the independent and dependent variables and compute $SS_{tot} = SS_{treat} + SS_{error}$.

Model	Film	Desensitization	Control
29	22	21	13
27	18	17	12
27	17	16	9
21	15	14	6

MEAN SQUARES AND DEGREES OF FREEDOM

After finding SS, the next step in an ANOVA is to find the mean squares. A mean square is simply a sum of squares divided by its degrees of freedom. It is an estimate of the population variance, σ^2, when the null hypothesis is true.

Each sum of squares has a particular number of degrees of freedom associated with it. In a one-way classification, the df are df_{tot}, df_{treat}, and df_{error}. Also, as in the relationship among sums of squares,

$$df_{tot} = df_{treat} + df_{error}$$

The formula for df_{tot} is $N_{tot} - 1$. The df_{treat} is the number of treatment groups minus one ($K - 1$). The df_{error} is the sum of the degrees of freedom for each group $[(N_1 - 1) + (N_2 - 1) + \cdots + (N_K - 1)]$. If there are equal numbers of scores in the K groups, the formula for df_{error} reduces to $K(N_t - 1)$. A little algebra will reduce this still further:

$$df_{error} = K(N_t - 1) = KN_t - K$$

However, $KN_t = N_{tot}$, so $df_{error} = N_{tot} - K$. This formula for df_{error} works whether the number in each group is the same or not.

To summarize,

$$df_{tot} = N_{tot} - 1$$
$$df_{treat} = K - 1$$
$$df_{error} = N_{tot} - K$$

ERROR DETECTION

$df_{tot} = df_{treat} + df_{error}$. Degrees of freedom are always positive.

Mean squares, then, can be found using the following formulas:[5]

$$MS_{treat} = \frac{SS_{treat}}{df_{treat}} \qquad \text{where} \quad df_{treat} = K - 1$$

$$MS_{error} = \frac{SS_{error}}{df_{error}} \qquad \text{where} \quad df_{error} = N_{tot} - K$$

For the data in Table 9.1,

$$df_{drugs} = K - 1 = 3 - 1 = 2$$
$$df_{error} = N_{tot} - K = 12 - 3 = 9$$

Check: $df_{tot} = df_{treat} + df_{error} = 2 + 9 = 11$

$$MS_{drugs} = \frac{SS_{drugs}}{df_{drugs}} = \frac{60.667}{2} = 30.333$$

$$MS_{error} = \frac{SS_{error}}{df_{error}} = \frac{24.25}{9} = 2.694$$

---

[5] MS_{tot} is not used in ANOVA; only MS_{treat} and MS_{error} are calculated.

Notice that, although $SS_{tot} = SS_{treat} + SS_{error}$ and $df_{tot} = df_{treat} + df_{error}$, mean squares are *not* additive; that is, $MS_{tot} \neq MS_{treat} + MS_{error}$.

CALCULATION AND INTERPRETATION OF F VALUES USING THE F DISTRIBUTION

The next two steps in an ANOVA are the calculation of a value for F and its interpretation using the F distribution.

■ *F* Value: Calculation

You learned earlier that F is a ratio of two estimates of the treatment population variance. MS_{treat} is an estimate based on the variability between groups. MS_{error} is an estimate based on the sample variances. An **F test** consists of dividing MS_{treat} by MS_{error} to obtain an F value:

$$F = \frac{MS_{treat}}{MS_{error}}$$

The null hypothesis is $H_0: \mu_1 = \mu_2 = \cdots = \mu_K$. When the null hypothesis is true, the expected value of F is about 1.00. Of course, sampling error will produce some F's greater than 1.00 and some less than 1.00.

There are two different degrees of freedom associated with any F value, one with the numerator and one with the denominator. For the data in Table 9.1,

$$F = \frac{MS_{drugs}}{MS_{error}} = \frac{30.333}{2.694} = 11.26 \qquad df = 2, 9$$

■ *F* Value: Interpretation

The next question is, What is the probability of obtaining $F = 11.26$ if all three samples come from populations with the same mean? If that probability is less than α, the null hypothesis should be rejected.

Turn now to **Table F** in Appendix B, which gives the critical values of F when $\alpha = .05$ and when $\alpha = .01$. Across the top of the table are degrees of freedom associated with the numerator (MS_{treat}). For Table 9.1, $df_{drugs} = 2$, so 2 is the column you want. Along the left side of the table are degrees of freedom associated with the denominator (MS_{error}). In this case, $df_{error} = 9$, so look for 9 along the side. The tabled value for 2 and 9 df is 4.26 at the .05 level (lightface type) and 8.02 at the .01 level (boldface type). Because the obtained F (11.26) is greater than the tabled F (8.02), you can reject the null hypothesis at the .01 level. The three samples do not have a common population mean. At least one of the drugs had a different effect on the number of psychotic episodes.

T A B L E 9.2 **Summary table of ANOVA for data in Table 9.1**

Source	SS	df	MS	F	p
Drugs	60.667	2	30.333	11.26	<.01
Error	24.250	9	2.694		
Total	84.917	11			

$F_{.01}$ (2, 9 df) = 8.02

At this point, your interpretation must stop. An ANOVA does not tell you which of the population means is greater than or less than the others. Such an interpretation requires more statistical analysis, which is a topic later in this chapter.

It is customary to present all of the calculated ANOVA statistics in a *summary table*. **Table 9.2** is an example. Look at the right side of the table under *p* (for probability). The notation "$p < .01$" is shorthand for "the probability is less than 1 in 100 of obtaining treatment means as different as the ones actually obtained, if, in fact, the samples all came from identical populations."

Sometimes Table F does not contain an *F* value for the *df* in your problem. For example, an *F* with 2, 35 *df* or 4, 90 *df* is not tabled. When this happens, be conservative; use the *F* value that is given for *fewer df* than you have. Thus, the proper *F* values for those two examples would be based on 2, 34 *df* and 4, 80 *df*, respectively.[6]

SCHEDULES OF REINFORCEMENT—A LESSON IN PERSISTENCE

Persistence is shown when you keep on trying even though the rewards are scarce or nonexistent. It is something that seems to vary a great deal from person to person, from task to task, and from time to time. What causes this variation? What has happened to lead to such differences in people, tasks, and times?

Persistence can often be explained if you know how frequently reinforcement (rewards) occurred in the past. I will illustrate with data produced by pigeons, but the principles illustrated are true for many other forms of life, including students and professors.

The data in **Table 9.3** are typical results from a study of *schedules of reinforcement*. A hungry pigeon is taught to peck at a disk on the wall. A peck produces food according to a schedule the experimenter has set up. For some pigeons every peck produces food—a continuous reinforcement schedule (crf). For some birds every other peck produces food—a fixed-ratio schedule of 2 : 1 (FR2). A third group of pigeons gets food after every fourth peck—an FR4 schedule. Finally, for a fourth group, eight pecks are required to produce food—an FR8 schedule. After all groups receive 100

[6] The *F* distribution may be used to test hypotheses about variances as well as hypotheses about means. To determine the probability that two sample variances came from the same population (or from populations with equal variances), form a ratio with the larger sample variance in the numerator. The resulting *F* value can be interpreted with Table F. The proper *df* are $N_1 - 1$ and $N_2 - 1$ for the numerator and denominator, respectively. For more information, see Kirk (1990, p. 410) or Ferguson and Takane (1989, p. 202).

TABLE 9.3 **Partial analysis of the data on the number of minutes to extinction after four schedules of reinforcement during learning**

	Schedule of reinforcement during learning			
	crf	*FR2*	*FR4*	*FR8*
	3	5	8	10
	5	7	12	14
	6	9	13	15
	2	8	11	13
	5	11	10	11
		10		
		6		
ΣX	21	56	54	63
ΣX^2	99	476	598	811
$\bar{X}$	4.2	8.0	10.8	12.6

$$\Sigma X_{tot} = 21 + 56 + 54 + 63 = 194$$

$$\Sigma X^2_{tot} = 99 + 476 + 598 + 811 = 1984$$

$$SS_{tot} = \Sigma X^2_{tot} - \frac{(\Sigma X_{tot})^2}{N_{tot}} = 1984 - \frac{(194)^2}{22} = 1984 - 1710.727 = 273.273$$

$$SS_{schedules} = \Sigma \left[\frac{(\Sigma X_t)^2}{N_t} \right] - \frac{(\Sigma X_{tot})^2}{N_{tot}} = \frac{(21)^2}{5} + \frac{(56)^2}{7} + \frac{(54)^2}{5} + \frac{(63)^2}{5} - \frac{(194)^2}{22} = 202.473$$

$$SS_{error} = \Sigma \left[\Sigma X_t^2 - \frac{(\Sigma X_t)^2}{N_t} \right]$$

$$= \left(99 - \frac{(21)^2}{5} \right) + \left(476 - \frac{(56)^2}{7} \right) + \left(598 - \frac{(54)^2}{5} \right) + \left(811 - \frac{(63)^2}{5} \right)$$

$$= 70.800$$

$$df_{tot} = N_{tot} - 1 = 22 - 1 = 21$$

$$df_{schedules} = K - 1 = 4 - 1 = 3$$

$$df_{error} = N_{tot} - K = 22 - 4 = 18$$

reinforcements, no more food is given (extinction begins). Under such conditions pigeons will continue to peck for a while and then stop. The dependent variable is the number of minutes a bird continues to peck (persist) after the food stops. As you can see in Table 9.3, three groups had five birds and one had seven.

Table 9.3 shows the raw data and the steps required to get the three *SS* figures and the three *df* figures. Work through **Table 9.3** now. When you finish, examine the summary table (**Table 9.4**), which continues the analysis by giving the mean squares, *F* test, and the probability of such an *F*.

When an experiment produces an *F* with a probability value less than .05, a conclusion is called for. Here it is: The schedule of reinforcement used during learning

TABLE 9.4 **Summary table of ANOVA analysis of the schedules of reinforcement study**

Source	SS	df	MS	F	p
Schedules	202.473	3	67.491	17.16	<.01
Error	70.800	18	3.933		
Total	273.273	21			

$F_{.05}$ (3, 18) = 3.16 $F_{.01}$ (3, 18) = 5.09

has a significant effect on persistence of responding during extinction. It is unlikely that the four samples have a common population mean.

You might feel unsatisfied by that conclusion. You might ask which groups differ from the others, or which groups are causing that "significant difference." Good questions. You will find answers in the next section, but first here are a few problems to reinforce what you learned in this section.

PROBLEMS

10. Suppose you obtained $F = 2.56$ with 5, 75 df. How many treatment groups are in the experiment? What are the critical values for the .05 and .01 levels? What conclusion should be reached?

11. Here are two more questions about those raw numbers you found sums of squares for in problem 7.
   a. Perform an F test and compose a summary table. Look up the critical value of F and make a decision about the null hypothesis. Write a sentence explaining what the analysis has shown.
   b. Perform an F test on the data from group 2 and group 3. Perform a t test on these same two groups.

*12. Perform an F test on the modernization and suicide data in problem 8 and write a conclusion.

*13. Perform an F test on the data from Bandura and colleagues' phobia treatment study (problem 9). Write a conclusion.   ■

COMPARISONS AMONG MEANS

A significant F by itself tells you only that it is not likely that the samples have a common population mean. Usually you want to know more, but exactly what you want to know often depends on the particular problem you are working on. For example, in the schedules of reinforcement study, many would ask which schedules are significantly different from others (all pairwise comparisons). Another question might be whether there are sets of schedules that do not differ among themselves but do differ from other sets. In the drug study, the focus of interest is the new experimental drug. Is it better than either of the standard drugs or is it better than the average of the standards? This is just a *sampling* of the kinds of questions that can be asked.

Questions such as these are answered by analyzing the data using specialized statistical tests. As a group, these tests are referred to as **multiple comparison tests**. Unfortunately, no one multiple comparison method is satisfactory for all questions. In fact, not even three or four methods will handle all questions satisfactorily.

The reason there are so many methods is that specific questions are best answered by tailor-made tests. Over the years, statisticians have developed many tests for specific situations, and each test has something to recommend it. For an excellent chapter on the tests and their rationale, see Howell (1992, "Multiple Comparisons Among Treatment Means").

With this background in place, here is the plan for this elementary statistics book. I will discuss the two major categories of tests subsequent to ANOVA (*a priori* and *post hoc*) and then cover one *post hoc* test (the Tukey HSD test).

■ *A Priori* and *Post Hoc* Tests

A priori **tests** require that you *plan* a limited number of tests before gathering the data. *A priori* means "based on reasoning, not on immediate observation." Thus, for *a priori* tests, you must choose during the design stage of the experiment the groups that will be compared.

Post hoc **tests** can be chosen after the data are gathered and you have examined the means. Any and all differences for which a story can be told are fair game for *post hoc* tests. *Post hoc* means "after this" or "after the fact."

The necessity for two kinds of tests will be evident to you if you think again about the 15 samples that were drawn from the same population. Before the samples are drawn, what would be your expectation that two randomly chosen means would be significantly different if tested with an independent-measures *t* test? About .05, I would suppose. Now, what if the data are drawn, the means compiled, and you then pick out the largest and the smallest means and test them with a *t* test? What is your expectation that these two would be significantly different? Much higher than .05, I would suppose. Thus, if you want to make comparisons after the data are gathered (keeping α at an acceptable level like .05), then larger differences should be required before a difference is declared "not due to chance."

The preceding paragraphs should give you a start on understanding that there is a good reason for more than one statistical test to answer questions subsequent to ANOVA. For this text, however, I have confined us to just one test, a *post hoc* test named the Tukey (Too´ Key) Honestly Significant Difference.

■ The Tukey Honestly Significant Difference

For many research problems, the principal question is, Are any of the treatments different from the others? John W. Tukey designed a test to answer this question when N is the same for all samples. A **Tukey Honestly Significant Difference** (HSD) test pairs each sample mean with every other mean, a procedure called *pairwise comparisons*.

For each pair, the Tukey HSD test tells you if one sample mean is significantly larger than the other.

When N_t is the same for all samples, the formula for HSD is

$$\text{HSD} = \frac{\bar{X}_1 - \bar{X}_2}{s_{\bar{X}}}$$

where $s_{\bar{X}} = \sqrt{\dfrac{MS_{\text{error}}}{N_t}}$

$N_t = $ the number in each treatment

The critical value against which the HSD is compared is found in Table G in Appendix B. Turn to **Table G**. Critical values for $\text{HSD}_{.05}$ are on the first page; those for $\text{HSD}_{.01}$ are on the second. For either page, enter the row that gives the *df* for the MS_{error} from the ANOVA. Go over to the column that gives K for the experiment. If the HSD you calculated from the data is equal to or larger than the number at the intersection, reject the null hypothesis and conclude that the two means are significantly different. Finally, describe that difference using the names of the independent and dependent variables.

I will illustrate the Tukey HSD by doing pairwise comparisons on the three groups in the drug experiment (Tables 9.1 and 9.2). The three separate pairwise comparisons can be done in any order, but there is a strategy that may save some arithmetic. The trick is to pick a difference between means of intermediate size and test that difference first. If this difference proves to be significant, then all larger differences also will be significant. If this difference fails to reach significance, then all smaller differences also will be not significant. This trick always works if the sample sizes are equal, but it *may not* work if the sample sizes are not equal, an exception pointed out to me by an undergraduate user of an earlier edition of this book.

In the drug study, there are three means and so there will be three comparisons. It will be easy to choose "a mean difference of intermediate size" because there is only one possibility—the difference between the experimental drug (C) and a standard drug (B):

$$\text{HSD} = \frac{\bar{X}_B - \bar{X}_C}{s_{\bar{X}}} = \frac{6.75 - 2.25}{\sqrt{\dfrac{2.694}{4}}} = \frac{4.50}{0.821} = 5.48$$

Because 5.48 is larger than the tabled value for HSD at the .01 level ($\text{HSD}_{.01} = 5.43$), you can conclude that the experimental drug (C) is significantly better than Drug B at reducing psychotic episodes ($p < .01$). Because the difference between Drug C and Drug A is even greater than the difference between Drugs C and B, you can also conclude that the experimental drug is significantly better than Drug A in reducing psychotic episodes ($p < .01$).

One comparison remains, the comparison between the two standard drugs, A and B:

$$\text{HSD} = \frac{7.25 - 6.75}{\sqrt{\dfrac{2.694}{4}}} = 0.61$$

Because $\text{HSD}_{.05} = 3.95$, you can conclude that this study does not provide evidence of the superiority of Drug B over Drug A.

Tukey HSD when N_t's are not equal.   The Tukey HSD was developed for studies in which N_t's are equal for all groups. Sometimes, despite the best of intentions, unequal N_t's occur. A modification in the denominator of HSD will produce a statistic that can be tested with the values in Table G. The modification, recommended by Kramer (1956) and shown to be accurate (Smith, 1971), is

$$s_{\bar{X}} = \sqrt{\frac{MS_{\text{error}}}{2}\left(\frac{1}{N_1} + \frac{1}{N_2}\right)}$$

■ Tukey HSD and ANOVA

In analyzing an experiment, it has been common practice to calculate an ANOVA and then, if the F value is significant, conduct Tukey HSD tests. Mathematical statisticians have shown, however, that this practice is overly conservative (Zwick, 1993). That is, if the *only* thing you want to know from your data is whether any *pairs* of treatments show a significant difference, then it is legitimate to run Tukey HSD tests without first showing that a one-way ANOVA produced a significant F. For the beginning statistics student, however, I recommend that you learn both procedures and apply them in the order of ANOVA and then HSD.

PROBLEMS

14. Distinguish between the two categories of tests subsequent to ANOVA: *a priori* and *post hoc*.
15. For Durkheim's modernization and suicide data, make all possible pairwise comparisons and write a conclusion telling what the data show.
16. For the schedules of reinforcement data, make all possible pairwise comparisons and write a conclusion telling what the data show.
*17. For the data on treatment of phobias, compare the best treatment with the second best, and the poorest treatment with the control group. Write a conclusion about the study. ■

ASSUMPTIONS OF THE ANALYSIS OF VARIANCE

In developing the analysis of variance and the F test, Fisher started with the following assumptions about the dependent-variable scores:

1. *Normality.* The dependent variable is assumed to be normally distributed in the populations from which samples are drawn. It is often difficult or impossible to demonstrate normality or lack of normality in the parent populations. Such a demonstration usually occurs only with very large samples. On the other hand, because

of extensive research, some populations are known to be skewed, and researchers in those fields may decide that ANOVA is not appropriate for their data analysis. Unless there is a reason to suspect that populations depart severely from normality, the inferences made from the F test will probably not be affected. ANOVA is robust. (It results in correct probabilities even when the populations are not exactly normal.) Where there is suspicion of a severe departure from normality, however, use the nonparametric method explained in Chapter 13.

2. *Homogeneity of variance.* The variances of the dependent-variable scores for each of the populations are assumed to be equal to each other. Figures 9.1, 9.2, and 9.3, which were used to illustrate the rationale of ANOVA, show populations whose variances are equal. Several methods of testing this assumption are presented in advanced texts such as Winer, Brown, and Michels (1991), Kirk (1995), and Keppel (1982). When variances are grossly unequal, the nonparametric method discussed in Chapter 13 may be called for. [See Wilcox (1987) for a discussion of this issue.]

3. *Random sampling.* ANOVA is based on the assumption that samples are taken randomly from the population. Most researchers cannot get random samples, but they are often able to use random assignment of subjects to groups. Random assignment accomplishes most of what random sampling accomplishes.

I hope these assumptions seem familiar to you. They are the same as those you learned for the t distribution in Chapter 8. This makes sense; t is the special case of F when there are two samples.

EFFECT SIZE

The important question, Could these differences be due to chance? can be answered with the hypothesis-testing F test. The equally important question, Are these differences large, medium, or small? requires an *effect size index*.

I'm sure you recall calculating an effect size index for the one-sample t test (symbolized d and discussed on pages 176–177) and for the two-sample t test (symbolized d and discussed on pages 202–204). For ANOVA, the effect size index is symbolized f. The formula is

$$f = \frac{\sigma_{\text{treat}}}{\sigma_{\text{error}}}$$

To find f, *estimators* of σ_{treat} and σ_{error} are required. The estimator of σ_{treat} is symbolized $\hat{s}_{\text{treat}}$ and is found using the formula[7]

$$\hat{s}_{\text{treat}} = \sqrt{\frac{K-1}{N_{\text{tot}}}(MS_{\text{treat}} - MS_{\text{error}})}$$

Likewise, the estimator of σ_{error} is symbolized $\hat{s}_{\text{error}}$, which is simply the square root of the error variance.

$$\hat{s}_{\text{error}} = \sqrt{MS_{\text{error}}}$$

[7] This formula is appropriate for an ANOVA with equal numbers of participants in each treatment group (Kirk, 1995, p. 181). For unequal N's, see Cohen (1988, pp. 359–362).

Here's an example of an effect size index for an ANOVA using the drug data in Table 9.1.

$$f = \frac{\sqrt{\dfrac{K-1}{N_{tot}}(MS_{treat} - MS_{error})}}{\sqrt{MS_{error}}} = \frac{\sqrt{\dfrac{3-1}{12}(30.333 - 2.694)}}{\sqrt{2.694}} = 1.31$$

The next step is to interpret an f value of 1.31. Jacob Cohen, who promoted the awareness of effect size and also contributed formulas, provided guidance for judging f values as small, medium, and large (Cohen, 1969, 1992).

Small $f = .10$
Medium $f = .25$
Large $f = .40$

Thus, an f value of 1.31 indicates that the effect size in the drug study was just huge. (This is not surprising because textbook authors often manufacture fictional data so that an example will produce a large effect.)

Here is the final set of problems on one-way ANOVA. All that you learn here will be of use to you in Chapter 10, which explains how to use ANOVA to analyze a more complex experiment.

PROBLEMS

18. List the three assumptions of the analysis of variance.
19. For the data from Bandura and colleagues' phobia treatment experiment (problems 9, 13, and 17), find f and the effect size index, and write a sentence of interpretation.
20. Hermann Rorschach, a Swiss psychiatrist, began to get an inkling of an idea in 1911. By 1921 he had developed, tested, and published his test, which measures motivation and personality. One of the several ways that the test is scored is to count the number of responses the test-taker gives (R scores). The data that follow are representative of those given by normals, schizophrenics, and depressives. Analyze the data with an F test and a set of Tukey HSDs. Write a conclusion about Dr. Rorschach's test.

Normals	Schizophrenics	Depressives
23	17	16
18	20	11
31	22	13
14	19	10
22	21	12
28	14	10
20	25	
22		

21. Almost all health professionals recommend that women practice monthly breast self-examination (BSE) as a way to detect breast cancer in its early stages (when treatment is much more effective). Several methods have been used to teach BSE. T. S. Spatz (1991) compared the method recommended by the American Cancer Society (ACS) to the 4MAT method, a method based on different learning styles that different people have. She also included an untrained control group. The dependent variable was the amount of information retained three months after training. Analyze the data in the table with an ANOVA, HSD tests, and an effect size index. Write a conclusion.

	ACS	4MAT	Control
ΣX	178	362	62
ΣX^2	2619	7652	1192
N	20	20	20

22. A researcher interested in the effects of stimulants on memory gave three groups of rats 20 trials of training to avoid a shock. Immediately after learning, one group of rats was injected with a small amount of strychnine (a stimulant), a second group was injected with saline as a control for the injection, and a third group was not injected. Seven days later, the researcher tested the rats' memory by training them until they reached a criterion of avoiding the shock five trials in a row. The number of trials to reach this criterion was recorded for each rat. (See McGaugh and Petrinovich, 1965.) Perform an ANOVA and write a sentence summary.

	Strychnine	Saline	No injection
ΣX	121	138	152
ΣX^2	1608	2464	3000
N	10	8	8

23. Look over the objectives at the beginning of the chapter. Can you do them?

10

Analysis of Variance: Factorial Design

OBJECTIVES FOR CHAPTER 10

After studying the text and working the problems in this chapter, you should be able to:

1. Define the terms *factorial design*, *factor*, *level*, and *cell*
2. Name the sources of variance in a factorial design
3. Compute sums of squares, degrees of freedom, mean squares, and *F* values in a factorial design
4. Determine whether *F* values are significant or not
5. Describe the concept of interaction
6. Interpret interactions and main effects
7. Make pairwise comparisons between means with Tukey HSD tests
8. List the conditions required for a factorial design ANOVA

In Chapter 9, you learned the basic concepts and procedures for conducting a simple one-way ANOVA. In this chapter, I will expand those concepts and procedures to cover a popular and more complex design: the factorial ANOVA. For many problems, a factorial ANOVA will give you a special kind of information that has not yet been discussed in this book. First, however, there are some new terms to learn and a review of the designs that lead up to a factorial ANOVA.

The first new term is **factor**, which is just a shorter word for independent variable. The **levels** *of a factor* are simply the different treatments that make up the independent variable. To illustrate, in Chapter 9 you analyzed data from a study on persistence that had four treatments. The factor in that study was schedule of reinforcement, and there were four levels of the factor. In Chapter 8, you analyzed data from a two-group study of wage discrimination. The factor was gender, and there were two levels of the factor, male and female. I will continue this summarizing by describing the designs you learned to analyze in the two previous chapters.

T A B L E 10.1 Illustration of a two-treatment design that can be analyzed with a *t* test

Factor *A*	
A_1	A_2
Scores on the dependent variable	Scores on the dependent variable
$\bar{X}_{A_1}$	$\bar{X}_{A_2}$

T A B L E 10.2 Illustration of a four-treatment design that can be analyzed with an *F* test

Factor *A*			
A_1	A_2	A_3	A_4
Scores on the dependent variable	Scores on the dependent variable	Scores on the dependent variable	Scores on the dependent variable
$\bar{X}_{A_1}$	$\bar{X}_{A_2}$	$\bar{X}_{A_3}$	$\bar{X}_{A_4}$

In Chapter 8, you learned to analyze data from a two-group experiment, schematically shown in **Table 10.1**. In Table 10.1, there is *one independent variable*, Factor *A*, with data from two levels, A_1 and A_2. For the *t* test, the question was whether $\bar{X}_{A_1}$ was significantly different from $\bar{X}_{A_2}$. You learned to determine this for both independent and correlated-samples designs.

In Chapter 9, you learned to analyze data from a two-or-more-treatment experiment, schematically shown in **Table 10.2** for a four-treatment design. In Table 10.2, there is *one independent variable*, Factor *A*, with data from four levels, A_1, A_2, A_3, and A_4. In the one-way ANOVA of Chapter 9, the question was whether the four treatment means could have come from populations that have a common mean. You learned to determine this for independent but not for correlated samples.

FACTORIAL DESIGN

In this chapter, you will learn to analyze data from a design in which there are *two independent variables* (factors), each of which may have two or more levels. **Table 10.3** illustrates an example of this design with one factor (Factor *A*) having three levels (A_1, A_2, and A_3) and another factor (Factor *B*) having two levels (B_1 and B_2). Such a design is called a **factorial design**.

A factorial design is one that has *two or more independent variables*. Each level of a factor is paired with all levels of the other factor(s). This chapter will cover only two-

T A B L E   10.3   Illustration of a 2 × 3 factorial design that can be analyzed with F tests

		Factor A			
		A_1	A_2	A_3	
Factor B	B_1	Cell A_1B_1 Scores on the dependent variable	Cell A_2B_1 Scores on the dependent variable	Cell A_3B_1 Scores on the dependent variable	$\bar{X}_{B_1}$
	B_2	Cell A_1B_2 Scores on the dependent variable	Cell A_2B_2 Scores on the dependent variable	Cell A_3B_2 Scores on the dependent variable	$\bar{X}_{B_2}$
		$\bar{X}_{A_1}$	$\bar{X}_{A_2}$	$\bar{X}_{A_3}$	

factor designs.[1] These two-factor designs will be for independent samples and not for correlated samples.

Factorial designs are identified with a shorthand notation such as "2 × 3" or "3 × 5." The general term is $R \times C$ (Rows × Columns). The first number tells you the number of levels of one factor; the second number tells you the number of levels of the other factor. The design in Table 10.3 is a 2 × 3 design. Assignment of a factor to a row or column is arbitrary; I could just as well have made Table 10.3 a 3 × 2 table.

In Table 10.3, there are six cells. Each **cell** represents a different way to treat participants. Participants in the upper left cell are given treatment A_1 *and* treatment B_1; that cell is identified as Cell A_1B_1. Participants in the lower right cell are given treatment A_3 and treatment B_2, and that cell is called Cell A_3B_2.

In this paragraph I will tell you what information a factorial ANOVA gives you.[2] Pay careful attention. In the first place, a factorial ANOVA gives you the same information that you can get from two separate one-way ANOVAs. Look again at **Table 10.3**. A factorial ANOVA will determine the probability of obtaining means such as $\bar{X}_{A_1}$, $\bar{X}_{A_2}$, and $\bar{X}_{A_3}$ from identical populations (a one-way ANOVA with three groups). The same factorial ANOVA will determine the probability of $\bar{X}_{B_1}$ and $\bar{X}_{B_2}$ coming from the same population (a second one-way ANOVA). In addition, a factorial ANOVA helps you decide whether the two factors *interact* on the dependent variable. The interaction test is the special kind of information that a factorial design provides. *An interaction*

---

[1] Advanced textbooks discuss the analysis of three-or-more-factor designs. See Howell (1992), Kirk (1995), or Winer, Brown, and Michels (1991).
[2] See Kirk (1995, p. 431) for both the advantages and the disadvantages of factorial ANOVA.

between two factors means that the effect that one factor has on the dependent variable depends on which level of the other factor is being administered. There would be an interaction in the factorial design in Table 10.3 if the difference between Level B_1 and Level B_2 depended on whether you looked at the A_1 column, the A_2 column, or the A_3 column. To expand on this, each of the three columns represents an experiment that compares B_1 to B_2. If there is *no* interaction (that is, Factor A does not interact with Factor B), then the difference between B_1 and B_2 will be the same for all three experiments. If there *is* an interaction, however, the difference between B_1 and B_2 will *depend* on whether you are examining the experiment at A_1, at A_2, or at A_3. [The effect of Factor B (B_1 or B_2) depends on Factor A (A_1, A_2, or A_3).]

Perhaps a couple of examples of interactions will help at this point. Suppose a group of friends were sitting in a dormitory lounge one Monday discussing the weather of the previous weekend. What would you need to know to predict each person's rating of the weather? The first thing you would probably want to know is what the weather was actually like. A second important variable would be the activity each person had planned for the weekend. For purposes of this little illustration, suppose that weather comes in one of two varieties, snow or no snow, and that our participants had planned only one of two activities, camping or skiing. This means there are two independent variables (weather and plans) and a dependent variable (rating of the weather). Now we have the ingredients for an interaction.

If plans called for camping, "no snow" is rated good, but if plans called for skiing, "no snow" is rated bad. To complete the possibilities, campers rated snow bad and skiers rated it good. Thus, the effect of the weather (Factor A) on the rating depends on plans (Factor B). **Table 10.4** summarizes this paragraph. Study it before going on.

Here is a similar example, but one with *no* interaction. Suppose you wanted to know how people would rate the weather, which again could be snow or no snow. This time, however, the people are divided into campers and rock climbers. For both groups, snow would rate as bad weather. You might make up a version of Table 10.4 that describes this second example.

In this example of no interaction, effect of the weather on a judgment about the weather does not depend on a person's plans for the weekend. A change from snow to no snow brings joy to the hearts of both groups. You can see this in the table you constructed.

T A B L E   10.4 **Illustration of a significant interaction— "good" and "bad" refer to ratings of the weather**

		Kind of weather	
		Snow	No snow
Plans for weekend	Camping	Bad	Good
	Skiing	Good	Bad

MAIN EFFECTS AND INTERACTION

A factorial ANOVA produces three statistical tests. Two of the tests are called **main effects** and one is called the **interaction**. A main effect is like the between-treatments test in a one-way ANOVA. With a factorial ANOVA, the data yield two of these analyses (a test for Factor A and a test for Factor B). The test of the interaction (symbolized AB) tells you if the scores for different levels of Factor A depend on the levels of Factor B.

Table 10.5 shows a 2×3 factorial ANOVA. The numbers in each of the six cells are cell means. Examine those six means and see for yourself what effect changing from A_1 to A_2 to A_3 has. Make the same examination for B_1 to B_2. (Incidentally, making this kind of preliminary examination is very informative for any set of data.)

I will use Table 10.5 to show you what goes on in a factorial ANOVA's analysis of the main effects and the interaction. Look at the margin means for Factor B ($B_1 = 30$, $B_2 = 70$). The B main effect will give the probability of obtaining means of 30 and 70 if both samples came from identical populations. Thus, with this probability you can decide whether the null hypothesis

$$H_0: \ \mu_{B_1} = \mu_{B_2}$$

should be rejected or not. The A main effect will give you the probability of getting means of A_1, A_2, and A_3 (30, 40, and 80) if $H_0: \mu_{A_1} = \mu_{A_2} = \mu_{A_3}$. Again, this probability can result in a rejection or a retention of the null hypothesis.

Notice that a comparison of B_1 to B_2 satisfies the requirements for an experiment (page 184) in that, except for getting either B_1 or B_2, the groups were treated alike. That is, Group B_1 and Group B_2 are alike in that both groups have equal numbers of participants who received A_1. Whatever the effect of receiving A_1, it will occur as much in the B_1 group as it does in the B_2 group; B_1 and B_2 differ only in their levels of Factor B. It would be worthwhile for you to apply this same line of reasoning to the question of whether comparisons of A_1, A_2, and A_3 satisfy the requirement for an experiment.

In Table 10.5 there is *no* interaction. The effect of changing from B_1 to B_2 is to increase the mean score 40 points and this is true at *each* level of A. A similar constancy

T A B L E   10.5   A 2×3 factorial design (The number in each cell represents the mean for all the subjects in the cell. N is the same for each cell.)

		Factor A			
		A_1	A_2	A_3	Factor B means
Factor B	B_1	10	20	60	30
	B_2	50	60	100	70
	Factor A means	30	40	80	Grand mean 50

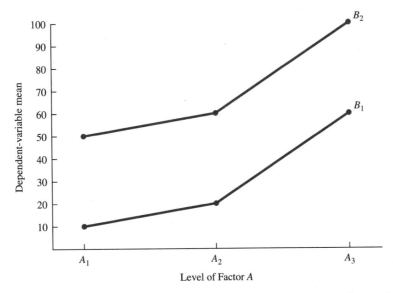

FIGURE 10.1 Graphic representation of data in Table 10.5; no interaction exists.

is seen in the rows. The effect of changing from A_1 to A_2 is to increase the mean 10 points, *regardless* of whether you are looking at the B_1 row or the B_2 row. In a similar way, the effect of moving from A_2 to A_3 is to increase scores by 40. Factor B has no effect on the change from A_2 to A_3; it is 40 for both levels. In Table 10.5 there is no interaction between Factor A and Factor B.

It is common to display an interaction or lack of one with a graph (with the dependent-variable scores on the ordinate). A graph is a good aid for writing an interpretation of an interaction. So, study each graph in this chapter until you understand it. In addition, I urge you to draw a graph of the interaction on the factorial problems you work. **Figure 10.1** is one graph of the data in Table 10.5. The two curves are parallel. Parallel curves mean that there is zero interaction between two factors.

The data in Table 10.5 can also be graphed with each curve representing a level of A. **Figure 10.2** is the result. Again, the parallel lines indicate that there is zero interaction. Interactions, like other statistics, are subject to sampling fluctuation. As a result, graphs of actual data almost never have exactly parallel curves, and this lack may indicate some degree of interaction. A statistical test, of course, is required to separate the effect of sampling fluctuation from the effect of a reliable, repeatable interaction.

Table 10.6 shows a 2 × 3 factorial design in which there *is* an interaction between the two independent variables. The main effect of Factor A is indicated by the margin means along the bottom. The average effect of a change from A_1 to A_2 to A_3 is to *reduce* the mean score by 10 (55 to 45 to 35). But look at the cells. For B_1, the effect of changing from A_1 to A_2 to A_3 is to *increase* the mean score by 10 points. For B_2, the effect is to *decrease* the score by 30 points. These data have an interaction because the effect of one factor *depends* on which level of the other factor you administer.

Let me describe this interaction in Table 10.6 another way. Look at the difference between Cell A_1B_1 and Cell A_2B_1—a difference of −10 (10 − 20). If there were no

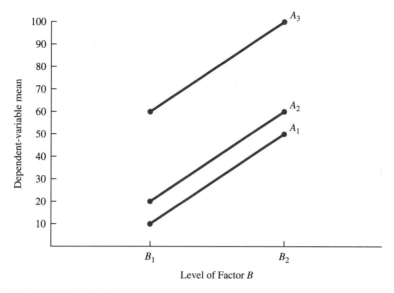

FIGURE 10.2 Graphic representation of data in Table 10.5; no interaction exists.

interaction, this same difference (-10) should have been found between Cells A_1B_2 and A_2B_2. But this latter difference is $+30$ ($100 - 70$); it is in the opposite direction. Something about B_2 reverses the effect of changing from A_1 to A_2 that was found under condition B_1.

This interaction is illustrated graphically in **Figure 10.3**. You can see that B_1 increases across the levels of Factor A. B_2, however, decreases as you go from A_1 to A_2 to A_3. The lines for the two levels of B are not parallel.

Finally, **Figure 10.4** shows a graph of the rating given the weather by skiers and campers. Again, the lines are not parallel.[3]

TABLE 10.6 A 2 × 3 factorial design with an interaction between factors (The numbers represent the mean of all scores within that cell.)

		Factor A			
		A_1	A_2	A_3	Factor B means
Factor B	B_1	10	20	30	20
	B_2	100	70	40	70
	Factor A means	55	45	35	Grand mean 45

[3] Students often have trouble with the concept of interaction. Usually, having the same idea presented in different words facilitates understanding. Three good references are Christensen and Stoup (1991, pp. 341–345), Howell (1995, pp. 330–332), and Minium, King, and Bear (1993, pp. 416–419).

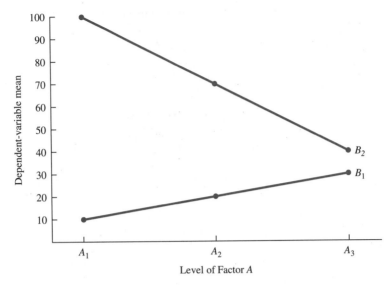

FIGURE 10.3 Graphic representation of the interaction of Factors A and B from Table 10.6.

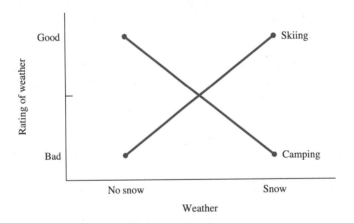

FIGURE 10.4 Graphic representation of the "rating of the weather" example indicating an interaction. Whether snow or no snow is rated higher depends on the plans of the raters: skiing or camping.

RESTRICTIONS AND LIMITATIONS

I have tried to emphasize throughout this book the limitations that go with each statistical test you learn. For the factorial analysis of variance presented in this chapter, the restrictions include the three that you learned for one-way ANOVA. If you cannot recall those three, refresh your memory by rereading the section near the end of Chapter 9, "Assumptions of the Analysis of Variance." In addition to the basic assumptions of ANOVA, the factorial ANOVA presented in this chapter requires the following:

1. The number of scores in each cell must be equal. For techniques dealing with unequal N's, see Kirk (1995), Howell (1992), or Ferguson and Takane (1989).
2. The cells must be independent. This is usually accomplished by randomly assigning a participant to only one of the cells. This restriction means that these formulas should not be used with any type of correlated-samples design. For factorial designs for correlated samples, see Howell (1992), or Winer, Brown, and Michels (1991).
3. The levels of both factors are chosen by the experimenter. The alternative is that the levels of one or both factors are chosen at random from several possible levels of the factor. The techniques of this chapter are used when the levels are *fixed* by the experimenter and not chosen randomly. For a discussion of fixed and random models of ANOVA, see Howell (1992), Winer, Brown, and Michels (1991), or Ferguson and Takane (1989).

PROBLEMS

1. Identify the following factorial designs, using $R \times C$ notation.
   a. Three methods of teaching Spanish were evaluated using either a grade report or a credit/no credit report.
   b. Four strains of mice were injected with three types of virus.
   c. Males and females worked on a complex problem at either 7 A.M. or 7 P.M.
   d. This one is a little different. Four different species of pine were grown with three different kinds of fertilizer in three different soil types.
2. What is a main effect?
3. What is an interaction?
4. Does a t test test for a main effect or for an interaction?
5. In a one-way ANOVA, are you testing for a main effect or for an interaction?
6. List the six requirements that must be met if data are to be analyzed by a factorial ANOVA using techniques outlined in this chapter.
7. A description of the dependent variable and a table with cell means are given for the five problems below. For each problem, do the following:
   a. Graph the means.
   b. Decide whether an interaction appears to be present. In these problems, you may state that there appears to be *no* interaction or that there appears to be one. Interactions, like main effects, are subject to sampling variations.
   c. Write a statement of explanation of the interaction or lack of one.
   d. Decide whether any main effects appear to be present.
   e. Write a statement of explanation of the main effects or lack of them.

i. The dependent variable for this ANOVA is attitude scores toward games with cards or games with computers (see following table).

Factor A (age)

		A_1 (young)	A_2 (old)
Factor B	B_1 (cards)	50	75
(games)	B_2 (computer)	75	50

ii. This is a study of the effect of dosage of a new drug on three types of schizophrenics. The dependent variable is scores on a competency test.

A (diagnosis)

		A_1 (disorganized)	A_2 (catatonic)	A_3 (paranoid)
B	B_1 (small)	5	10	70
(dose)	B_2 (medium)	20	25	85
	B_3 (large)	35	40	100

iii. The dependent variable in this one is an improvement score measured after a 90-day period of therapy.

A (therapy)

		A_1 (Psychoanalytic)	A_2 (Cognitive)	A_3 (Behavior modification)
B verbal ability	B_1 (low)	15	60	30
	B_2 (high)	75	120	90

iv. In this problem, the dependent variable is the score on a test that measures self-consciousness.

A (grade)

		A_1 (5th grade)	A_2 (9th grade)	A_3 (12th grade)
B	B_1 (short)	10	30	20
(height)	B_2 (tall)	5	65	35

v. Here, the dependent variable is attitude toward lowering the taxes on profits from investments.

A (socioeconomic status)

		A_1 (low)	A_2 (middle)	A_3 (high)
B	B_1 (men)	10	35	60
(gender)	B_2 (women)	15	35	55

■

A SIMPLE EXAMPLE OF A FACTORIAL DESIGN

As you read the following story, identify the dependent variable and the two independent variables (factors).

> The term was over. A dozen students who had all been in the same class were talking about what a bear of a test the comprehensive final exam had been. Competition surfaced.
>
> "Of course, I was ready. Those of us who major in natural science disciplines just prepare more than you humanities types."
>
> "Poot, poot, and balderdash," exclaimed one of the humanities types, who was a sophomore.
>
> A third student said, "Well, there are some of both types in this group; let's just gather some data. How many hours did each of you study for that final?"
>
> "Wait a minute," exclaimed one of younger students. "I'm a natural science type, but this is my first term in college. I didn't realize how much time I would need to cover the readings. I don't want my score to pull down my group's tally."
>
> "Hmmm," mused the data-oriented student. "Let's see. Look, some of you are in your first term and the rest of us have had some college experience. We'll just take experience into account. Maybe the dozen of us will divide up evenly."

This story establishes the conditions for a 2 × 2 factorial ANOVA. One of the factors is area of interest; the two levels here are natural science and humanities. The other factor is previous experience in college; the two levels are none and some. The dependent variable is hours of study for a final examination.

Not surprisingly, when I created a data set to go with this story, I found that the dozen did divide up evenly—there were three students in each of the four cells. These data, along with cell means and the computation of the sums of squares, are illustrated in **Table 10.7**.

Look at the margin means in Table 10.7. Compare the number of hours of study for natural science and humanities. What about the effect of experience in college (the other margin means)? Is the difference as large or larger there? Now, look at the four cell means. Does there appear to be an interaction?

This kind of preliminary inspection of the data will improve your understanding of a study and will prepare you to detect gross computational errors. For example, in Table 10.7, the difference in the experience variable is greater than the difference in the area of study variable. This will be reflected in an *F* value for experience that is larger than the *F* value for area of study.

T A B L E 10.7 **Cell means and computation of sums of squares for the final exam data**

		Previous college experience (Factor A)			
		Some (A_1)		None (A_2)	
		X	X^2	X	X^2
	Humanities (B_1)	12	144	8	64
		11	121	7	49
		10	100	7	49

Major (Factor B)

$\Sigma X_{\text{cell}} = 33$ $\Sigma X_{\text{cell}} = 22$ $\Sigma X_{B_1} = 55$
$\Sigma X^2_{\text{cell}} = 365$ $\Sigma X^2_{\text{cell}} = 162$ $\Sigma X^2_{B_1} = 527$
$\bar{X}_{\text{cell}} = 11.00$ $\bar{X}_{\text{cell}} = 7.3333$ $\bar{X}_{B_1} = 9.1667$

Natural science (B_2)	12	144	10	100
	10	100	9	81
	9	81	8	64

$\Sigma X_{\text{cell}} = 31$ $\Sigma X_{\text{cell}} = 27$ $\Sigma X_{B_2} = 58$
$\Sigma X^2_{\text{cell}} = 325$ $\Sigma X^2_{\text{cell}} = 245$ $\Sigma X^2_{B_2} = 570$
$\bar{X}_{\text{cell}} = 10.3333$ $\bar{X}_{\text{cell}} = 9.00$ $\bar{X}_{B_2} = 9.6667$

$\Sigma X_{A_1} = 64$ $\Sigma X_{A_2} = 49$ $\Sigma X_{\text{tot}} = 113$
$\Sigma X^2_{A_1} = 690$ $\Sigma X^2_{A_2} = 407$ $\Sigma X^2_{\text{tot}} = 1097$
$\bar{X}_{A_1} = 10.6667$ $\bar{X}_{A_2} = 8.1667$ $\bar{X}_{\text{tot}} = 9.4167$

$$SS_{\text{tot}} = \Sigma X^2_{\text{tot}} - \frac{(\Sigma X_{\text{tot}})^2}{N_{\text{tot}}} = 1097 - \frac{(113)^2}{12} = 1097 - 1064.0833 = 32.9167$$

$$SS_{\text{cells}} = \Sigma\left[\frac{(\Sigma X_{\text{cell}})^2}{N_{\text{cell}}}\right] - \frac{(\Sigma X_{\text{tot}})^2}{N_{\text{tot}}} = \frac{(33)^2}{3} + \frac{(22)^2}{3} + \frac{(31)^2}{3} + \frac{(27)^2}{3} - \frac{(113)^2}{12}$$
$$= 363 + 161.3333 + 320.3333 + 243 - 1064.0833 = 23.5833$$

$$SS_{\text{experience}} = \frac{(\Sigma X_{A_1})^2}{N_{A_1}} + \frac{(\Sigma X_{A_2})^2}{N_{A_2}} - \frac{(\Sigma X_{\text{tot}})^2}{N_{\text{tot}}} = \frac{(64)^2}{6} + \frac{(49)^2}{6} - \frac{(113)^2}{12}$$
$$= 682.6667 + 400.1667 - 1064.0833 = 1082.8334 - 1064.0833$$
$$= 18.7501$$

$$SS_{\text{majors}} = \frac{(\Sigma X_{B_1})^2}{N_{B_1}} + \frac{(\Sigma X_{B_2})^2}{N_{B_2}} - \frac{(\Sigma X_{\text{tot}})^2}{N_{\text{tot}}} = \frac{(55)^2}{6} + \frac{(58)^2}{6} - \frac{(113)^2}{12}$$
$$= 504.1667 + 560.6667 - 1064.0833 = 1064.8334 - 1064.0833 = 0.7501$$

$$SS_{AB} = N_{cell}[(\bar{X}_{A_1B_1} - \bar{X}_{A_1} - \bar{X}_{B_1} + \bar{X}_{tot})^2 + (\bar{X}_{A_2B_1} - \bar{X}_{A_2} - \bar{X}_{B_1} + \bar{X}_{tot})^2$$
$$+ (\bar{X}_{A_1B_2} - \bar{X}_{A_1} - \bar{X}_{B_2} + \bar{X}_{tot})^2 + (\bar{X}_{A_2B_2} - \bar{X}_{A_2} - \bar{X}_{B_2} + \bar{X}_{tot})^2]$$
$$= 3[(11.00 - 10.6667 - 9.1667 + 9.4167)^2 + (7.3333 - 8.1667 - 9.1667 + 9.4167)^2$$
$$+ (10.3333 - 10.6667 - 9.6667 + 9.4167)^2 + (9.00 - 8.1667 - 9.6667 + 9.4167)^2]$$
$$= 4.0836$$

Check: $SS_{AB} = SS_{cells} - SS_A - SS_B = 23.5833 - 18.7501 - 0.7501 = 4.0831$

$$SS_{error} = \Sigma\left[\Sigma X_{cell}^2 - \frac{(\Sigma X_{cell})^2}{N_{cell}}\right]$$
$$= \left(365 - \frac{(33)^2}{3}\right) + \left(162 - \frac{(22)^2}{3}\right) + \left(325 - \frac{(31)^2}{3}\right) + \left(245 - \frac{(27)^2}{3}\right)$$
$$= (365 - 363) + (162 - 161.3333) + (325 - 320.3333) + (245 - 243)$$
$$= 2 + 0.6667 + 4.6667 + 2 = 9.3333$$

Check : $SS_{tot} = SS_{cells} + SS_{error}$
$$32.9167 = 23.5833 + 9.3333$$

■ Sources of Variance and Sums of Squares

Remember that in Chapter 9 you identified three sources of variance in the one-way analysis of variance. They were (1) the total variance, (2) the between-treatments variance, and (3) the error variance. In a factorial design with two factors, the same sources of variance can be identified. However, the between-treatments variance, which is called the between-cells variance in factorial ANOVA, is partitioned into *three* components. These are the two *main effects* and the *interaction*. Thus, of the variability among the means of the four cells in Table 10.7, some can be attributed to the *A* main effect, some to the *B* main effect, and the rest to the *AB* interaction.

Total sum of squares.   Calculating the total sum of squares will be easy for you because it is the same as SS_{tot} in the one-way analysis. It is defined as $\Sigma(X - \bar{X}_{tot})^2$, or the sum of the squared deviations of all the scores in the experiment from the grand mean of the experiment. To actually compute SS_{tot}, use the formula

$$SS_{tot} = \Sigma X_{tot}^2 - \frac{(\Sigma X_{tot})^2}{N_{tot}}$$

For the final exam data, you get

$$SS_{tot} = 1097 - \frac{(113)^2}{12} = 32.9167$$

Between cells sum of squares.   To find the main effects and interaction, you should first find the between-cells variability and then partition it into its component parts. SS_{cells} is defined $\Sigma[N_{cell}(\bar{X}_{cell} - \bar{X}_{tot})^2]$. A "cell" in a factorial ANOVA is a group of participants treated alike; for example, humanities majors with no previous college experience constitute a cell.

The computational formula for SS_{cells} is similar to that for a one-way analysis:

$$SS_{cells} = \Sigma\left[\frac{(\Sigma X_{cell})^2}{N_{cell}}\right] - \frac{(\Sigma X_{tot})^2}{N_{tot}}$$

For the final exam data, you get

$$SS_{cells} = \frac{(33)^2}{3} + \frac{(22)^2}{3} + \frac{(31)^2}{3} + \frac{(27)^2}{3} - \frac{(113)^2}{12} = 23.5833$$

After SS_{cells} is obtained, it is partitioned into its three components: the A main effect, the B main effect, and the AB interaction.

The sum of squares for each main effect is somewhat like a one-way ANOVA. The sum of squares for Factor A ignores the existence of Factor B and considers the deviations of the Factor A means from the grand mean. Thus,

$$SS_A = N_{A_1}(\bar{X}_{A_1} - \bar{X}_{tot})^2 + N_{A_2}(\bar{X}_{A_2} - \bar{X}_{tot})^2$$

where N_{A_1} is the total number of scores in the A_1 cells.

For Factor B, you get

$$SS_B = N_{B_1}(\bar{X}_{B_1} - \bar{X}_{tot})^2 + N_{B_2}(\bar{X}_{B_2} - \bar{X}_{tot})^2$$

Computational formulas for the main effects are like formulas for SS_{treat} in a one-way design:

$$SS_A = \frac{(\Sigma X_{A_1})^2}{N_{A_1}} + \frac{(\Sigma X_{A_2})^2}{N_{A_2}} - \frac{(\Sigma X_{tot})^2}{N_{tot}}$$

And for the final exam data,

$$SS_{experience} = \frac{(64)^2}{6} + \frac{(49)^2}{6} - \frac{(113)^2}{12} = 18.7501$$

The computational formula for the B main effect simply substitutes B for A in the previous formula. Thus,

$$SS_B = \frac{(\Sigma X_{B_1})^2}{N_{B_1}} + \frac{(\Sigma X_{B_2})^2}{N_{B_2}} - \frac{(\Sigma X_{tot})^2}{N_{tot}}$$

For the final exam data,

$$SS_{majors} = \frac{(55)^2}{6} + \frac{(58)^2}{6} - \frac{(113)^2}{12} = 0.7501$$

To find the sum of squares for the interaction, use this formula:

$$SS_{AB} = N_{\text{cell}}[(\bar{X}_{A_1B_1} - \bar{X}_{A_1} - \bar{X}_{B_1} + \bar{X}_{\text{tot}})^2$$
$$+ (\bar{X}_{A_2B_1} - \bar{X}_{A_2} - \bar{X}_{B_1} + \bar{X}_{\text{tot}})^2$$
$$+ (\bar{X}_{A_1B_2} - \bar{X}_{A_1} - \bar{X}_{B_2} + \bar{X}_{\text{tot}})^2$$
$$+ (\bar{X}_{A_2B_2} - \bar{X}_{A_2} - \bar{X}_{B_2} + \bar{X}_{\text{tot}})^2]$$

For the final exam data,

$$SS_{AB} = 3[(11.00 - 10.6667 - 9.1667 + 9.4167)^2$$
$$+ (7.3333 - 8.1667 - 9.1667 + 9.4167)^2$$
$$+ (10.3333 - 10.6667 - 9.6667 + 9.4167)^2$$
$$+ (9.00 - 8.1667 - 9.6667 + 9.4167)^2]$$
$$= 4.0836$$

Because SS_{cells} contains only the components SS_A, SS_B, and the interaction SS_{AB}, you can also obtain SS_{AB} by subtraction:

$$SS_{AB} = SS_{\text{cells}} - SS_A - SS_B$$

For the final exam data,

$$SS_{AB} = 23.5833 - 18.7501 - 0.7501 = 4.0831$$

Error sum of squares.   As in the one-way analysis, the error variability is due to the fact that participants treated alike differ from one another on the dependent variable. Because all were treated the same, this difference must be due to uncontrolled variables or random variation. SS_{error} for a 2 × 2 design is defined as

$$SS_{\text{error}} = \Sigma(X_{A_1B_1} - \bar{X}_{A_1B_1})^2 + \Sigma(X_{A_2B_1} - \bar{X}_{A_2B_1})^2 + \Sigma(X_{A_1B_2} - \bar{X}_{A_1B_2})^2$$
$$+ \Sigma(X_{A_2B_2} - \bar{X}_{A_2B_2})^2$$

In words, SS_{error} is the sum of the sums of squares for each *cell* in the experiment. The computational formula is

$$SS_{\text{error}} = \Sigma\left[\Sigma X_{\text{cell}}^2 - \frac{(\Sigma X_{\text{cell}})^2}{N_{\text{cell}}}\right]$$

For the final exam data,

$$SS_{\text{error}} = \left(365 - \frac{(33)^2}{3}\right) + \left(162 - \frac{(22)^2}{3}\right)$$
$$+ \left(325 - \frac{(31)^2}{3}\right) + \left(245 - \frac{(27)^2}{3}\right) = 9.3333$$

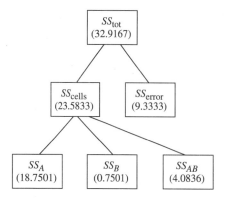

F I G U R E 10.5 The partition of sums of squares in a factorial ANOVA problem

As in a one-way ANOVA, the total variability in a factorial ANOVA is composed of two components: SS_{cells} and SS_{error}. Thus,

$$32.9167 = 23.5833 + 9.3333$$

As you read earlier, SS_{cells} can be partitioned among SS_A, SS_B, and SS_{AB} in a factorial ANOVA. Thus,

$$23.5833 = 18.7501 + 0.7501 + 4.0836$$

These relationships among sums of squares in a factorial ANOVA are shown graphically in **Figure 10.5**.

I will interrupt the analysis of the final exam data so that you may practice what you have learned about the sums of squares in a factorial ANOVA.

ERROR DETECTION

One computational check for a factorial ANOVA is the same as that for the one-way classification: $SS_{tot} = SS_{treat} + SS_{error}$. A more complete version is $SS_{tot} = SS_A + SS_B + SS_{AB} + SS_{error}$. As before, this check will not catch errors in ΣX or ΣX^2.

PROBLEMS

*8. You know that women live longer than men. What about right-handed and left-handed people? Coren and Halpern's (1991) review included data on both gender and handedness, as well as age at death. An analysis of the age-at-death numbers in this problem will produce conclusions like those reached by Coren and Halpern.

	Gender	
	Women	Men
Left-handed	76	67
	74	61
	69	58
Right-handed	82	76
	78	72
	74	68

a. Identify the design, using $R \times C$ notation.
b. Name the independent variables and the dependent variable.
c. Calculate SS_{tot}, SS_{cells}, SS_{gender}, $SS_{handedness}$, SS_{AB}, and SS_{error}.

*9. An educational psychologist was interested in whether different teacher responses affected boys and girls differently. Three classrooms with both boys and girls were used in the experiment. In one classroom, the teacher's response was one of "neglect." The children were not even observed during the time they were working arithmetic problems. In a second classroom, the teacher's response was "reproof." The children were observed and errors were corrected and criticized. In the third classroom, children were "praised" for correct answers and incorrect answers were ignored. The summary numbers in the table are based on the number of errors the children made on a comprehensive examination of arithmetic achievement given at the end of the experiment.

a. Identify the design, using $R \times C$ notation.
b. Name the independent variables.
c. Name the dependent variable.
d. Calculate SS_{tot}, SS_{cells}, $SS_{response}$, SS_{gender}, SS_{AB}, and SS_{error}. ($N = 10$ per cell)

Teacher response

		Neglect (A_1)	Reproof (A_2)	Praise (A_3)
Gender	Boys (B_1)	$\Sigma X = 169$ $\Sigma X^2 = 3129$ $\bar{X} = 16.9$	$\Sigma X = 136$ $\Sigma X^2 = 2042$ $\bar{X} = 13.6$	$\Sigma X = 118$ $\Sigma X^2 = 1620$ $\bar{X} = 11.8$
	Girls (B_2)	$\Sigma X = 146$ $\Sigma X^2 = 2244$ $\bar{X} = 14.6$	$\Sigma X = 167$ $\Sigma X^2 = 2955$ $\bar{X} = 16.7$	$\Sigma X = 109$ $\Sigma X^2 = 1291$ $\bar{X} = 10.9$

*10. Many experiments have established the concept of *state-dependent memory.* Participants learn a task in a particular state—say, under the influence of a drug or not under the influence of the drug. They recall what they have learned in the same state or in the other state. The dependent variable is recall score (memory). The data in the table are designed to illustrate the phenomenon of state-dependent memory. Calculate SS_{tot}, SS_{cells}, SS_{learn}, SS_{recall}, SS_{AB}, and SS_{error}.

During learning

			Drug	No drug
During recall	Drug	ΣX	43	20
		ΣX^2	400	100
		N	5	5
	No drug	ΣX	25	57
		ΣX^2	155	750
		N	5	5

11. Examine the following summary statistics and explain why the techniques of this chapter cannot be used for this factorial design.

		A_1	A_2
B_1	ΣX	405	614
	ΣX^2	8503	15,540
	N	20	25
B_2	ΣX	585	246
	ΣX^2	12,060	3325
	N	30	20

Degrees of Freedom, Mean Squares, and *F* Tests

Now that you are skilled at calculating sums of squares, you can proceed with the analysis of the final exam data. Mean squares, as before, are found by dividing the sums of squares by their appropriate degrees of freedom. Degrees of freedom for the sources of variance are

In general:

$$df_{tot} = N_{tot} - 1$$
$$df_A = A - 1$$
$$df_B = B - 1$$
$$df_{AB} = (A - 1)(B - 1)$$
$$df_{error} = N_{tot} - (A)(B)$$

For the final exam data:

$$df_{tot} = 12 - 1 = 11$$
$$df_{experience} = 2 - 1 = 1$$
$$df_{majors} = 2 - 1 = 1$$
$$df_{AB} = (1)(1) = 1$$
$$df_{error} = 12 - (2)(2) = 8$$

In these equations, A and B stand for the number of levels of Factor A and Factor B, respectively.

ERROR DETECTION

$$df_{tot} = df_A + df_B + df_{AB} + df_{error}$$

$$df_{cells} = df_A + df_B + df_{AB}$$

Mean squares for the final exam data are

$$MS_{experience} = \frac{SS_{experience}}{df_{experience}} = \frac{18.7501}{1} = 18.7501$$

$$MS_{majors} = \frac{SS_{major}}{df_{major}} = \frac{0.7501}{1} = 0.7501$$

$$MS_{AB} = \frac{SS_{AB}}{df_{AB}} = \frac{4.0831}{1} = 4.0831$$

$$MS_{error} = \frac{SS_{error}}{df_{error}} = \frac{9.3333}{8} = 1.1667$$

F is computed by dividing MS_{error} into each of the other three mean squares:

$$F_{experience} = \frac{MS_{experience}}{MS_{error}} = \frac{18.7501}{1.1667} = 16.07$$

$$F_{majors} = \frac{MS_{major}}{MS_{error}} = \frac{0.7501}{1.1667} = 0.64$$

$$F_{AB} = \frac{MS_{AB}}{MS_{error}} = \frac{4.0831}{1.1667} = 3.50$$

Results of a factorial ANOVA are usually presented in a summary table. **Table 10.8** is a summary table for the final exam data.

You now have three F values from the data. Each has 1 degree of freedom in the numerator and 8 degrees of freedom in the denominator. To find the probabilities of obtaining such F values if the null hypothesis is true, use Table F. Table F yields the following critical values: $F_{.05}(1,8) = 5.32$ and $F_{.01}(1,8) = 11.26$.

The next step is to compare the obtained F's to the critical value F's and then tell what your analysis reveals about relationships between the independent variables and the dependent variable. With factorial ANOVAs, you should interpret the interaction F first and then proceed to the F's for the main effects. (As you continue this chapter, be alert for the explanation of *why* you should use this order.)

The interaction F_{AB} (3.50) is less than 5.32, so the interaction between experience and major is not significant. Although the effect of experience was to increase study time by 50 percent for humanities majors and only about 15 percent for natural science

TABLE 10.8 ANOVA summary table for the final exam data

Source	SS	df	MS	F	p
A (experience)	18.7501	1	18.7501	16.07	< .01
B (majors)	0.7501	1	0.7501	0.64	> .05
AB	4.0831	1	4.0831	3.50	> .05
Error	9.3333	8	1.1667		
Total	32.9167	11			

majors, this interaction difference was not statistically significant. When *the interaction is not significant*, you can interpret each main effect as if it came from a one-way ANOVA.

The main effect for experience ($F_{experience}$) is significant beyond the .01 level (16.07 > 11.26). The null hypothesis, $\mu_{some} = \mu_{none}$, can be rejected. By examining the margin means (8.17 and 10.67), you can conclude that experienced students study significantly more than those without experience. The F_{majors} was less than 1, and F values less than 1 are never significant. Thus, these data do not provide evidence that either major studies more than the other.

PROBLEMS

12. For the longevity data in problem 8, compute *df*, *MS*, and *F* values. Arrange these in a summary table that footnotes appropriate critical values. Plot the cell means. Tell what the analysis shows.

*13. For the data on educational practices in problem 9, compute *df*, *MS*, and *F* values. Arrange these in a summary table and note the appropriate critical values. Plot the cell means. Tell what the analysis shows.

*14. For the data on state-dependent memory in problem 10, compute *df*, *MS*, and *F* values. Arrange these in a summary table and note the appropriate critical values. Plot the cell means. ∎

ANALYSIS OF A 3 × 3 DESIGN

Your next problem is to analyze a 3 × 3 design. The procedures are exactly like those for the other designs you have analyzed, but this section will emphasize the interpretation of main effects when the interaction is significant.

Two researchers were interested in the Gestalt principle of closure—the drive to have things finished, or closed. A result of this drive is that people often see a circle with a gap as closed, even if they are looking for the gap. These researchers thought that the strength of the closure drive in an anxiety-arousing situation would depend on the subjects' general anxiety level. Thus, the researchers hypothesized an interaction between the anxiety of a person and the kind of situation he or she is in.

The independent variables for this experiment were (1) anxiety level of the participant[4] and (2) kind of situation the person was in—that is, whether it was anxiety arousing or not. As you probably realized from the title of this section, there were three levels for each of these independent variables. The dependent variable was a measure of the closure drive.

First, the researchers administered the Taylor Manifest Anxiety Scale (Taylor, 1953) to a large group of college students. From this large group, they selected the 15 lowest scorers, 15 of the middle scorers, and the 15 highest scorers as participants in the study. The first factor in the experiment, then, was anxiety, with three levels: low (A_1), middle (A_2), and high (A_3).

The second factor was the kind of situation. The three kinds were dim illumination (B_1), normal illumination (B_2), and very bright illumination (B_3). The researchers believed that dim and bright illumination would create more anxiety than normal illumination.

Participants viewed 50 circles projected onto a screen. Ten of the circles were closed, ten contained a gap at the top, ten a gap at the bottom, ten a gap on the right, and ten a gap on the left. Participants simply stated whether the gap was at the top, bottom, right, or left, or whether the circle was closed. The researchers recorded as the dependent variable the number of circles reported as closed by each participant.

The hypothetical data and their analysis are reported in **Table 10.9**. Here is my suggestion for working through this complex table. Reread the description of the experiment in the text. In the top portion of Table 10.9 (the data portion), examine the headers and the descriptive statistics (sums, cell means, and margin means). Finally, work through the ANOVA analysis in the lower portion of the table.

Table 10.10 is the ANOVA summary table for the anxiety and illumination study. Note that the F value for each main effect has 2 and 36 degrees of freedom. The interaction, however, has 4 and 36 degrees of freedom [$df_{AB} = (A - 1)(B - 1)$]. This change in df results in a smaller critical value for the interaction F than for the main effects F.

The F test for the interaction is significant (4.07 > 3.89). A significant interaction requires interpretation, which is facilitated by a graph of the cell means, such as that shown in **Figure 10.6**.[5] The low-anxiety curve and the middle-anxiety curve appear relatively parallel, but the high-anxiety curve is very different in a particular way. Compared to the other seven groups, highly anxious participants in dim illumination and highly anxious participants in bright illumination reported many more circles as closed. You may remember that the researchers expected the dim and the bright illumination to be anxiety arousing, and, thus, the significant interaction is statistical confirmation that the closure drive is greater in highly anxious persons who are in anxiety-arousing situations.

---

[4] For an experiment that manipulated only this variable, see Calhoun and Johnston (1968).

[5] The interpretation of many interactions requires knowledge of techniques that are not covered in this book. The "examine a graph of the cell means" technique that I use in this chapter will produce adequate interpretations for some data sets. The best thing about this method, however, is that it serves as a fairly understandable introduction to interactions. A more sophisticated method is examining cell *residuals* (what is left in a cell after both main effects are removed). An article by Rosnow and Rosenthal (1989) explains how to use residuals to interpret interactions (and also includes some surprising statistics about how frequently authors of journal articles are confused about interactions).

T A B L E 10.9 ANOVA of hypothetical study of effects of anxiety and illumination on closure drive

		Anxiety (Factor A)						Summary values
		Low (A_1)		Middle (A_2)		High (A_3)		
		X_{A_1}	$X_{A_1}^2$	X_{A_2}	$X_{A_2}^2$	X_{A_3}	$X_{A_3}^2$	
Dim (B_1)		17	289	14	196	21	441	$\Sigma X_{B_1} = 206$
		15	225	14	196	19	361	$\Sigma X_{B_1}^2 = 3048$
		13	169	12	144	17	289	$\bar{X}_{B_1} = 13.7333$
		10	100	10	100	16	256	
		8	64	7	49	13	169	
	Sum	63	847	57	685	86	1516	
	Means	12.60		11.40		17.20		
Normal (B_2)		14	196	12	144	12	144	$\Sigma X_{B_2} = 158$
		13	169	11	121	10	100	$\Sigma X_{B_2}^2 = 1708$
		11	121	11	121	10	100	$\bar{X}_{B_2} = 10.5333$
		11	121	9	81	9	81	
		9	81	8	64	8	64	
	Sum	58	688	51	531	49	489	
	Means	11.60		10.20		9.80		
Bright (B_3)		12	144	11	121	18	324	$\Sigma X_{B_3} = 178$
		11	121	11	121	17	289	$\Sigma X_{B_3}^2 = 2228$
		10	100	10	100	15	225	$\bar{X}_{B_3} = 11.8667$
		10	100	9	81	14	196	
		9	81	9	81	12	144	
	Sum	52	546	50	504	76	1178	
	Means	10.40		10.00		15.20		
Summary values		$\Sigma X_{A_1} = 173$		$\Sigma X_{A_2} = 158$		$\Sigma X_{A_3} = 211$		$\Sigma X_{\text{tot}} = 542$
		$\Sigma X_{A_1}^2 = 2081$		$\Sigma X_{A_2}^2 = 1720$		$\Sigma X_{A_3}^2 = 3183$		$\Sigma X_{\text{tot}}^2 = 6984$
		$\bar{X}_{A_1} = 11.5333$		$\bar{X}_{A_2} = 10.5333$		$\bar{X}_{A_3} = 14.0667$		$\bar{X}_{\text{tot}} = 12.0444$

Illumination (Factor B)

Sums of squares

$$SS_{tot} = \Sigma X_{tot}^2 - \frac{(\Sigma X_{tot})^2}{N_{tot}} = 6984 - \frac{(542)^2}{45} = 455.9111$$

$$SS_{cells} = \Sigma\left[\frac{(\Sigma X_{cell})^2}{N_{cell}}\right] - \frac{(\Sigma X_{tot})^2}{N_{tot}} = \frac{(63)^2}{5} + \frac{(57)^2}{5} + \frac{(86)^2}{5} + \frac{(58)^2}{5} + \frac{(51)^2}{5} + \frac{(49)^2}{5}$$

$$+ \frac{(52)^2}{5} + \frac{(50)^2}{5} + \frac{(76)^2}{5} - \frac{(542)^2}{45} = 263.9111$$

$$SS_A = \frac{(\Sigma X_{A_1})^2}{N_{A_1}} + \frac{(\Sigma X_{A_2})^2}{N_{A_2}} + \frac{(\Sigma X_{A_3})^2}{N_{A_3}} - \frac{(\Sigma X_{tot})^2}{N_{tot}} = \frac{(173)^2}{15} + \frac{(158)^2}{15} + \frac{(211)^2}{15} - \frac{(542)^2}{45} = 99.5111$$

$$SS_B = \frac{(\Sigma X_{B_1})^2}{N_{B_1}} + \frac{(\Sigma X_{B_2})^2}{N_{B_2}} + \frac{(\Sigma X_{B_3})^2}{N_{B_3}} - \frac{(\Sigma X_{tot})^2}{N_{tot}} = \frac{(206)^2}{15} + \frac{(158)^2}{15} + \frac{(178)^2}{15} - \frac{(542)^2}{45} = 77.5111$$

$$SS_{AB} = N_{cell}[(\bar{X}_{A_1B_1} - \bar{X}_{A_1} - \bar{X}_{B_1} + \bar{X}_{tot})^2 + (\bar{X}_{A_1B_2} - \bar{X}_{A_1} - \bar{X}_{B_2} + \bar{X}_{tot})^2$$
$$+ (\bar{X}_{A_1B_3} - \bar{X}_{A_1} - \bar{X}_{B_3} + \bar{X}_{tot})^2 + \cdots + (\bar{X}_{A_3B_3} - \bar{X}_{A_3} - \bar{X}_{B_3} + \bar{X}_{tot})^2]$$

$$= 5[(12.60 - 11.5333 - 13.7333 + 12.0444)^2$$
$$+ (10.40 - 11.5333 - 11.8667 + 12.0444)^2 + (11.60 - 11.5333 - 13.7333 + 12.0444)^2$$
$$+ (10.20 - 10.5333 - 13.7333 + 12.0444)^2 + (11.40 - 10.5333 - 13.7333 + 12.0444)^2$$
$$+ (10.00 - 10.5333 - 11.8667 + 12.0444)^2$$
$$+ (17.20 - 14.0667 - 13.7333 + 12.0444)^2 + (9.80 - 14.0667 - 10.5333 + 12.0444)^2$$
$$+ (15.20 - 14.0667 - 11.8667 + 12.0444)^2] = (5)(17.3778) = 86.8890$$

(*continued*)

TABLE 10.9 (continued)

$Check:$ $SS_{AB} = SS_{cells} - SS_A - SS_B = 263.9111 - 99.5111 - 77.5111 = 86.8889$

$$SS_{error} = \Sigma \left[\Sigma X^2_{cell} - \frac{(\Sigma X_{cell})^2}{N_{cell}} \right]$$

$$= \left(847 - \frac{(63)^2}{5} \right) + \left(688 - \frac{(58)^2}{5} \right) + \left(546 - \frac{(52)^2}{5} \right) + \left(685 - \frac{(57)^2}{5} \right)$$

$$+ \left(531 - \frac{(51)^2}{5} \right) + \left(504 - \frac{(50)^2}{5} \right) + \left(1516 - \frac{(86)^2}{5} \right) + \left(489 - \frac{(49)^2}{5} \right) + \left(1178 - \frac{(76)^2}{5} \right) = 192$$

$Check:$ $455.9111 = 263.9111 + 192$

Degrees of freedom

$df_A = A - 1 = 3 - 1 = 2$

$df_B = B - 1 = 3 - 1 = 2$

$df_{AB} = (A - 1)(B - 1) = (2)(2) = 4$

$df_{error} = N_{tot} - (A)(B) = 45 - (3)(3) = 36$

$df_{tot} = N_{tot} - 1 = 45 - 1 = 44$

Mean squares

$$MS_A = \frac{SS_A}{df_A} = \frac{99.51}{2} = 49.76$$

$$MS_B = \frac{SS_B}{df_B} = \frac{77.51}{2} = 38.76$$

$$MS_{AB} = \frac{SS_{AB}}{df_{AB}} = \frac{86.89}{4} = 21.72$$

$$MS_{error} = \frac{SS_{error}}{df_{error}} = \frac{192}{36} = 5.33$$

F values

$$F_A = \frac{MS_A}{MS_{error}} = \frac{49.76}{5.33} = 9.33$$

$$F_B = \frac{MS_B}{MS_{error}} = \frac{38.76}{5.33} = 7.27$$

$$F_{AB} = \frac{MS_{AB}}{MS_{error}} = \frac{21.72}{5.33} = 4.07$$

TABLE 10.10 ANOVA summary table for the closure study

Source	SS	df	MS	F	p
A (anxiety)	99.51	2	49.76	9.33	< .01
B (illumination)	77.51	2	38.76	7.27	< .01
AB	86.89	4	21.72	4.07	< .01
Error	192.00	36	5.33		
Total	455.91	44			

$$F_{.01}(2, 36 \ df) = 5.25 \qquad F_{.01}(4, 36 \ df) = 3.89$$

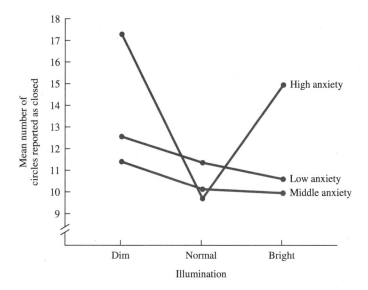

FIGURE 10.6 Graph of cell means, illustrating an interaction effect for the closure study

■ Interpreting Main Effects When the Interaction Is Significant

After interpreting a significant interaction, the next step is to deal with main effects. Such a step would take us from our topic of elementary statistics to a more advanced level, so I will just indicate what the problems are. The anxiety main effect is significant. From **Table 10.9**, you can see that the margin means for anxiety are: 14.07 for high, 10.53 for medium, and 11.53 for low. A main effects analysis would produce a conclusion such as, "Highly anxious people have a greater closure drive than low- and middle-anxiety people." However, look at the graph of cell means (**Figure 10.6**) and attend to the normal illumination means. This interpretation makes no sense. The high-anxiety people saw *fewer* circles as closed. Thus, an interpretation of this main effect (when the interaction is significant) is misleading.

Main effects are part of any factorial design. They give you information about one factor (one set of margin means) *by ignoring the other factor* in the experiment. When the interaction is not significant, main effects are useful. However, a significant interaction means that both factors are having an effect on the data in some combined way. When this is the case, you *should not ignore* one of the factors. Thus, when the interaction is significant, a main effect analysis is just too simple. The situation is complicated; main effects are useful only in simpler situations.

Here's another example of the problem of interpreting main effects when the interaction is significant. Reexamine Figure 10.4, which I used to illustrate an interaction. The two factors in this illustration are weather and plans. The main effect for *weather* would compare two means, the average snow rating and the average no snow rating. Both of these rating scores are halfway between good and bad (the tick mark on the figure). Of course, the F value for two identical sample means would be zero. In the same way, the main effect for *plans* would also produce an $F = 0$; neither main effect is significant, but the interaction is.

Does an interpretation of these nonsignificant main effects lead to better understanding? Let's see. The interpretation of *plans* is that we don't have evidence that a person's plans has an effect on rating the weather. Of course, this is not correct. Plans are most important; it's just that the effect of a person's plans on the rating depends on what the weather actually is. So, once again, with a significant interaction, the interpretation of a main effect is not helpful.

Fortunately, there are techniques that allow you to extract more information from factorial experiments that have significant interactions. Techniques such as testing for *treatment-contrast interactions*[6] or *decomposing cell means*[7] are available. In addition, there are times when the data are such that the interpretation of a main effect is correct, even though the interaction is significant (see Howell, 1995, pp. 330–332). I constructed **Table 10.11** with cell means that illustrate Howell's point. Note that the data appear to have an interaction (diverging curves). The story for the main effect of anxiety (the margin means at the bottom) is "the greater the anxiety, the larger the score." This story is true at all three levels of the other factor. Similarly, the story for the main effect of illumination (the brighter the illumination, the larger the score) is true at all levels of anxiety.

At this stage of your education, what should you do about interpreting a factorial design that has a significant interaction? I think you should seek the advice of your

T A B L E 10.11 Hypothetical data that show a significant interaction and main effects that can be interpreted

		Anxiety			
		Low	Middle	High	$\bar{X}$
	Dim	5	10	15	10
Illumination	Normal	20	30	40	30
	Bright	50	80	110	80
	$\bar{X}$	25	40	55	

[6] For information on treatment-contrast interactions, see Kirk (1995, pp. 383–389).
[7] For an article explaining the decomposition of cell means, see Rosnow and Rosenthal (1989).

instructor. In the case of my own students, I ask them to interpret only the interaction (and to recognize that further analysis requires knowledge they don't yet have).

PROBLEMS

15. The following three problems are designed to help you learn to *interpret* the results of a factorial experiment. For each problem (**a**) identify the independent and dependent variables, (**b**) fill in the rest of the summary table, and (**c**) interpret the results.

i. A clinical psychologist has been investigating the relationship between humor and aggression. The participant's task in these experiments is to think up as many captions as possible for four cartoons. The test lasts eight minutes, and the psychologist simply counts the number of captions produced. In the psychologist's latest experiment, a participant is either insulted or treated in a neutral way by an experimenter. Next, the participant responds to the cartoons at the request of either the same experimenter or a different experimenter. The cell means and the summary table are shown in the tables.

Participant was

Experimenter was		Insulted	Treated neutrally
	Same	31	18
	Different	19	17

Source	df	MS	F	p
A (between treatments)	1	675.00		
B (between experimenters)	1	507.00		
AB	1	363.00		
Error	44	74.31		

ii. An educational psychologist was interested in response biases, such as making a decision on the basis of an irrelevant stimulus. For example, a response bias occurs if the grade an English teacher puts on a theme written by a tenth grader is influenced by the student's name or by the occupation of his father. For this particular experiment, the psychologist picked a "high-prestige" name (David) and a "low-prestige" name (Elmer). In addition, she made up two biographical sketches that differed only in the occupation of the student's father (research chemist or unemployed). The participants in this experiment were given the task of grading a theme on a scale of 50 to 100. They had been told the name of the writer and had read his biographical sketch. The same theme was given to each participant in the experiment.

Name

	David	Elmer
Research chemist	86	81
Unemployed	80	88

Occupation of father

Source	df	MS	F	p
A (between names)	1	22.50		
B (between occupations)	1	2.50		
AB	1	422.50		
Error	36	51.52		

iii. A social psychologist in charge of a large class in community psychology invited 20 outside speakers who held community service jobs to give short talks. Ten of these speakers were men, ten women. Five of the men had been at their jobs two years or longer, and the other five had been at their jobs less than six months. This same job-experience variable divided the women into two groups of five. The psychologist wanted to determine if the two variables (gender and job experience) were related to the amount of attention the speakers received. She arranged a wide-angle camera to take a picture of the class when the speaker had completed the third minute of his or her talk. The dependent variable was the number of persons who were looking directly at the speaker when the picture was taken. Fortunately, the same number of people were in class on every occasion. The cell means and the summary table are shown here.

Gender of speaker

	Female	Male
Less than 6 months	23	28
Longer than 2 years	31	39

Job experience

Source	df	MS	F	p
A (between gender)	1	211.25		
B (between experience)	1	451.25		
AB	1	11.25		
Error	16	46.03		

16. Write an interpretation of your analysis of the data from the state-dependent memory experiment (problems 10 and 14). ■

COMPARING LEVELS WITHIN A FACTOR

You can compare levels within a factor when you have a factorial ANOVA that has an interaction that is *not significant*. Each main effect in a factorial ANOVA is like a one-way ANOVA; the question that remains after testing a main effect is, Are any of the levels of this factor significantly different from any others? You will be pleased to learn that the solution you learned for one-way ANOVA also applies to factorial ANOVA.

When there are more than two levels of one factor in a factorial design, a Tukey HSD test can be used to test for significant differences between pairs of levels. There is one restriction, however, which is that the Tukey HSD test is appropriate *only if the interaction is **not** significant*.[8] To keep you from getting bored while doing the same test you did in the last chapter, I will show you a way to arrange the arithmetic that will save you some time when you do Tukey tests.

The HSD formula for a factorial ANOVA is the same as that for a one-way ANOVA:

$$\text{HSD} = \frac{\bar{X}_1 - \bar{X}_2}{s_{\bar{X}}}$$

where $s_{\bar{X}} = \sqrt{\dfrac{MS_{\text{error}}}{N}}$

Let me explain what $\bar{X}$'s and what N you need for the HSD formula. The $\bar{X}$'s are margin means and N is the number of scores used to calculate the $\bar{X}$ (a margin N). I'll illustrate using the data that analyzed arithmetic errors of boys and girls who were taught by different methods. In that study (problem 13), the interaction was not significant, so Tukey HSD tests are appropriate. The actual data are shown in problem 9 (page 255). Look at that data table now. To compare the effect of *neglect* to that of *praise* with a Tukey HSD, N is 20 (10 from Cell A_1B_1 and 10 from Cell A_1B_2). The margin $\bar{X}$'s are 17.75 and 11.35 (315/20 and 227/20). (For a comparison of levels of teacher response, the factor of gender is ignored. When one factor is ignored, statisticians say that the design is *collapsed* over the ignored factor.)

ERROR DETECTION

The N in $\sqrt{\dfrac{MS_{\text{error}}}{N}}$ is the N that each $\bar{X}$ in the numerator is based on. This rule applies to one-way ANOVA and to factorial ANOVAs.

---

[8] A nonsignificant interaction is a condition required by many *a priori* and *post hoc* tests.

When Tukey HSDs are calculated on means that are all based on equal N's (as is the case for all factorial ANOVAs in this chapter), there is an arithmetic arrangement that is more efficient than that used in Chapter 9. Start by calculating a product, $(HSD_{.05})(s_{\bar{X}})$. $HSD_{.05}$ comes from Table G and $s_{\bar{X}}$ comes from the data.[9] Any difference between means that is larger than this product is significant. If a mean difference is smaller, it is not significant. (You can play with the formula above and see how this works.) For the neglect-reproof-praise study,

$$(HSD_{.05})(s_{\bar{X}}) = (3.44)\sqrt{\frac{19.89}{20}} = 3.43$$

Arithmetic on the cell data produces the following margin means and differences:

Neglect and Reproof $15.75 - 15.15 = 0.60$
Neglect and Praise $15.75 - 11.35 = 4.40$
Reproof and Praise $15.15 - 11.35 = 3.80$

Because 4.40 and 3.80 are larger than 3.43, the conclusion is that a teacher response of praise results in significantly fewer arithmetic errors than does a response of neglect or reproof ($p < .05$). Responses of neglect or reproof are not significantly different.

Of course, for some main effects you do not need a test subsequent to ANOVA; the F test on that main effect gives you the answer. This is true for *any main effect that involves only two levels*. Just like a t test, a significant F tells you that the two means are from different populations. Thus, if you are analyzing a 2 × 2 factorial ANOVA, you do not need a test subsequent to ANOVA for *either* factor.

PROBLEMS

17. Suppose you were asked to determine whether there is a significant difference between the low-anxiety scores and the high-anxiety scores in the study of the effects of anxiety and illumination on closure drive. Could you use a Tukey HSD test?

18. Two social psychologists asked freshman and senior college students to write an essay supporting recent police action on campus. (The students were known to be against the police action.) The students were given 50 cents, one dollar, five dollars, or ten dollars for their essay. Later, each student's attitude toward the police was measured on a scale of 1 to 20. Perform an ANOVA, fill out a summary table, and write an interpretation of the analysis at this point. (See Brehm and Cohen, 1962, for a similar study of Yale students.) If appropriate, calculate HSDs for all pairwise comparisons. Write an overall interpretation of the results of the study.

---

[9] Because all comparisons are based on equal N's, the $s_{\bar{X}}$ for every comparison will be the same.

Reward

			$10	$5	$1	50¢
Students	Freshmen	ΣX	73	83	99	110
		ΣX^2	750	940	1290	1520
		N	8	8	8	8
	Seniors	ΣX	70	81	95	107
		ΣX^2	820	910	1370	1610
		N	8	8	8	8

19. Eckhard Hess used a scientific experiment to show that a person's pupil diameter changes as emotional arousal changes. (Hess reports that his wife first noted this phenomenon while watching her husband leaf through a magazine—a magazine with pictures of birds.) The data in the table illustrate some of Hess's findings. Men and women were shown one picture and their pupil size was recorded (large numbers indicate more arousal). Analyze the data with a factorial ANOVA and, if appropriate, a set of Tukey HSDs. Write a conclusion.

Pictures

			Nude man	Nude woman	Infant	Landscape
Gender	Women	ΣX	22	9	23	15
		ΣX^2	166	29	179	77
		N	3	3	3	3
	Men	ΣX	11	26	9	14
		ΣX^2	41	226	29	70
		N	3	3	3	3

20. Review the objectives at the beginning of the chapter. ■

Analysis of Variance:
One-Factor Correlated Measures

OBJECTIVES FOR CHAPTER 11

After studying the text and working the problems in this chapter, you should be able to:

1. Distinguish between correlated-samples and independent-samples designs and identify the statistical tests that go with each design

2. Describe the characteristics of a one-factor correlated-measures analysis of variance (ANOVA)

3. For a one-factor correlated-measures ANOVA, identify the sources of variance, calculate their sums of squares, degrees of freedom, mean squares, and the *F* value

4. Interpret the *F* value from a one-factor correlated-measures ANOVA

5. Use a Tukey Honestly Significant Difference test to make pairwise comparisons

6. Distinguish between Type I and Type II errors

7. List and explain the advantages and limitations of a one-factor correlated-measures ANOVA

8. List and explain the assumptions of a one-factor correlated-measures ANOVA

There are three chapters in this book on analysis of variance (ANOVA). No other inferential statistical test gets as much space from me or as much time from you.

This heavy emphasis on ANOVA in an introductory statistics book does make sense; among behavioral scientists, ANOVA is the most widely used inferential statistics technique. When I surveyed 20 recent empirical articles in *Psychological Science*, I found that 60 percent of the articles used some form of analysis of variance. Of those articles that used analysis of variance, two-thirds of them used a correlated-measures ANOVA at least once in analyzing the data. It seems likely (if my sample is a clue to the future) that you will have opportunities to use what you are learning in this chapter as you continue your education.

You have already covered material that prepares you to answer the question, What makes an analysis of variance a **correlated-measures ANOVA**? That material was in

TABLE 11.1 Identification of statistical tests based on kind of design

Independent variables and number of levels	Design of experiment	
	Correlated samples	Independent samples
One independent variable, 2 levels	Correlated-samples t test	Independent-samples t test
One independent variable, 2 or more levels	Correlated-measures ANOVA	One-way ANOVA
Two independent variables, each with 2 or more levels		Factorial ANOVA

Chapter 8, where you learned to distinguish between an independent-samples t test and a correlated-samples t test. As you probably recall, you looked at a data set and asked if there was some reason why the top two scores in each group were paired. If there was a reason [such as natural pairs, matched pairs, or same subjects (repeated measures)], then the design was one that required a correlated-samples t test. If there was no reason for two scores to be paired side by side, an independent-samples t test was called for.

The experimental designs in Chapter 9 (one-way ANOVA) and Chapter 10 (factorial ANOVA) both require that the analyzed scores be independent. This requirement is often satisfied by randomly assigning different participants to the different treatments.[1]

All the statistical tests covered in Chapters 8, 9, 10, and 11 are given in **Table 11.1**, which shows which tests are appropriate for correlated-samples designs and which are appropriate for independent-samples designs. In this chapter, you return to the analysis of data from a correlated-samples design. (Now is the time for you to decide for yourself whether you need to review the distinction between correlated-samples and independent-samples designs. If the answer is yes, reread pages 189–192.)

The designs that can be analyzed using correlated-measures ANOVA are the same designs that use a correlated t: natural pairs, matched pairs, and repeated measures.[2] One other reminder may be helpful: The "one-factor" in the chapter title means that just one independent variable is involved. The number of levels of this variable can be two or more. For this chapter, all the problems you do will have three or more levels of the independent variable. (When you have two levels, use a t test.)

EXAMINING THE NUMBERS

Look at the data in **Table 11.2**. There are three levels of the one independent variable (named simply X at this point). The mean score for each level is shown at the bottom of

---

[1] To understand this chapter, you must have completed Chapter 9, "Analysis of Variance: One-Way Classification." It isn't necessary (though it might be helpful) for you to have also studied Chapter 10, "Analysis of Variance: Factorial Design."

[2] Some writers use the name *randomized blocks* for this design. In particular contexts, it is also referred to as *subjects × treatments*, *subjects × trials*, and *split plot*.

TABLE 11.2 **A data set**

| Subjects | Levels of independent variable, X | | |
	X_1	X_2	X_3
S_1	57	60	64
S_2	71	72	74
S_3	75	76	78
S_4	93	92	96
$\bar{X}$	74	75	78

its column. Using the skills you learned in Chapter 9, estimate whether an F ratio for these data would be significant. Remember that $F = MS_{treat}/MS_{error}$. Go ahead, stop reading and make your estimate. (If you also write the steps in your thinking, you can compare them to the analysis that follows.)

Here's one line of thinking that might be expected from a person who remembers the lessons of Chapter 9:

"Well, OK, let's start with the means of the three treatments. They are pretty close together; 74, 75, and 78. Not much variability there. How about *within* each treatment? Uh-huh, scores range from 57 to 93 in the first column, so there is a good bit of variability there. Same story in the second and the third columns, too. There's lots of variability within the groups, so MS_{error} will be large. Nope, I don't think the F ratio would be very big. My guess is that F won't be significant—that the means are not significantly different."

Most of us who teach statistics would say that this quote shows good understanding of what goes into the F value in a one-way ANOVA. With a correlated-measures ANOVA, however, there is an *additional* consideration.

The additional consideration for the data in Table 11.2 is that *each subject*[3] contributed a score to *each level* of the independent variable. To explain, look at the second column, X_2, which has scores 60, 72, 76, 92. The numbers are quite variable. But notice that some of this variability is *predictable* from the X_1 scores. That is, low scores (such as 60) are associated with S_1 (who had a low score of 57 on X_1). Notice, too, that in column X_3, S_1 had a score of 64, the lowest in the X_3 condition. Thus, knowing that a score was made by S_1, you can predict that the score will be low. Please do a similar analysis on the scores for the person labeled S_4.

In a correlated-measures ANOVA, then, some of the variability of the scores within a level of the independent variable is predictable if you know which subject contributed the score. If we could *remove* the variability that goes with the differences between the subjects, we could *reduce* the variability within a level of the independent variable.

One way to reduce the variability between subjects is to use subjects who are all alike. However, although this would work well, finding people who are all alike is

---

[3] "Participant" is usually preferable to "subject" when referring to people whose scores are analyzed in a behavioral science study. Statistical methods, however, are used by all manner of disciplines; dependent variable scores come from agricultural plots, production records, and *Rattus norvegicus* as well as from people. Thus, statistics books continue to use the more general term, *subjects*.

almost impossible. Fortunately, a correlated-measures ANOVA provides a statistical alternative that accomplishes the same goal.

ONE-FACTOR CORRELATED-MEASURES ANOVA: THE RATIONALE

Here's the paragraph that gives the overview of the topic of this chapter. A correlated-measures ANOVA allows you to calculate and then discard the variability that comes from the differences among the subjects. The remaining variability in the data set is then partitioned into two components: one that is due to the differences between treatments (between-treatment variance) and another that is due to the inherent variability in the measurements (the error variance). These two variances are put into a ratio; the between-treatments variance is divided by the error variance. This ratio is an F value. This F ratio is a sensitive measure of any differences that may exist among the treatments because the variability produced by the differences between individuals has been removed.

AN EXAMPLE PROBLEM

The numbers in Table 11.2 can be used to illustrate the procedures for calculating a one-factor correlated-measures ANOVA. To include the important step of telling what the analysis shows, however, you must know what the independent and dependent variables are. The numbers in the table (the dependent variable) are scores on multiple-choice tests. The independent variable is the strategy the test-taker used.[4]

The strategies used by subjects in this experiment were:

1. Always return to your first choice if you become doubtful about an item, and
2. Write notes and comments about the items on the test itself.

Do you use either of these strategies? Do you have an opinion about the effect of using either of these tactics? Perhaps you would be willing to express your opinion by deciding which of the columns in Table 11.2 represent which of the two strategies and which column represents the control condition (taking the test without instructions about strategy). (If you want to guess, do it now, because I'm going to label the columns in the following paragraph.)

The numbers in the X_2 column represent typical multiple-choice test performances (control condition). The mean is 75. The X_1 column shows scores when test takers follow strategy 1 (go with your initial choice). The scores in column X_3 represent performance when test takers make notes on the test itself (though these notes are not graded in any way). Does either strategy have a significant effect on scores? A correlated-measures ANOVA will give you an answer.

[4] By the way, should you be one of those people who find yourself taking multiple-choice tests from time to time, the analysis of these numbers will reveal a strategy for improving your score on such exams.

TABLE 11.3 A correlated-measures ANOVA comparing two test-taking strategies with a control condition. The data set is the same as Table 11.2.

Subjects	Use First Choice	Control	Write Notes	ΣX_s
	Treatments			
S_1	57	60	64	181
S_2	71	72	74	217
S_3	75	76	78	229
S_4	93	92	96	281
ΣX_t	296	300	312	908
$\bar{X}$	74	75	78	

$\Sigma X_{\text{tot}} = 296 + 300 + 312 = 908$

$\Sigma X_{\text{tot}}^2 = 57^2 + 60^2 + \cdots + 92^2 + 96^2 = 70{,}460$

$$SS_{\text{tot}} = \Sigma X_{\text{tot}}^2 - \frac{(\Sigma X_{\text{tot}})^2}{N_{\text{tot}}} = 70{,}460 - 68{,}705.333 = 1754.667$$

$$SS_{\text{subjects}} = \frac{181^2}{3} + \frac{217^2}{3} + \frac{229^2}{3} + \frac{281^2}{3} - \frac{908^2}{12}$$
$$= 70{,}417.333 - 68{,}705.333 = 1712.00$$

$$SS_{\text{strategies}} = \frac{296^2}{4} + \frac{300^2}{4} + \frac{312^2}{4} - \frac{908^2}{12}$$
$$= 68{,}740.000 - 68{,}705.333 = 34.667$$

$$SS_{\text{error}} = SS_{\text{tot}} - SS_{\text{subjects}} - SS_{\text{strategies}}$$
$$= 1754.667 - 1712.000 - 34.667 = 8.000$$

$df_{\text{tot}} = N_{\text{tot}} - 1 = 12 - 1 = 11$

$df_{\text{subjects}} = N_s - 1 = 4 - 1 = 3$

$df_{\text{strategies}} = N_t - 1 = 3 - 1 = 2$

$df_{\text{error}} = (N_s - 1)(N_t - 1) = (3)(2) = 6$

The calculations for this analysis are shown in Table 11.3 (which starts with the same data set as in Table 11.2, except that labels and sums have been added). Three subscripts (explained below) will be new to you. Examine **Table 11.3** now.

■ Sums of Squares

In the calculations that follow the data set in Table 11.3, note the familiar symbols, ΣX_{tot} (the sum of all the scores) and ΣX_{tot}^2. I have illustrated the calculation of ΣX_{tot}^2

(the sum of all the squared scores) without using a column for X^2, as I did in the two previous chapters. The formula for SS_{tot} is the same as that for all ANOVA problems.

$$SS_{tot} = \Sigma X_{tot}^2 - \frac{(\Sigma X_{tot})^2}{N_{tot}}$$

For the data in Table 11.3, $SS_{tot} = 1754.667$.

$SS_{subjects}$ is the variability due to differences among subjects. As you know, people differ in their abilities when it comes to multiple-choice exams. These differences are reflected in the different row totals. The symbol for the score of a subject is X_s. The sum of one subject's scores (a row total) is symbolized ΣX_s, and N_s indicates the number of scores contributed by one subject. The general formula for $SS_{subjects}$ is

$$SS_{subjects} = \Sigma \left[\frac{(\Sigma X_s)^2}{N_s} \right] - \frac{(\Sigma X_{tot})^2}{N_{tot}}$$

Work through the calculations that lead to the conclusion: $SS_{subjects} = 1712.000$. You may have already noted that 1712 is a big portion of the total sum of squares.

SS_{treat} is the variability that is due to differences among the strategies the subjects used; that is, SS_{treat} is the variability due to the independent variable. A particular score is symbolized X_t; the sum of the scores for one treatment (one level of the independent variable) is ΣX_t. N_t tells you the number of scores that received that treatment. The general formula for SS_{treat} is

$$SS_{treat} = \Sigma \left[\frac{(\Sigma X_t)^2}{N_t} \right] - \frac{\Sigma (X_{tot})^2}{N_{tot}}$$

For the analysis in Table 11.3, the generic term *treatments* has been replaced with the term specific to these data: *strategies*. As you can see, $SS_{strategies} = 34.667$.

SS_{error} is the variability that remains in the data when the effects of the other identified sources have been removed. To find SS_{error},

$$SS_{error} = SS_{tot} - SS_{subjects} - SS_{treat}$$

For these data, $SS_{error} = 8.000$.

PROBLEM

1. In Chapter 9, page 225 (and in Chapter 10, page 254), the text included a graphic showing how SS_{tot} is partitioned into its components. Construct a graphic that shows the partition of the components of sums of squares in a one-factor correlated-measures ANOVA. ∎

■ Degrees of Freedom, Mean Squares, and F

After calculating sums of squares, the next step is to determine their degrees of freedom. Degrees of freedom for one-way correlated-measures ANOVA follow the general rule: number of observations minus one.

In general:

$$df_{tot} = N_{tot} - 1$$
$$df_{subjects} = N_s - 1$$
$$df_{treat} = N_t - 1$$
$$df_{error} = (N_s - 1)(N_t - 1)$$

For the strategy study:

$$df_{tot} = 12 - 1 = 11$$
$$df_{subjects} = 4 - 1 = 3$$
$$df_{strategies} = 3 - 1 = 2$$
$$df_{error} = (3)(2) = 6$$

As is always the case for ANOVAs, mean squares are found by dividing a sum of squares by its df. With mean squares in hand, you can form any appropriate F ratios. For a reason to be explained later, the only F ratio you will calculate is one that determines if there are significant differences among the *treatments*. Because you will not test the significance of the factor *subjects*, there is no need to calculate $MS_{subjects}$.

$$MS_{strategies} = \frac{SS_{strategies}}{df_{strategies}} = \frac{34.667}{2} = 17.333$$

$$MS_{error} = \frac{SS_{error}}{df_{error}} = \frac{8.000}{6} = 1.333$$

To find F, divide $MS_{strategies}$ by the error term:

$$F = \frac{MS_{strategies}}{MS_{error}} = 13.00$$

The now familiar ANOVA summary table is the conventional way to present the results of an analysis of variance. **Table 11.4** shows the analysis that assessed strategies for taking multiple-choice exams.

■ Interpretation of F

The F from the data (13.00) is larger than the F_{01} from the table (10.92), which allows you to conclude that at least one of the strategies differs from the others. Such a

TABLE 11.4 ANOVA summary table for the exam strategy study

Source	SS	df	MS	F
Subjects	1712.000	3		
Strategies	34.667	2	17.333	13.00
Error	8.000	6	1.333	
Total	1754.667	11		
	$F_{.05}(2, 6) = 5.14$		$F_{.01}(2, 6) = 10.92$	

statement, of course, doesn't answer the question that any intelligent, eager student would ask: "Should I use or avoid either of these strategies?"

With this question in mind, look back at **Table 11.3** for the means of each of the strategies. What pairwise comparison is the most interesting to you?

TUKEY HSD TESTS

The one pairwise comparison that I'll do for you is the comparison between the Write Notes strategy and the control group. The formula for a Tukey HSD is the same as that you used previously:

$$\text{HSD} = \frac{\bar{X}_1 - \bar{X}_2}{s_{\bar{X}}}$$

where $s_{\bar{X}} = \sqrt{\dfrac{MS_{\text{error}}}{N_t}}$

N_t = the number of scores in each treatment

Applying the formula to the data for the Write Notes strategy and the control group:

$$\text{HSD}_{\text{write v. cont}} = \frac{78 - 75}{\sqrt{\dfrac{1.333}{4}}} = \frac{3}{.577} = 5.20 \qquad p < .05$$

The critical value of HSD at the .05 level for a study with three groups and with a df_{error} of 6 is 4.34. Because the data-produced HSD is larger than the critical value of HSD, you can conclude that making notes on a multiple-choice exam will produce a statistically significantly better score.[5]

OK, now it is time for you to put into practice what you have been learning.

PROBLEMS

2. Many studies have shown that interspersing rest between practice sessions (spaced practice) produces better performance than a no-rest condition (massed practice). Most of these studies used short rests, such as minutes or hours. Bahrick et al. (1993), however, studied the effect of weeks of rest. Thirteen practice sessions of learning the English equivalents of foreign words were interspersed with rests of two weeks, four weeks, or eight weeks. The data for all four participants in this study (for percent of recall) are shown here. Analyze the data with a one-factor correlated-measures ANOVA, construct a summary table, perform Tukey HSD tests, and write an interpretation.

---

[5] This comparison is based on a conversation with Wilbert J. McKeachie.

	Rest interval (weeks)		
Subject	2	4	8
1	40	50	58
2	58	56	65
3	44	70	69
4	57	61	74

3. Perform a Tukey HSD test using the data in Table 11.3. Compare the strategy of using the first choice to the control group.[6] ■

TYPE I AND TYPE II ERRORS

In problem 2 you reached the conclusion that an 8-week rest between practice sessions is better than a 2-week rest. Could this conclusion be in error? Of course it could be, although it isn't very likely. But suppose the conclusion is erroneous. What kind of a statistical error would it be: a Type I error or a Type II error?

In problem 3 you compared the strategy of *return to your initial choice* to the *control* condition. You did not reject the null hypothesis. If this conclusion is wrong, then you've made an error. Is it a Type I error or a Type II error?

If you are uncertain about the definitions of Type I and Type II errors, don't feel too bad. They are easy to forget. Maybe what you need is spaced practice?

A Type I error is rejecting the null hypothesis when it is true. A Type II error is failing to reject a false null hypothesis. For a review of these concepts, see pages 171–173 and 204–205.

SOME BEHIND-THE-SCENES INFORMATION ABOUT CORRELATED-MEASURES ANOVA

Now that you can analyze data from a one-factor correlated-measures ANOVA, you are ready to appreciate something about what is really going on when you perform this statistical test. I'll begin by addressing *the fundamental issue* in any ANOVA: *partitioning the variance*.

Think about SS_{tot}. You may recall that the definitional formula for SS_{tot} is $\Sigma(X - \bar{X}_{tot})^2$. When you calculate SS_{tot}, you get a number that measures how variable the scores are. If the scores are all close together, SS_{tot} will be small. (Think this through. Do you agree?) If the scores are quite different, SS_{tot} will be large. (Agree?)

Regardless of the amount of total variance, an ANOVA divides it into separate, independent parts. Each part is associated with a different source.

[6] This comparison is based on data presented by Benjamin, Cavell, and Shallenberger (1984).

Look again at the raw scores in **Table 11.3**. Some of the variability in those scores is associated with different subjects. You can see this in the *row totals*—there is a lot of variability from S_1 to S_2 to S_3 to S_4; that is, there is variability *between the subjects*.

Now, please focus on the variability *within the subjects*. That is, look at the variability within S_1. S_1's scores vary from 57 to 60 to 64. S_2's scores vary from 71 to 72 to 74. Where does this variability within a subject come from? Clearly, *treatments* are implicated because, as treatments vary from the first to second to third column, scores increase.

But the change in treatments doesn't account for all the variation. That is, for S_1 the change in treatments produced a 3-point or 4-point increase in the scores, but for S_2 the change in treatments produced only a 1-point or 2-point increase, going from the first to the third treatment. Thus, the change from the first treatment to the third is not consistent; although an increase always comes with the next treatment, the increase isn't the same for each subject. This means that some of the variability within a subject is accounted for by the treatments, but not all. The variability that remains in the data in Table 11.3 is used to calculate an error term. It is sometimes called the **residual variance**.

The analysis presented in this chapter for correlated-measures ANOVA is appropriate if (1) the levels of the treatments were chosen by the researcher (a fixed-effects model) rather than being selected randomly, and (2) the subjects are randomly assigned to rows. These conditions are common in behavioral science experiments. For other situations, see Howell (1992, Chapter 14) or Kirk (1995, Chapter 7).

■ Advantages of Correlated-Measures ANOVA

Recall that correlated-measures ANOVA is used for designs in which there is a reason why the scores in each row belong together. That reason might be that the subjects were matched before the experiment began, that the subjects were "natural triplets" or "natural quadruplets," or that the same subject contributed all the scores in one row.

The case where the same subjects are used for all levels of the independent variable has the advantage of saving time and effort. To illustrate, it takes time to recruit participants, give explanations, and provide debriefing. So, if your study has three levels of the independent variable, and you can run each participant in all three conditions, you can probably save two-thirds of the time it takes to recruit, explain, and debrief. Clearly, a correlated-measures design can be more *efficient*.

What about statistical *power*?[7] Any procedure that reduces the size of the error term (MS_{error}) will produce a larger F ratio. By removing the between-subject variability from the analysis, the error term is reduced, making a correlated-measures design more powerful. To summarize, two of the advantages of a correlated-measures ANOVA are *efficiency* and *power*.

---

[7] You may recall that the more powerful the statistical test, the more likely it will reject a false null hypothesis; see pages 204–205.

■ Cautions about Correlated-Measures ANOVA

The previous section, titled "Advantages of . . . " leads good readers to respond internally with, "OK, *disadvantages* is next." Well, this section will give you two cautions that go with correlated-measures ANOVAs. To address the first, work the problem that follows.

PROBLEM

4. One of the following experiments does not lend itself to a correlated-measures design using the same participants in every treatment. Figure out which one and explain why.
   a. Suppose you want to design an experiment to find out the physiological effects of meditation. Your plan is to measure oxygen consumption before, during, and after a meditation session.
   b. A friend, interested in the treatment of depression, wants to compare insight therapy, drug therapy, and a combination of the two, using Beck Depression Inventory scores as a dependent variable. ■

The first caution about correlated-measures ANOVA is that it often is not useful if the effect of one level of the independent variable continues to affect the subject's response in the next treatment condition (a *carryover* effect). Problem 4b is an example of such a situation. To use a correlated-measures ANOVA, some way of matching three patients who could then be assigned randomly to one of the three treatment conditions would be needed. In the oxygen consumption experiment in problem 4a, however, there is no reason to expect that measuring consumption before meditation will carry over and affect consumption at a later time.

Please stop reading and apply the issue of carryover effects to the test strategies problem you worked at the beginning of this chapter.

If scores in the S_1 row were all from the same person, then carryover effects would be expected. That is, the effect of practicing one strategy probably lasts long enough to carry over and have an effect in the next phase of the experiment. A carryover effect would make it difficult to interpret the outcome. However, if S_1 represents three participants who were matched before the experiment, or S_1 represents some type of natural triplets, then no carryover effects would be expected. With no carryover effects, any differences among the column means can be attributed solely to treatments.

Having cautioned you about carryover effects, I must mention that *sometimes* you want and expect carryover effects. For example, to evaluate a workshop, you might give the participants a pretest, a test after the workshop, and a follow-up test six months later. The whole issue is carryover effects, and a correlated-measures ANOVA is the statistical test to use to measure these effects.

The second caution that goes with correlated-measures ANOVA, a caution that was discussed earlier, deals with the issue of fixed and random effects. Giving you an understandable explanation of this issue involves the *statistical model* that ANOVA is based on, and this book has not provided you with the background information that you need.[8]

Don't worry about your ignorance at this point; you can eliminate it by taking an intermediate course in statistics, which usually covers the topic of statistical models. In the meantime, if you are faced with analyzing data not in this textbook, do what many who analyze data do: Ask for advice from someone knowledgeable about statistics.

ASSUMPTIONS OF CORRELATED-MEASURES ANOVA

If the probability figure you get from Table F is to be accurate, then the data used to calculate the *F* value must have certain characteristics. Mathematical statisticians refer to these characteristics as the "assumptions of the statistical test." Most of the assumptions about the data you use to calculate a correlated-measures ANOVA will be familiar to you; they are similar to the assumptions you read about in the previous three chapters.

One assumption of ANOVA is that the population scores for every treatment are distributed as a normal curve. A second assumption is that sampling from the treatment populations was random. To have complete confidence in the probability figures from a one-factor correlated-measures ANOVA, the data must meet an additional assumption of covariance matrix sphericity (Kirk, 1995). As was the case with statistical models, you have reached a point where the simplified approach I have taken with this book does not give you the background for an advanced topic.

PROBLEMS

5. What are two cautions about using a correlated-measures ANOVA?
6. List the assumptions required for assurance that the *F* distribution will give accurate probabilities for a set of data.
*7. Most college students have some familiarity with meditational exercises. One question about meditation that has been answered is, "Are there any physiological effects?" The data that follow are representative of answers to this question. These data on oxygen consumption (in cc^3 per minute) are based on Wallace and Benson's article in *Scientific American* (1972). Analyze the data. Include HSD tests and an interpretation that describes the effect of meditation on oxygen consumption.

[8] Writers of introductory textbooks must choose a framework that beginners can use. Unfortunately, in statistics, a simplified framework that works fine for elementary statistics is not adequate for more advanced concepts. Fortunately, learning a more complex framework when the time comes doesn't lead to too much confusion.

Subjects	Oxygen consumption (cc^3/min)		
	Before meditation	During meditation	After meditation
S_1	175	125	180
S_2	320	290	315
S_3	250	210	250
S_4	270	215	260
S_5	220	190	240

8. In problem 7 you drew some conclusions, each of which could be wrong. For each conclusion, identify what type of error is possible.

9. Most of the problems you have worked in this book have asked you to do problems that were already "set up" (as they are in problem 7). Researchers, however, usually have to begin by arranging the data into a usable form. As you have probably suspected by now, your first step in this problem will be to arrange the data.

About 10–20 percent of the population experience a period of major depression at some point in their lives. Several kinds of therapy are available. Many studies have tried to answer questions of whether one kind is better than others and whether an untreated control group improves as much as people who receive therapy. Robinson, Berman, and Neimeyer (1990) combined the results from 73 studies to reach the conclusions that are built into the data that follow. For these data, I have given you characteristics that allow you to match up a group of four subjects for each row. The dependent variable is Beck Depression Inventory scores (BDI) recorded at the end of treatment. Before you begin arranging, identify the independent variable.

Female, age 20, BDI = 10, drug therapy

Female, age 45, BDI = 16, drug therapy

Female, age 20, BDI = 8, cognitive therapy

Female, age 45, BDI = 10, cognitive therapy

Male, age 30, BDI = 15, cognitive therapy

Female, age 45, BDI = 16, behavior therapy

male, age 30, BDI = 9, behavior therapy

female, age 20, BDI = 11, behavior therapy

female, age 20, BDI = 18, no therapy

male, age 30, BDI = 10, drug therapy

female, age 45, BDI = 24, no therapy

male, age 30, BDI = 21, no therapy

10. Look over the objectives at the beginning of the chapter. Can you do them? ■

So far in your practice of hypothesis testing, you have used three different sampling distributions. The normal curve is appropriate when you know σ (Chapter 6). When you don't know σ, you can use the t distribution or the F distribution (Chapters 6–11) if the populations you are sampling from have certain characteristics (such as being normally distributed).

In the next two chapters, you will learn about statistical tests that require neither knowledge of σ nor that the data have the characteristics needed for t and F tests. These tests have sampling distributions that will be new to you. They do, however, have the same purpose as those you have been working with—providing you with the probability of obtaining the observed sample results, *if the null hypothesis is true*.

In Chapter 12, "The Chi Square Distribution," you will learn to analyze frequency count data. These data occur when observations are classified into categories and the frequencies in each category are counted. In Chapter 13, "Nonparametric Statistics," you will learn four techniques for analyzing scores that are ranks or can be reduced to ranks.

The techniques in Chapters 12 and 13 are often described as "less powerful." This means that *if* the populations you are sampling from have the characteristics required for t and F tests, then a t or F test is more likely than a nonparametric test to reject H_0 if it should be rejected. To put this same idea another way, t and F tests have a smaller probability of a Type II error if the population scores have the characteristics the tests require.

12

The Chi Square Distribution

OBJECTIVES FOR CHAPTER 12

After studying the text and working the problems in this chapter, you should be able to:

1. Identify the kind of data that requires a chi square test for hypothesis testing
2. Distinguish between problems that require goodness-of-fit tests and those that require tests of independence
3. For goodness-of-fit tests: State the null hypothesis and calculate and interpret chi square values
4. For tests of independence: State the null hypothesis and calculate and interpret chi square values
5. Calculate a chi square value from a 2 × 2 table using the shortcut method
6. Describe the issues associated with chi square tests based on small N's

A sociology class was deep into a discussion of methods of population control. As the class ended, the topic was abortion; it was clear that there was strong disagreement about using abortion as a method of population control. Both sides in the argument had declared that "educated people" supported their side. Two empirically minded students left the class determined to get actual data on attitudes toward abortion. They solicited the help of a friendly psychology professor, who encouraged them to distribute a questionnaire to his General Psychology class. The questionnaire asked, "Do you consider abortion to be an acceptable or unacceptable method of population control?" The questionnaire also asked the respondent's gender. The results were presented to the sociology class in the following tabular form:

	Abortion is	
	Acceptable	Unacceptable
Females	59	29
Males	15	37
Σ	74	66
	53%	47%

The general conclusion of the class was that college students were pretty evenly divided on this issue, 53 percent to 47 percent.

The attitude data, then, did not seem to indicate that "educated people" were clearly on one side or the other on this issue. The class discussion resumed, with personal opinions dominating, until an observant student said, "Look! Look at that table! The women are for abortion and the guys are against it. Look, 59 of the 88 women say 'acceptable'—that's 67 percent. But only 15 of the 52 guys say 'acceptable'—that's just 29 percent. There is a big gender difference here!"

"Maybe," said another student, "and maybe not. That looks like a big difference, but maybe both groups really have the same attitude and the difference we see here is just the usual variation you find in samples."

Do the two positions in the story sound familiar? Indeed, the issue seems like one to resolve by comparing the sample results to a sampling distribution that shows what happens when the null hypothesis is true. In this case, the null hypothesis is that females and males do not differ in their attitudes. If this is true, how likely is it that two samples would produce proportions of .67 and .29?[1] A chi square analysis will give you a probability figure that will help you decide. Thus, chi square is another inferential statistics technique—one that leads you to reject or retain a null hypothesis.

Chi square (pronounced "ki," as in *kind*, "square," and symbolized χ^2) is a sampling distribution that gives probabilities about frequencies. Frequencies like those in the abortion–gender table are distributed approximately as chi square, so the chi square distribution will provide the probabilities needed for decision making about the difference between the attitudes of females and males.

The characteristic that distinguishes chi square from the techniques in previous chapters is the *kind of data* you get when you make an observation. For the *t* test and ANOVA, the data are *scores* on a quantitative variable. That is, the subject of the experiment is measured on a quantitative variable and scores, such as IQs, attitudes, time, or errors result.

With chi square, the data are *frequency counts*. The researcher begins by identifying the categories for the study. Each subject in the study is observed and placed in one category. The frequency of observations in a category is counted and the analysis is performed on the frequency counts. What a chi square analysis does is compare the observed frequencies of a category to frequencies that would be expected if the null hypothesis is true.

Karl Pearson (1857–1936), of Pearson product-moment correlation coefficient fame, published the first article on chi square in 1900. Pearson wanted a way to measure the "fit" between data generated by a theory and data obtained from observations. Until chi square was dreamed up, theory testers presented theoretical predictions and empirical data side by side, followed by a declaration such as "good fit" or "poor fit."

The most popular theories at the end of the 19th century predicted that data would be distributed as a normal curve. Many data gatherers had adopted Quételet's position that measurements of almost any social phenomenon would be normally distributed if the number of cases was large enough. Pearson (and others) thought this was not true, and they proposed other curves. By inventing chi square, Pearson provided everyone with a quantitative method of *choosing* the curve of best fit.

---

[1] Casting the chi square problems in this chapter into proportions will help you in *interpreting* the results. However, this book does not cover chi square tests of proportions (see Kirk, 1990, pp. 549–552).

Chi square turned out to be very versatile, being applicable to many problems besides curve-fitting ones. As a test statistic, it is used by psychologists, sociologists, health professionals, biologists, educators, political scientists, economists, foresters, and others. In addition to its use as a test statistic, the chi square *distribution* has come to occupy an important place in theoretical statistics. As further evidence of the importance of Pearson's chi square, the editors of *Science 84* chose it as one of the 20 most important discoveries of the 20th century (Hacking, 1984).

Pearson, too, turned out to be very versatile, contributing data and theory to both biology and statistics. He is credited with naming the standard deviation, giving it the symbol σ, and coining the word *biometry*. He (and Galton and Weldon) founded the journal *Biometrika* to publicize and promote the marriage between biology and mathematics.[2] In addition to his scientific efforts, Pearson was an advocate of women's rights and the father of an eminent statistician, Egon Pearson.

As a final note, Pearson's overall goal was not to be a biologist or a statistician. His goal was a better life for the human race. An important step in accomplishing this was to "develop a methodology for the exploration of life" (Walker, 1968, pp. 499–500).[3]

THE CHI SQUARE DISTRIBUTION

The **chi square distribution** is a theoretical distribution, just as t and F are. Like them, as the number of degrees of freedom increases, the shape of the distribution changes. **Figure 12.1** shows how the shape of the chi square distribution changes as degrees of freedom change from 1 to 5 to 10. Note that χ^2, like the F distribution, is a positively skewed curve.

Critical values for χ^2 are given in Table E in Appendix B for α levels of .10, .05, .02, .01, and .001. Look at **Table E**.

The design of the χ^2 table is similar to that of the t table: α levels are listed across the top and each row shows a different *df* value. Notice, however, that with chi square, as degrees of freedom increase (looking down columns), larger and larger values of χ^2 are required to reject the null hypothesis. *This is just the opposite of what occurs in the t and F distributions*. This will make sense if you examine the three curves in **Figure 12.1**.

Once again, I will symbolize a critical value by giving the statistic (χ^2), the α level, *df*, and the critical value. Here's an example (that you will see again): $\chi^2_{.05}\,(1\;df) = 3.84$.

To calculate a χ^2 to compare to a critical value from Table E, use the formula

$$\chi^2 = \Sigma\left[\frac{(O - E)^2}{E}\right]$$

where O = observed frequency
E = expected frequency

[2] In 1901 when *Biometrika* began, The Royal Society, the principal scientific society in Great Britain, accepted articles on biology and articles on mathematics, but they would not permit papers that combined the two. (See Galton, 1901, for a thinly disguised complaint against the establishment.)
[3] For a short biography of Pearson, see Helen M. Walker (1968).

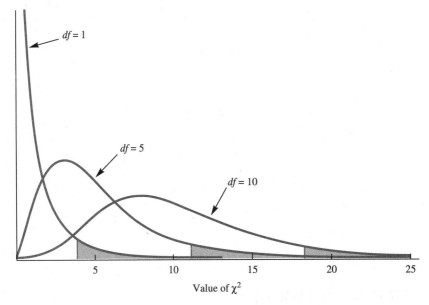

Value of χ^2

F I G U R E 12.1 Chi square distribution for three different degrees of freedom. Rejection regions at the .05 level are shaded.

Obtaining the observed frequency is simple enough—count the events in each category. Finding the **expected frequency** is a bit more complex. Two methods can be used to determine the expected frequency. One method is used if the problem is one of "goodness of fit" and the other method is used if the problem is a "test of independence."

CHI SQUARE AS A TEST FOR GOODNESS OF FIT

Chi square can be used to determine whether or not there is a good "fit" between a set of data [actual (observed) frequencies] and predictions made by a hypothesis, theory, or model [theoretical (expected) frequencies]. In testing for **goodness of fit**, the *null hypothesis is that the actual data fit the expected data*. A rejected H_0 means that the data do *not* fit the model—that is, the model is inadequate. A retained H_0 means that *the data are not at odds with the model*. A retained H_0 does not prove that the model, theory, or hypothesis is true because other models may also predict such results. A retained H_0 does, however, lend support for the model.

Where do hypotheses, theories, and models come from? At a very basic level, they come from our nature as human beings. Humans are always trying to understand things, and this often results in a guess at how things are. When these guesses are developed, supported, and discussed (and perhaps, published), they earn the more sophisticated label of hypothesis, theory, or model. Hypotheses, theories, and models that make quantitative predictions can be evaluated using empirical data, which in some cases require a chi square test.

The chi square goodness-of-fit test is used frequently in population genetics, where Mendelian laws predict ratios such as $3:1$, $1:2:1$, and $9:3:3:1$. For example, the law (model) might predict that crossing two pea plants will result in three times as many seeds in the smooth category as in the wrinkled category ($3:1$ ratio). If you perform the crosses, you might get 316 smooth seeds and 84 wrinkled seeds, for a total of 400. How well do these actual frequencies fit the expected frequencies of 300 and 100? Chi square will give you a probability figure that will help you decide if the data you have are consistent with the theory.

Although the mechanics of χ^2 require you to manipulate raw frequency counts (such as 316 and 84), you can best understand this test by thinking of proportions. Thus, 316 and 84 out of 400 represent proportions of .79 and .21. How likely are such proportions if the population proportions are .75 and .25? Chi square provides you with the probability of obtaining the observed proportions if the theory is true. As a result, you will have a quantitative way to decide "good fit" or "poor fit."

In the genetics example, the expected frequencies were predicted by a theory. Sometimes, in a goodness-of-fit test, the expected frequencies are predicted by chance. In such a case, a rejected H_0 means that something besides chance is at work. When you interpret the analysis, you should identify what that "something" is. Here is an example.

Suppose you were so interested in sex stereotypes and job discrimination that you conducted the following experiment, which is modeled after that of Mischel (1974). You made up four one-page résumés and four fictitious names—two female, two male. Names and résumés were randomly combined, and each subject was asked to read the résumés and "hire" one of the four "applicants" for a management trainee position. The null hypothesis tested is a "no discrimination" hypothesis. The hypothesis says that gender is *not* being used as a basis for hiring and therefore equal numbers of men and women will be hired. Thus, if the data cause the null hypothesis to be rejected, the hypothesis of "no discrimination" will be rejected.

Suppose you had 120 participants in your study, and they "hired" 75 men and 45 women. Is there statistical evidence that sex stereotypes are leading to discrimination? That is, given the hypothesized result of 60 men and 60 women, how good a fit are the observed data of 75 men and 45 women?

Applying the χ^2 formula, you get

$$\chi^2 = \Sigma\left[\frac{(O-E)^2}{E}\right] = \frac{(75-60)^2}{60} + \frac{(45-60)^2}{60} = 3.75 + 3.75$$
$$= 7.50$$

The number of degrees of freedom for this problem is the number of categories minus one. There are two categories here—hired men and hired women. Thus, $2 - 1 = 1$ *df*. Looking in Table E, you will find in row 1 that, if $\chi^2 \geq 6.64$, the null hypothesis may be rejected at the .01 level. Thus, the model may be rejected.

The last step in data analysis (and perhaps the most important one) is to write a conclusion. Begin by returning to descriptive statistics of the original data. Because 62.5 percent of the hires were men and 37.5 percent were women, you can conclude that discrimination in favor of men was demonstrated.

The term *degrees of freedom* implies that there are some restrictions. In the case of χ^2, one restriction is always that the sum of the expected events must be equal to the sum of the observed events; that is, $\Sigma E = \Sigma O$. If you manufacture a set of expected frequencies from a model, their sum must be equal to the sum of the observed frequencies. In our sex discrimination example, $\Sigma O = 75 + 45 = 120$, and $\Sigma E = 60 + 60 = 120$.

ERROR DETECTION

The sum of the expected frequencies must be equal to the sum of the observed frequencies: $\Sigma E = \Sigma O$.

PROBLEMS

1. Using the data on wrinkled and smooth pea seeds on page 291, test how well the data fit a 3 : 1 hypothesis.

2. This problem illustrates work of behaviorist John B. Watson (1879–1958) on the question of whether there are any basic, inherited emotions. Watson (1924) thought there were three: fear, rage, and love. He suspected that the wide variety of emotions experienced by adults was learned over the years. Proving his suspicion would be difficult because "unfortunately there are no facilities in maternity wards for keeping mother and child under close observation for years." So Watson attacked the apparently simpler problem of showing that the emotions of fear, rage, and love could be distinguished in infants. (He recognized the difficulty of proving that these were the *only* basic emotions, and he made no such claim.) Fear could be elicited by dropping the child onto a soft feather pillow (but not by the dark, dogs, white rats, or a pigeon fluttering its wings in the baby's face). Rage could be elicited by holding the child's arms tightly at its sides, and love by "tickling, shaking, gentle rocking, and patting" among other things.

    Here is an experiment reconstructed from the conclusions Watson described in his book. A child was stimulated in a way designed to elicit fear, rage, or love. An observer then looked at the child and judged the emotion the child was experiencing. Each judgment was scored as correct or incorrect—correct meaning that the judgment (say, fear) corresponded to the stimulus (say, dropping). Sixty observers made one judgment each, with the results shown in the table. To find the expected frequencies, think about the chance of being correct or incorrect when there are three possible outcomes. Analyze the data with χ^2 and write your conclusion in a sentence.

Correct	Incorrect
32	28

CHI SQUARE AS A TEST OF INDEPENDENCE

Another use of χ^2 is to test the *independence* of two variables on which frequency data are available. **Table 12.1**, the work of empirically minded students, shows the abortion questionnaire data. Such a table is called a *contingency table*; here, the question is whether a person's attitude toward abortion is contingent upon gender. *The null hypothesis is that attitudes and gender in the population are independent*—that knowing a person's gender gives you *no* clue to his or her attitude and vice versa. Rejection of the null hypothesis will support the alternate hypothesis, which is that gender and attitude are not independent but related—that knowing a person's gender does help you to predict his or her attitude and that attitude is contingent upon gender.

Table 12.1 gives the observed frequencies, with the expected frequencies in parentheses. Pay careful attention to the logic behind the calculation of the expected frequencies.

Let's start with an explanation of the expected frequency in the upper left corner, 46.51, the expected number of females who think abortion is acceptable. Of all the 140 subjects, 88 (row total) were female, so if we chose a subject at random, the probability that the person would be female is 88/140 = .6286.

In a similar way, of the 140 subjects, 74 (column total) thought that abortion was acceptable. Thus, the probability that a randomly chosen person would think that abortion is acceptable is 74/140 = .5286.

T A B L E 12.1 Hypothetical data on attitudes toward abortion (Expected frequencies are in parentheses.)

	Abortion is		
	Acceptable	Unacceptable	Σ
Females	59 (46.51)	29 (41.49)	88
Males	15 (27.49)	37 (24.51)	52
Σ	74	66	140

[4] This is like determining the chances of obtaining two heads in two tosses of a coin. For each toss, the probability of a head is $\frac{1}{2}$. The probability of two heads in two tosses is obtained by multiplying the two probabilities: $(\frac{1}{2})(\frac{1}{2}) = \frac{1}{4} = .25$. See the section, "A Binomial Distribution" in Chapter 5 for a review of this topic.

If we now ask what the probability is that a person chosen at random is a female who thinks that abortion is acceptable, we can find the answer by multiplying together the probability of the two separate events.[4] Thus, $(.6286)(.5286) = .3323$.

Finally, we can find the expected frequency of such people by multiplying the probability of such a person by the total number of people. Thus, $(.3323)(140) = 46.52$. Notice what happens to the arithmetic when these steps are combined:

$$\left(\frac{88}{140}\right)\left(\frac{74}{140}\right)(140) = \frac{(88)(74)}{140} = 46.51$$

The slight difference in the two answers (46.52 vs. 46.51) is due to rounding. The second answer is the correct one even though I carried four decimal places in our first answer. (This serves as a lesson for those of you who value precision.)

Thus, the formula for the expected value of a cell is its row total, multiplied by its column total, divided by N.

In a similar way, the expected frequency of males with an attitude that abortion is unacceptable is the probability of a male times the probability of an attitude of unacceptable times the number of subjects. For each of the four cells, here are the calculations of the expected values:

$$\frac{(88)(74)}{140} = 46.51 \qquad \frac{(88)(66)}{140} = 41.49$$

$$\frac{(52)(74)}{140} = 27.49 \qquad \frac{(52)(66)}{140} = 24.51$$

Once the expected values are calculated, a table like **Table 12.2** can be constructed, which leads to a χ^2 value of 19.14.

To determine the df for any $R \times C$ (rows by columns) table such as Table 12.1, use the formula $(R - 1)(C - 1)$. In this case, $(2 - 1)(2 - 1) = 1$. Perhaps having just 1 degree of freedom for four cells of data surprises you. ("Well, yes, it does! Before, with two categories, there was 1 df and now with four, there is still just 1 df? What's going on?") To understand this, remember the "freedom to vary" explanation.

For any $R \times C$ table, the *marginal* totals of the expected frequencies are fixed so that they equal the observed frequency.[5] In a 2×2 table, this leaves only one cell free to vary. Once a value is fixed for one cell, the other cell values are determined if you are

TABLE 12.2   Calculation of χ^2 for the data in Table 12.1

O	E	$O - E$	$(O - E)^2$	$\dfrac{(O - E)^2}{E}$
59	46.51	12.49	156.00	3.35
15	27.49	−12.49	156.00	5.67
29	41.49	−12.49	156.00	3.76
37	24.51	12.49	156.00	6.36
				$\chi^2 = 19.14$

[5] In Table 12.1, the marginal total of 88 must be the sum of the two expected frequencies, 46.51 and 41.49, as well as the sum of the obtained frequencies, 59 and 29.

to maintain the marginal totals. The general rule again is $df = (R - 1)(C - 1)$, where R and C refer to the number of rows and columns.

To determine the significance of $\chi^2 = 19.14$ with 1 df, look at Table E in Appendix B. In the first row, you will find that, if $\chi^2 = 10.83$, the null hypothesis may be rejected at the .001 level of significance. Because our χ^2 exceeds this, you can conclude that the attitudes toward abortion are influenced by gender; that is, gender and attitudes toward abortion in the population are not independent but related. By examining the proportions (.67 and .29), you can conclude that females are significantly more likely to consider abortions to be acceptable.

A test of independence is essentially a test of whether there is an interaction between the two variables—in this case, gender and attitude. A significant χ^2 means that the proportions you find for the different levels of one variable *depend* on which level of the other variable you are considering. For the example you just worked, the number (proportion) of people who consider abortion acceptable depends on whether the person is a female or a male.

SHORTCUT FOR ANY 2 × 2 TABLE

When you have a 2 × 2 table to analyze and a calculator, the following shortcut will save you time. With this shortcut, you do not have to calculate the expected frequencies, which reduces calculating time by as much as one-half. The general case of a 2 × 2 table is

		Row totals
A	B	$A + B$
C	D	$C + D$

| Column totals | $A + C$ | $B + D$ | N |

To calculate χ^2 from a 2 × 2 table, use the formula:

$$\chi^2 = \frac{N(AD - BC)^2}{(A + B)(C + D)(A + C)(B + D)}$$

The term in the numerator, $AD - BC$, is the difference between the cross-products. The denominator is the product of the four marginal totals.

To illustrate the equivalence of the shortcut method and the method explained in the previous section, I will calculate χ^2 for the data in Table 12.1. Translating cell letters into cell totals produces $A = 59$, $B = 29$, $C = 15$, $D = 37$, and $N = 140$. Thus,

$$\chi^2 = \frac{(140)[(59)(37) - (29)(15)]^2}{(88)(52)(74)(66)} = 19.14$$

Both methods yield $\chi^2 = 19.14$.

PROBLEMS

3. In the late 1930s and early 1940s an important sociology experiment took place in the Boston area. At that time 650 boys (median age $= 10\frac{1}{2}$ years) participated in the Cambridge-Somerville Youth Study (named for the two economically depressed communities where the boys lived). The participants were *randomly* assigned to either a delinquency-prevention program or a control group. Boys in the delinquency-prevention program had a counselor and experienced several years of opportunities for enrichment. At the end of the study, police records were examined for evidence of delinquency among all 650 boys. Analyze the data in the table and write a conclusion.

	Received program	Control
Police record	114	101
No police record	211	224

4. A student of psychology was interested in the effect of group size on the likelihood of joining an informal group. On one part of the campus, he placed a group of *two* people looking intently up into a tree. On a distant part of the campus, a group of *five* stood looking up into a tree. Single passersby were classified as joiners or nonjoiners. If a passerby stopped and looked up for five seconds or longer or made some comment to the group, he or she was classified as a joiner. The data in the table were obtained. Use a χ^2 test to determine whether group size had an effect on joining. (See Milgram, 1969.) Use the calculation method you did not choose for problem 3. Write a conclusion.

	Group size 2	Group size 5
Joiners	9	26
Nonjoiners	31	34

5. You have probably heard that salmon return to the stream in which they hatched. According to the story, after years of maturing in the ocean, the salmon arrive at the mouth of a river and swim upstream, choosing at each fork the stream that leads to the pool in which they hatched. Arthur D. Hasler did a number of interesting experiments on this homing instinct [reported in Hasler (1966) and Hasler and Scholz (1983)]. Here are two sets of data that he gathered.
   a. These data help answer the question of whether salmon really do make consistent choices at the fork of a stream in their return upstream. Forty-six salmon were captured from the Issaquah Creek (just east of Seattle, Washington) and another 27 from its East Fork. All salmon were marked and

released below the confluence of these two streams. All of the 46 captured in the Issaquah were recaptured there. Of the 27 originally captured in the East Fork, 19 were recaptured from the East Fork and eight from the Issaquah. Use χ^2 to determine if salmon make consistent choices at the confluence of two streams. Write a sentence explanation of the results.

b. Hasler believed that the salmon were making the sequence of choices at each fork on the basis of olfactory (smell) cues. He thought that young salmon become imprinted on the particular mix of dissolved mineral and vegetable molecules in their home stream. As adults, they simply make choices at forks on the basis of where the smell of home is coming from. To test this, he captured 70 salmon from the two streams, plugged their nasal openings, and released them below the confluence of the two streams. The fish were recaptured above the fork in one stream or the other. Compute χ^2 and write a sentence about Hasler's hypothesis.

Recapture site

Capture site		Issaquah	East Fork
	Issaquah	39	12
	East Fork	16	3

6. Here are some data that I analyzed for an attorney. Of the 4200 white applicants at a large manufacturing facility, 390 were hired. Of the 850 African-American applicants, 18 were hired. Analyze the data and write a conclusion. *Note:* You will have to work on the data some before you set up the table. ■

CHI SQUARE WITH MORE THAN ONE DEGREE OF FREEDOM

The χ^2 values you have found so far have been evaluated by comparing them to a chi square distribution with 1 *df*. In this section the problems will require chi square distributions with more than 1 *df*. Some of these will be goodness-of-fit problems and some will be tests of independence.

Here's a question for you. Suppose that for six days next summer you get a free vacation, all expenses paid. Which location would you choose for your vacation?

__City
__Mountains
__Seashore

I gave that questionnaire to 78 students and found, not surprisingly, that different people chose different locations.

A chi square analysis can test the hypothesis that all three of these locations are preferred equally. In this case, the null hypothesis is that, among those in the population, all three locations are equally likely as a first choice. Using sample data and a chi square test, you can reject or retain this hypothesis.

Think for a moment about the expected value for each cell in this problem. If each of the three locations is equally likely, then the expected value for a location is $\frac{1}{3}$ times the number of respondents; that is, $(\frac{1}{3})(78) = 26$.

The results I got from my sample are as follows:

	Choice		
City	Mountains	Seashore	Total
9	27	42	78

The arrangement of the arithmetic for this χ^2 problem is a simple extension of what you have already learned. The analysis is shown in **Table 12.3**.

What is the *df* for the χ^2 value in Table 12.3? The marginal total of 78 is fixed, and, because ΣE must equal ΣO, only two of the expected cell frequencies are free to vary. Once two are determined, the third is restricted to whatever value will make $\Sigma E = \Sigma O$. Thus, there are $3 - 1 = 2$ *df*. For this design, the number of *df* is the number of categories minus 1. From Table E, for $\alpha = .001$, χ^2 with 2 *df* = 13.82. Therefore, reject the hypothesis of independence and conclude that the students were responding in a nonchance manner to the questionnaire.

One advantage of χ^2 is its *additive* nature. Each $(O - E)^2/E$ value is a measure of deviation of the data from the model, and the final χ^2 is simply the sum of these measures. Because of this, you can examine the $(O - E)^2/E$ values and see which deviations are contributing the most to χ^2. For the vacation preference data, it is a greater number of seashore choices and fewer city choices that make the final χ^2 significant. The mountain choices are about what would be predicted by the "equal likelihood" model.

Suppose you showed the vacation preference data to a few of your friends, and one of them (a senior psychology major) said, "That's typical of data in psychology—it's based only on college students! If you do any studies of your own, why don't you find out what the noncollege population thinks? Try some high school kids; try some business types downtown."

T A B L E   12.3   Calculation of the vacation location preference data

Locations	O	E	$O - E$	$(O - E)^2$	$\dfrac{(O - E)^2}{E}$
City	9	26	−17	289	11.115
Mountains	27	26	1	1	0.038
Seashore	42	26	16	256	9.846
					$\chi^2 = 20.999$

With this admonition in mind, you embark on a study of family planning. Your questionnaire reads, "A couple just starting a family this year should plan to have how many children?"

__0 or 1
__2 or 3
__4 or more

As respondents fill out your questionnaire, you classify them as high school, college, or business types. Suppose you obtained the 275 responses that are shown in **Table 12.4**.

For this problem, you do not have a theory or model to tell you what the theoretical frequencies should be. The question is whether there is a relationship between attitudes toward family size and group affiliation. A chi square test of independence may answer this question. The null hypothesis is that there is *no* relationship—that is, that recommended family size and group affiliation are independent.

To get the expected frequency for each cell, assume independence (H_0) and apply the reasoning explained earlier about multiplying probabilities.

$$\frac{(92)(104)}{275} = 34.793 \qquad \frac{(92)(130)}{275} = 43.491 \qquad \frac{(92)(41)}{275} = 13.716$$

$$\frac{(117)(104)}{275} = 44.247 \qquad \frac{(117)(130)}{275} = 55.309 \qquad \frac{(117)(41)}{275} = 17.444$$

$$\frac{(66)(104)}{275} = 24.960 \qquad \frac{(66)(130)}{275} = 31.200 \qquad \frac{(66)(41)}{275} = 9.840$$

These expected frequencies are incorporated into **Table 12.5**, which shows the calculation of the χ^2 value.

The *df* for a contingency table such as this is obtained from the formula

$$df = (R - 1)(C - 1)$$

Thus, $df = (R - 1)(C - 1) = (3 - 1)(3 - 1) = 4$.

For $df = 4$, $\chi^2 = 18.46$ is required to reject H_0 at the .001 level. The obtained χ^2 exceeds this, so reject H_0 and conclude that attitudes toward family size are related to group affiliation.[6]

T A B L E 12.4 **Recommended family size by those in high school, college, and business**

Subjects	0–1	2–3	4 or more	Σ
		Number of children		
High school	26	57	9	92
College	61	38	18	117
Business	17	35	14	66
Σ	104	130	41	275

[6] The statistical analysis of these data is straightforward; the null hypothesis is rejected. However, a flaw in the experimental design has left two variables confounded, leaving the interpretation clouded. Because the three groups differ in both age and education, we are unsure whether the differences in attitude are related only to age, only to education, or to both of these variables. A better design would have groups that differed only on the age or only on the education variable.

T A B L E   12.5   Calculation of χ^2 for the family planning data

O	E	$O - E$	$(O - E)^2$	$\dfrac{(O - E)^2}{E}$
26	34.793	−8.793	77.317	2.222
57	43.491	13.509	182.493	4.196
9	13.716	−4.716	22.241	1.622
61	44.247	16.753	280.663	6.343
38	55.309	−17.309	299.602	5.417
18	17.444	0.556	0.309	0.018
17	24.960	−7.960	63.362	2.539
35	31.200	3.800	14.440	0.463
14	9.840	4.160	17.306	1.759
				$\chi^2 = 24.579$

By examining the right-hand column in **Table 12.5**, you can see that 6.343 and 5.417 constitute a large portion of the final χ^2 value. By working backward on those rows, you will discover that college students chose the 0–1 category more often than expected and the 2–3 category less often than expected. The interpretation then is that college students think that families should be smaller than high school students and businesspeople do.

PROBLEMS

*7. Professor Stickler always grades "on the curve." "Yes, I see to it that my grades are distributed as 7 percent A's, 24 percent B's, 38 percent C's, 24 percent D's, and 7 percent flunks," he explained to a young colleague. The colleague, convinced that the professor was really a softer touch than he sounded, looked at Professor Stickler's grade distribution for the past year. He found frequencies of 20, 74, 120, 88, and 38, respectively, for the five categories, A through F. Run a χ^2 test for the colleague. What would *you* conclude about Professor Stickler?

8. Is problem 7 a problem of goodness of fit or of independence?

*9. "Snake-eyes," a local gambler, let a group of sophomores know they would be welcome in a dice game. After three hours the sophomores went home broke. However, one sharp-minded, sober lad had recorded the results on each throw. He decided to see if the results of the evening would fit an "unbiased dice" model. Conduct the test and write a sentence summary of your conclusions.

Number of spots	Frequency
6	195
5	200
4	220
3	215
2	190
1	180

10. Identify problem 9 as a test of independence or goodness of fit.

*11. Remember the controversy, described in Chapter 1, over the authorship of 12 of *The Federalist Papers*? The question was whether they were written by Alexander Hamilton or James Madison. Mosteller and Wallace (1989) selected 48 1000-word passages known to have been written by Hamilton and 50 1000-word passages known to have been written by Madison. In each passage they counted the frequency of certain words. The results for the word *by* are shown in the table. Is *by* a word that the two writers used with significantly different frequency? Explain how these results would help in deciding about the 12 disputed papers.

Rate per 1000 words	Hamilton	Madison
0–6	21	5
7–12	27	31
13–18	0	14

12. Is problem 11 one of independence or goodness of fit? ■

SMALL EXPECTED FREQUENCIES

The theoretical chi square distribution, which comes from a mathematical formula, is a continuous function that can have any positive numerical value. Chi square test statistics calculated from frequencies do not work this way. They change in discrete steps and, when the expected frequencies are very small, the discrete steps are quite large. It is clear to mathematical statisticians that, as the expected frequencies approach zero, the theoretical chi square distribution becomes a less and less reliable way to estimate probabilities. In particular, the fear is that such chi square analyses will reject the null hypothesis more often than warranted.

The question for researchers who hope to analyze their data with chi square has always been, Those expected frequencies—how small is too small? For years the usual recommendation was that if an expected frequency was less than 5, a chi square analysis was suspect. A number of studies, however, have led to a revision of the answer.

■ When *df* = 1

Several studies used a computer to draw thousands of random samples from a known population for which the null hypothesis was true. Each sample was analyzed with a chi square, and the proportion of rejections of the null hypothesis (Type I errors) calculated. (If this proportion were approximately equal to α, the chi square analysis would be clearly appropriate.) The general trend of these studies has been to show that the theoretical chi square distribution gives accurate conclusions even when the expected frequencies are considerably less than 5.[7] (See Camilli and Hopkins, 1978, for an analysis of a 2 × 2 test of independence.)

When $df > 1$

When $df > 1$, the same uncertainty exists if one or more expected frequencies is small. Is the chi square test still advisable? Bradley and colleagues (1979) addressed some of the cases in which $df > 1$. Again, they used a computer to draw thousands of samples and analyzed each sample with a chi square test. The proportion of cases in which the null hypothesis was mistakenly rejected was calculated. Nearly all were in the .03 to .06 range but a few fell outside this range, especially when the sample sizes were small ($N = 20$). Once again, the conclusion was that the chi square test gives fairly accurate probabilities, even with sample sizes that are considered small.

An Important Consideration

Now, let's back away from these trees we've been examining and visualize the forest that you are trying to understand. The big question, as always, is, What is the nature of the population? The question is answered by studying a sample, which may lead you to reject or to retain the null hypothesis. Either decision could be in error. The concern of this section so far has been whether the chi square test is keeping the percentage of mistaken rejections (Type I errors) near an acceptable level (.05).

Because the big question is, "What is the nature of the population?" you should keep in mind the other error you could make: retaining the null hypothesis when it should be rejected (a Type II error). How likely is a Type II error when expected frequencies are small? Overall (1980) has addressed this question and his answer is, "very."

Overall's warning is that if the effect of the variable being studied is not just huge, then small expected frequencies have a high probability of dooming you to make the mistake of failing to discover a real difference. You may recall that this issue was discussed in Chapter 8 under the topic of power.

How can this mistake be avoided? Use large N's. The larger the N, the more power you have to detect any real differences. And, in addition, the larger the N, the more confident you are that the actual probability of a Type I error is your adopted α level.

Combining Categories

Combining categories is a technique that allows you to ensure that your chi square analysis will be more accurate. (It reduces the probability of a Type II error and keeps

---

[7] **Yates' correction** used to be recommended for 2 × 2 tables that had 1 expected frequency less than 5. The effect of this was to reduce the size of the obtained chi square. According to current thinking, Yates' correction overcorrected, resulting in α levels that were much smaller than the investigator claimed to be using. Therefore, I do not recommend Yates' correction for the kinds of problems found in most research. (See Camilli and Hopkins, 1978.)

the probability of a Type I error close to α.) I will illustrate this technique with some data from Chapter 5 on the normal distribution.

In Chapter 5, I claimed that IQ scores are distributed normally and offered as evidence a frequency polygon, Figure 5.7. I asked you simply to look at it and note that this empirical distribution appeared to fit the normal curve. (I was using the pre-1900 method of presenting evidence.) Now you are in a position to use a Pearson goodness-of-fit test to determine whether the scores are distributed as a normal curve. For that set of 261 IQ scores, the mean was 101 and the standard deviation was 13.4. The question this goodness-of-fit test will answer is, Will a theoretical normal curve with a mean of 101 and a standard deviation of 13.4 fit the data graphed in Figure 5.7?

To find the expected frequency of each class interval, I constructed **Table 12.6**. The first three columns show the observed scores arranged into a grouped frequency distribution and the lower limit of each class interval. The z-score column shows the z score for the lower limit of the class interval, based on a mean of 101 and a standard deviation of 13.4. The next column was obtained by entering the normal curve table at the z score and finding the proportion of the curve above the z score. From this proportion I subtracted the proportion of the curve associated with class intervals with higher scores. (For example, from the proportion above $z = 1.75$, .0401, I subtracted .0166 to get .0235.) Finally, the proportion in the interval was multiplied by 261, the number of students, to get the expected frequency.

Now you have the ingredients necessary for a χ^2 test. The observed frequencies are in the second column. The frequencies expected if the data are distributed as a normal curve with a mean of 101 and a standard deviation of 13.4 are seen in the last column. I could combine these observed and expected frequencies into the now familiar

T A B L E 12.6 IQ scores and expected frequencies for 261 fifth-grade students

Class interval	Observed frequency	Lower limit of class interval	z Score	Proportion of normal curve within class interval	Expected frequency (261 × proportion)
130 and up	4	129.5	2.13	.0166	4.3
125–129	5	124.5	1.75	.0235	6.1
120–124	18	119.5	1.38	.0437	11.4
115–119	19	114.5	1.01	.0724	18.9
110–114	25	109.5	0.63	.1081	28.2
105–109	28	104.5	0.26	.1331	34.7
100–104	36	99.5	−0.11	.1464	38.2
95–99	41	94.5	−0.49	.1441	37.6
90–94	31	89.5	−0.86	.1172	30.6
85–89	28	84.5	−1.23	.0856	22.3
80–84	14	79.5	−1.60	.0545	14.2
75–79	7	74.5	−1.98	.0309	8.1
70–74	4	69.5	−2.35	.0145	3.8
65–69	1	64.5	−2.72	.0061	1.6
64 and below	0		below −2.72	.0033	0.9
Sum	261			1.0000	260.9

$(O - E)^2/E$ formula, except that several categories have suspiciously small expected frequencies.

The solution to this problem is to combine some adjacent categories. If the top category is 125 and up, the expected frequency becomes 10.4 (4.3 + 6.1). In a similar fashion, the bottom category becomes 74 and below and has an expected frequency of 6.3 (0.9 + 1.6 + 3.8).[8] These combinations are shown in **Table 12.7**, a familiar χ^2 table with 12 categories. The χ^2 value is 7.98.

What is the *df* for this problem? This is a case in which *df* is not equal to the number of categories minus 1. The reason is that there is more than our usual one restriction on the expected frequencies. Usually, the requirement that $\Sigma E = \Sigma O$ accounts for the loss of 1 *df*. For this problem, however, I also required that the mean and standard deviation of the expected frequency be equal to the observed mean and observed standard deviation. Thus, *df* in this case is the number of categories minus 3, and $12 - 3 = 9$ *df*.

By consulting Table E in Appendix B, you can see that a χ^2 of 16.92 would be required to reject the null hypothesis at the .05 level. The null hypothesis for a goodness-of-fit test is that the data *do* fit the model. Thus, to reject H_0 in this case is to show that the data are not normally distributed. However, because the obtained $\chi^2 = 7.98$ is less than the critical value of 16.92, the conclusion is that these data do not differ significantly from a normal curve model in which $\mu = 101$ and $\sigma = 13.4$.

There are other situations in which small expected frequencies may be combined. For example, in an opinion survey, the response categories might be "strongly agree," "agree," "no opinion," "disagree," and "strongly disagree." If there are few respondents who check the "strongly" categories, those who do might simply be combined with their "agree" or "disagree" neighbors—reducing the table to three cells instead of five. The basic rule on combinations is that they must "make sense." It

TABLE 12.7 Testing the goodness-of-fit of IQ data and a normal curve

Class interval	O	E	$O - E$	$(O - E)^2$	$\dfrac{(O - E)^2}{E}$
125 and up	9	10.4	−1.4	1.96	0.189
120–124	18	11.4	6.6	43.56	3.821
115–119	19	18.9	0.1	0.01	0.001
110–114	25	28.2	−3.2	10.24	0.363
105–109	28	34.7	−6.7	44.89	1.294
100–104	36	38.2	−2.2	4.84	0.127
95–99	41	37.6	3.4	11.56	0.307
90–94	31	30.6	0.4	0.16	0.005
85–89	28	22.3	5.7	32.49	1.457
80–84	14	14.2	−0.2	0.04	0.003
75–79	7	8.1	−1.1	1.21	0.149
74 and below	5	6.3	−1.3	1.69	0.268
Σ	261	260.9			$\chi^2 = 7.984$

[8] I have to confess that the guidelines I used for how much combining to do was the old notion about not having expected frequencies of less than 5.

wouldn't make sense to combine the "strongly agree" and the "no opinion" categories in this example.

WHEN YOU MAY USE CHI SQUARE

1. Chi square, like all the methods of statistical inference that you have studied, is based on the assumption that you have random (or at least representative) samples from the population to which you wish to generalize.
2. Chi square is appropriate when the data are frequency counts. Chi square is not appropriate to test a difference between means or a difference between medians.
3. The sampling distribution of χ^2 can be used to test for (a) goodness of fit and (b) independence of two variables. In addition, χ^2 has other applications, which you are likely to encounter if you take a second course in statistics.
4. The procedures presented in this book require that each event be **independent** of the other events. *Independent* here means uncorrelated; *independent* is used in the same sense as it was in the assumption of independent samples for t and F. As one example, if each subject contributed two or more observations to the study, the observations will *not* be independent. In addition, if the outcome of one event depends on the outcome of an already recorded event, the events are not independent. For example, if people are asked for their choice of vacation spots in the presence of others who have already responded, their stated choice is not independent of the other choices.

PROBLEMS

13. An early hypothesis about schizophrenia (a form of severe mental illness) was that it has a simple genetic cause. In accordance with the theory, one-fourth (a $1:3$ ratio) of the offspring of a certain selected group of parents would be expected to be diagnosed as schizophrenic. Suppose that of 140 offspring, 19.3 percent were schizophrenic (and 80.7 percent were not schizophrenic). Use this information to test the goodness of fit of a $1:3$ model. Remember that a χ^2 analysis is performed on frequency counts.

14. What is the difference between the goodness-of-fit test and the test of independence?

15. You may recall from Chapter 7 that I bought Doritos tortilla chips and Snickers candy bars to have data for problems. For this chapter I bought M&Ms. In a "fun size" package I got 4 browns, 6 reds, 8 yellows, 2 greens, 4 oranges, and 0 blues. Mars, Inc., the manufacturer, claims that the population breakdown is 30 percent browns, 20 percent reds, 20 percent yellows, 10 percent greens, 10 percent oranges, and 10 percent blues (M&M/Mars, 1993). Use chi square to test the fit of the data to the expected values.

16. Suppose a friend of yours decides to gather his own data on sex stereotypes and job discrimination. He makes up ten résumés and ten fictitious names, five male and five female. He asks his 24 subjects each to "hire" five of the candidates rather than just one. The data show that 65 men and 55 women are hired. He asks you for advice on χ^2. What would you say to your friend?

17. As an assignment for a political science class, students were asked to gather data to establish a "voter profile" for backers of each of the two candidates for senator (Hill and Dale). One student stationed herself at a busy intersection and categorized the cars turning left that had bumpers stickers for each of the candidates, according to whether they signaled the turn or not. After two hours, she left with the frequencies shown in the table. Test these data with χ^2 to see whether there is a story to be told about the backers of the two candidates. Carefully phrase the student's contribution to the voter profile.

Signaled turn	Sticker	Frequency
No	Hill	11
No	Dale	2
Yes	Hill	57
Yes	Dale	31

18. Another friend, who knows that you are almost finished with this book, comes to you for advice on the statistical analysis of the following data. He wants to know whether there is a significant difference between males and females. What do you tell your friend?

	Males	Females
Mean score	123	206
N	18	23

19. Our political science student from problem 17 has launched a new experiment—to determine whether affluence is related to voter choice. This time she looks for yard signs for the two candidates and then classifies the houses as brick or one-story frame. Her reasoning is that the one-story frame homes represent less affluent voters in her city. Houses that do not fit the categories are ignored. After three hours of driving and three-fourths of a tank of gas, the frequency counts are as shown in the table. Test the hypothesis that candidate preference and affluence (as measured by style of home) are independent. Carefully write a statement of the results that can be used by the political science class in the voter profile.

Type of house	Yard signs	Frequency
Brick	Hill	17
Brick	Dale	88
Frame	Hill	59
Frame	Dale	51

20. When using χ^2 what is the proper *df* for the following tables?
    a. 1×4 **b.** 4×5 **c.** 2×4 **d.** 6×3

21. Please reread the chapter objectives. Can you do each one?

13

Nonparametric Statistics

OBJECTIVES FOR CHAPTER 13

After studying the text and working the problems in this chapter, you should be able to:

1. Describe the rationale of nonparametric statistical tests
2. Determine when a Mann-Whitney U test is appropriate, perform the test, and interpret the results
3. Determine when a Wilcoxon matched-pairs signed-ranks T test is appropriate, perform the test, and interpret the results
4. Determine when a Wilcoxon-Wilcox multiple-comparisons test is appropriate, perform the test, and interpret the results
5. Calculate a Spearman r_s correlation coefficient and determine the probability that the coefficient came from a population with a correlation of zero

Two child psychologists were talking shop over coffee one morning. (Much research begins with just such sessions.) The topic was intensive early training in athletics. Both psychologists were convinced that such training made the child less sociable as an adult, but one psychologist went even further. "I think that really intensive training of young kids is ultimately detrimental to their performance in the sport. Why, I'll bet that, among the top ten men's singles tennis players, those with intensive early training are not in the highest ranks."

"Well, I certainly wouldn't go that far," said the second psychologist. "I think all that early intensive training is quite helpful."

"Good. In fact, great. We disagree and we may be able to decide who is right. Let's get the ground rules straight. For tennis players, how early is early and what is intensive?"

"Oh, I'd say early is starting by age 7 and intensive is playing every day for two or more hours."[1]

---

[1] Because the phrase *intensive early training* can mean different things to different people, the first psychologist has provided the second with an operational definition. An **operational definition** is a definition that specifies a concrete meaning for a term. Concrete meanings are ones that everyone understands, such as "seven years old" and "two or more hours of practice every day."

"That seems reasonable. Now, let's see, our population is 'excellent tennis players' and these top ten will serve as our representative sample."

"Yes, indeed. What we would have among the top ten players would be two groups to compare. One had intensive early training, and the other didn't. The dependent variable is the player's rank. What we need is some statistical test that will tell us whether the difference in average ranks of the two groups is statistically significant."[2]

"Right. Now, a *t* test won't give us an accurate probability figure because *t* tests assume that the population of dependent variable scores is normally distributed. A distribution of ranks is rectangular with each score having a frequency of 1."

"I think there is a nonparametric test that is the proper one for ranks."

Here is a new category of tests that can be used to analyze experiments in which the dependent variable is ranks. This category of tests is called **non-parametric statistics**. As you will see, there are data other than ranks for which nonparametric tests are appropriate. To begin, study the rationale of these tests until you understand the ideas they are based on.

THE RATIONALE OF NONPARAMETRIC TESTS

Many of the reasoning steps you go through in doing nonparametric statistics are already familiar to you—they are the hypothesis-testing steps. By this time in your study of statistics, you should be able to write a coherent paragraph about the null hypothesis, the alternative hypothesis, gathering data, using a sampling distribution to find the probability of such results when the null hypothesis is true, making a decision about the null hypothesis, and telling a story that the data support.

The steps in this rationale that are unique to this chapter are those on sampling distributions of ranks. Here's an explanation of how a sampling distribution based on ranks might be constructed.

Suppose you drew two samples of equal size (for example, $N_1 = N_2 = 10$) from the same population.[3] You then arranged all the scores from both samples into one overall ranking, from 1 to 20. Because the samples are from the same population, the sum of the ranks of one group should be equal to the sum of the ranks of the second group. In this case, the expected sum for each group is 105. (With a little figuring, you can prove this for yourself now, or you can wait for an explanation later in the chapter.)

Although the expected sum is 105, actual sampling from the population would also produce sums greater than 105 and less than 105. After repeated sampling, all the sums could be arranged into a sampling distribution, and a sampling distribution would allow you to determine the likelihood of any sum (if both samples come from the same population).

Even if the two sample sizes are unequal, the same logic will work. A sampling distribution can be constructed that will show the expected variation in sums of ranks for one of the two groups.

---

[2] This is how the experts convert vague questionings into comprehensible ideas that can be communicated to others—they identify the independent and the dependent variables.

[3] As always, drawing two samples from one population is statistically the same as starting with two identical populations and drawing a random sample from each.

TABLE 13.1 The function of some nonparametric and parametric tests

Nonparametric test	Function	Parametric test
Mann-Whitney U	Tests for a significant difference between two independent samples	Independent-samples t test
Wilcoxon matched-pairs signed-ranks T or Z	Tests for a significant difference between two correlated samples	Correlated-samples t test
Wilcoxon-Wilcox multiple-comparisons	Tests for significant differences among all possible pairs of independent samples	One-way ANOVA and Tukey HSDs
Spearman r_s	Describes the degree of correlation between two variables	Pearson product-moment correlation coefficient, r

Handwritten annotations: ← pretest/post test; natural pairs; matched — does a difference exist? ← unequal variance ← does a relationship exist?

With this rationale in mind, you are ready to learn four new techniques. The first three are hypothesis-testing methods that determine whether samples came from the same population. The fourth technique is a correlation coefficient for ranked data, symbolized r_s. You will learn to calculate r_s and then test the hypothesis that a sample r_s came from a population with a correlation coefficient of .00.

The four nonparametric techniques in this chapter and their functions are listed in Table 13.1. In earlier chapters, you studied parametric tests that have similar functions, which are listed on the right side of the table. Study **Table 13.1** carefully now.

There are many other nonparametric statistical methods beside the four that you will study in this chapter. Peter Sprent (1989) has written a superb handbook, *Applied Nonparametric Statistical Methods*, that covers many of these methods. Sprent's book describes applications in fields such as road safety, space research, trade, and medicine, as well as applications in traditional academic disciplines.

COMPARISON OF NONPARAMETRIC AND PARAMETRIC TESTS

In what ways are nonparametric tests similar to parametric tests (t tests and ANOVA), and in what ways are they different? The following are similarities:

1. The logic of hypothesis testing is the same for both parametric and nonparametric tests. Thus, both tests yield the probability of the observed data *when the null hypothesis is true*. As you will see, though, the null hypotheses are different for the two kinds of tests.
2. Both parametric and nonparametric tests require you to have random samples (or at least assign participants randomly to subgroups).

As for differences, the t test and ANOVA require assumptions about the populations that nonparametric tests do not require. For example, a one-way ANOVA requires that you assume that the populations are normally distributed and have equal variances, but nonparametric tests make no such assumption. Also, with parametric tests, the null hypothesis is that the population *means* are the same (H_0: $\mu_1 = \mu_2$). In nonparametric tests, the null hypothesis is that the population *distributions* are the same. Because distributions can differ in form, variability, central tendency, or all three, the interpretation of a rejection of the null hypothesis may not be quite so clear-cut with a nonparametric test.

Recommendations on how to choose between a parametric and a nonparametric test have varied over the years. Two of the issues involved in the debate have been the scale of measurement and the power of the tests.

In the 1950s and after, some texts recommended that nonparametric tests be used if the scale of measurement was nominal or ordinal. After a period of controversy [see Chapter 2 of Kirk (1972) or Gardner (1975)], this consideration was dropped. Later, it resurfaced. [See the titles of the articles by Gaito (1980) and Townsend and Ashby (1984) to get the flavor.] Michell (1986), who describes the clash as one of underlying theories, may have an analysis that will lead to resolution.

The other issue, power, is also unresolved. Power, you may recall, comes up when the null hypothesis *should* be rejected. A test's power is a measure of how likely the test is to reject a false null hypothesis. It is clear to mathematical statisticians that if the populations being sampled from are normally distributed and have equal variances, then parametric tests are more powerful than nonparametric ones.

If the populations are not normal or do not have equal variances, it is less clear what to recommend. Early work on this question showed that parametric tests were robust, meaning that they gave approximately correct probabilities even though populations were not normal or did not have equal variances. Now, however, this robustness is being questioned. For example, Blair, Higgins, and Smitley (1980) show that a nonparametric test (Mann-Whitney U) is generally more powerful than its parametric counterpart (t test) for the nonnormal distribution they tested. Blair and Higgins (1985) arrived at a similar conclusion when they compared the Wilcoxon matched-pairs signed ranks test to the t test. I am sorry to leave these issues without giving you more specific advice, but no simple rule of thumb about superiority is possible except for one: If the data are ranks, use a nonparametric test.

Finally, there is not even a satisfactory name for these tests. Besides nonparametric, they are also referred to as **distribution-free statistics**. Although nonparametric and distribution-free mean different things to statisticians, the two words are used almost interchangeably by research workers. I will use the term *nonparametric tests* and illustrate them in the order given in Table 13.1.

THE MANN-WHITNEY *U* TEST

The **Mann-Whitney *U* test** is used to determine whether two sets of data based on two *independent* samples came from the same population. Thus, it is the appropriate test for the child psychologists to use to test the difference in ranks of tennis players. The

Mann-Whitney U test is identical to the **Wilcoxon rank-sum test**. Wilcoxon published his test first (1945). However, when Mann and Whitney (1947) independently published a test based on the same logic, they provided tables *and a name* for the statistic (U). Currently, the Mann-Whitney U test appears to be referred to more often than the Wilcoxon rank-sum test.

The Mann-Whitney U test produces a statistic, U, that is evaluated by consulting the sampling distribution of U. Like all the distributions you have encountered that can be used in the analysis of small samples, the sampling distribution of U depends on sample size.

When sample sizes are relatively small, critical values of U can be found in Table H in Appendix B. Use Table H if $N_1 = 20$ or less *and* $N_2 = 20$ or less. If the number of scores in one of the samples is greater than 20, the statistic U is distributed approximately as a normal curve. In this case, a z score is calculated, and the values of ± 1.96 and ± 2.58 are used as critical values for $\alpha = .05$ and $\alpha = .01$.

■ Mann-Whitney U Test for Small Samples

To provide data to illustrate the Mann-Whitney U test, I invented the information in Table 13.2 about the intensive early training of the top ten male singles tennis players.

In **Table 13.2** the players are listed by initials in the left column; the right column indicates that they had intensive early training ($N_{yes} = N_1 = 4$) or that they did not ($N_{no} = N_2 = 6$). Each player's rank is shown in the middle column. At the bottom of the table, the ranks of the *yes* players are summed (27), as are the ranks of the *no* players (28).

T A B L E 13.2 Early training of top ten male singles tennis players

Players	Rank	Intensive early training
Y.O.	1	No
U.E.	2	No
X.P.	3	No
E.C.	4	Yes
T.E.	5	No
D.W.	6	Yes
O.R.	7	No
D.S.	8	Yes
H.E.	9	Yes
R.E.	10	No

$\Sigma R_{yes} = 4 + 6 + 8 + 9 = 27$
$\Sigma R_{no} = 1 + 2 + 3 + 5 + 7 + 10 = 28$

The sums of the ranks are used to calculate two **U values**. The smaller of the two U values is used to enter Table H, which yields a probability figure. For the *yes* group, the U value is

$$U = (N_1)(N_2) + \frac{N_1(N_1 + 1)}{2} - \Sigma R_1$$

$$= (4)(6) + \frac{(4)(5)}{2} - 27 = 7$$

For the *no* group, the U value is

$$U = (N_1)(N_2) + \frac{N_2(N_2 + 1)}{2} - \Sigma R_2$$

$$= (4)(6) + \frac{(6)(7)}{2} - 28 = 17$$

A convenient way to check your calculation of U values is to know that the sum of the two U values is equal to $(N_1)(N_2)$. For this example, $7 + 17 = 24 = (4)(6)$.

Now, please examine **Table H**. It appears on two pages. On the first page of the table, the *lightface* type gives the critical values for α levels of .01 for a one-tailed test and .02 for a two-tailed test. The numbers in *boldface* type are critical values for $\alpha = .005$ for a one-tailed test and $\alpha = .01$ for a two-tailed test. In a similar way, the second page gives larger α values for both one- and two-tailed tests. The commonly used two-tailed test with $\alpha = .05$ is on this second page (boldface type). To enter the table, the value for N_1 gives the correct column and N_2 gives the correct row. Each intersection of N_1 and N_2 has two critical values. Now you are almost ready to enter Table H to find a probability figure for a U value of 7, the smaller of the two U values.

From the conversation of the two child psychologists, it is clear that a two-tailed test is appropriate; they would be interested in knowing whether intensive, early training *helps* or *hinders* players. Because an α level was not discussed, we will do what they would do—see if the difference is significant at the .05 level, and if it is, see if it is also significant at some smaller α level. Thus, in Table H we begin by looking for the critical value of U for a two-tailed test with $\alpha = .05$. This number is on the second page in boldface type at the intersection of $N_1 = 4$, $N_2 = 6$. The critical value is 2.

Note that the *smaller* the U value, the more different the two samples are. This is just the opposite of the situation with t tests, ANOVA, and χ^2.

Because our obtained value of U is 7, we must *retain* the null hypothesis and conclude that there is no evidence from our sample that the distribution of players trained early and intensively is significantly different from the distribution of those without such training.

Although you can easily find a U value using the preceding method and quickly go to Table H and reject or retain the null hypothesis, it would help your understanding of this test to think about small values of U. Under what conditions would you get a small U value? What kinds of samples would give you a U value of zero?

By examining the formula for U, you can see that $U = 0$ when the members of one sample all rank lower than every member of the other sample. Under such conditions, rejecting the null hypothesis seems reasonable. By playing with numbers in this manner, you can move from the rote memory level to the understanding level.

■ Assigning Ranks and Tied Scores

Sometimes you may choose a nonparametric test for data that are not already in the form of ranks.[4] In such cases, you will have to rank the scores. Two questions often arise. Is the largest or the smallest score ranked 1, and what should I do about the ranks for scores that are tied?

You will find the answer to the first question to be very satisfactory. For the Mann-Whitney test, it doesn't make any difference whether you call the largest or the smallest score 1. (This is not true for the Wilcoxon T test.)

Ties are handled by giving all tied scores the same rank. This rank is the mean of the ranks the tied scores would have if no ties had occurred. For example, if a distribution of scores was 12, 13, 13, 15, and 18, the corresponding ranks would be 1, 2.5, 2.5, 4, 5. The two scores of 13 would have been 2 and 3 if they had not been tied, and 2.5 is the mean of 2 and 3. As a slightly more complex example, the scores 23, 25, 26, 26, 26, 29 would have ranks of 1, 2, 4, 4, 4, 6. Ranks of 3, 4, and 5 average out to be 4.

Ties do not affect the value of U if they are in the same group. If several ties involve both groups, a correction factor may be advisable.[5]

ERROR DETECTION

Assigning ranks is tedious. Practice (with feedback) leads to fewer errors. Dealing with ties is tedious *and* troublesome. Pay careful attention to the examples.

■ Mann-Whitney U Test for Larger Samples

If one sample (or both samples) has 21 scores or more, the normal curve should be used to assess probability. The z value is obtained by the formula

$$z = \frac{(U + c) - \mu_U}{\sigma_U}$$

where $c = 0.5$, a correction factor explained later

$$\mu_U = \frac{(N_1)(N_2)}{2}$$

$$\sigma_U = \sqrt{\frac{(N_1)(N_2)(N_1 + N_2 + 1)}{12}}$$

Here c is a correction for continuity. It is used because the normal curve is a continuous function but the values of z obtained in this test are discrete. U, as before, is the smaller of the two possible U values.

Once again, the formula is familiar: the difference between a statistic based on data $(U + c)$ and the expected value of a parameter (μ_U) divided by a measure of variability.

---

[4] Severe skewness or populations with very unequal variances are often reasons for such a choice.
[5] See Kirk (1990, p. 572) for the correction factor.

After you have obtained z, the decision rules are the ones you have used in the past. For a two-tailed test, reject H_0 if $z \geqslant 1.96$ ($\alpha = .05$). For a one-tailed test, reject H_0 if $z \geqslant 1.65$ ($\alpha = .05$). The corresponding values for $\alpha = .01$ are $z \geqslant 2.58$ and $z \geqslant 2.33$.

Here is a problem for which the normal curve is necessary. An undergraduate psychology major was devoting a year to the study of memory. The principal independent variable was gender. Among her several experiments was one in which she asked the students in a General Psychology class to write down everything they remembered that was unique to the previous day's class, during which a guest had lectured. Students were encouraged to write down every detail they remembered. This class was routinely videotaped so it was easy to check each recollection for accuracy and uniqueness. Because the samples indicated that the population data were severely skewed, the psychology major chose a nonparametric test. (A plot of the scores in Table 13.3 will show this skew.)

The scores, their ranks, and the statistical analysis are presented in **Table 13.3**.

The z scores of -2.61 led to rejection of the null hypothesis, so the psychology major returned to the original data in order to interpret the results. Because the mean rank of the females, 15 ($258 \div 17$), is higher than that of the males, 25 ($603 \div 24$), and because higher ranks (those closer to 1) mean more recollections, she concluded that females recalled significantly more items than the males did.

Her conclusion is one that singles out central tendency for emphasis. On the average, females did better than males. The Mann-Whitney test, however, compares distributions. What our undergraduate has done is what most researchers who use the Mann-Whitney do: assume that the two populations have the same form but differ in central tendency. Thus, when a significant U value is found, it is common to attribute it to a difference in central tendency.

ERROR DETECTION

Here are two checks you can easily make. First, the last rank will be the sum of the two N's. In Table 13.3, $N_1 + N_2 = 41$, which is the lowest rank. Second, when ΣR_1 and ΣR_2 are added together, they will equal $N(N + 1)/2$, where N is the total number of scores. In Table 13.3, $603 + 258 = 861 = (41)(42)/2$.

From the information in the error detection box, you can see how I found the expected sum of 105 in the section on the rationale of nonparametric tests. There were 20 scores, so the overall sum of the ranks is

$$\frac{(20)(21)}{2} = 210$$

Half of this total should be found in each group, so the *expected* sum of ranks of each group, both of which come from the same population, is 105.

T A B L E 13.3 **Number of items recalled by males and females, ranks, and a Mann-Whitney U test**

Males ($N = 24$)		Females ($N = 17$)	
Items recalled	Rank	Items recalled	Rank
70	3	85	1
51	6	72	2
40	9	65	4
29	13	52	5
24	15	50	7
21	16.5	43	8
20	18.5	37	10
20	18.5	31	11
17	21	30	12
16	22	27	14
15	23	21	16.5
14	24.5	19	20
13	26.5	14	24.5
13	26.5	12	28.5
11	30.5	12	28.5
11	30.5	10	33
10	33	10	33
9	35.5		$\Sigma R_2 = 258$
9	35.5		
8	37.5		
8	37.5		
7	39		
6	40		
3	41		
	$\Sigma R_1 = 603$		

$$U = (N_1)(N_2) + \frac{N_1(N_1 + 1)}{2} - \Sigma R_1 = (24)(17) + \frac{(24)(25)}{2} - 603 = 105$$

$$\mu_U = \frac{(N_1)(N_2)}{2} = \frac{(24)(17)}{2} = 204$$

$$\sigma_U = \sqrt{\frac{(N_1)(N_2)(N_1 + N_2 + 1)}{12}} = \sqrt{\frac{(24)(17)(42)}{12}} = 37.79$$

$$z = \frac{(U + c) - \mu_U}{\sigma_U} = \frac{105 + 0.5 - 204}{37.79} = -2.61$$

PROBLEMS

1. In an experiment to determine the effects of estrogen on dominance, seven rats were given injections of that female hormone. The control group ($N = 8$) was injected with sesame oil, which is inert. Each rat engaged in a series of contests in a narrow runway. The rat that pushed the other one out was considered the more dominant of the two (see Work and Rogers, 1972). The number of bouts each won is given. Those marked with asterisks are the estrogen-injected rats. Analyze the results with a Mann-Whitney test and write a conclusion about the effects of estrogen.

   14, 13, 12, 12, *11, 9, 8, *7, *6, 4, 3, *2, *2, *1, *0

2. Grackles, commonly referred to as blackbirds, are hosts for a number of parasites. One variety of parasite is a thin worm that lives in the tissue around the brain. To see if the incidence of this parasite was changing, 24 blackbirds were captured one winter from a pine thicket, and the number of brain parasites in each bird was recorded. These data were compared with the infestation of 16 birds that had been captured from the same pine thicket 10 years earlier. Analyze the data with a Mann-Whitney test and write a conclusion. Be careful and systematic in assigning ranks. Errors here are quite frustrating.

   Present day

20	16	12	11	51	8	23	68	23	44	0	78
0	28	53	20	44	20	36	32	64	16	101	0

   Ten years earlier

16	19	43	90	16	72	29	62
103	39	70	29	110	32	87	57

*3. A friend of yours is trying to convince a mutual friend that the ride in an automobile built by (Ford, Chrysler [you choose]) is quieter than the ride in an automobile built by (Chrysler, Ford [no choice this time; you had just 1 degree of freedom—once the first choice was made, the second was determined]). This friend has arranged to borrow six fairly new cars—three made by each company—and to drive your blindfolded mutual friend around in the six cars (labeled A–F) until a stable ranking for quietness is achieved. So convinced is your friend that he insists on adopting $\alpha = .01$, "so that only the recalcitrant will not be convinced."

   As the statistician in the group, you decide to do a Mann-Whitney U test on the results. However, being the type who thinks through the statistics *before* gathering any experimental data, you look at the appropriate subtable in Table H and find that the experiment, as designed, is doomed to retain the null hypothesis. Write an explanation for your friend, explaining why the experiment must be redesigned.

4. Suppose your friend talked three more persons out of their cars for an afternoon, labeled the cars A–I, changed α to .05, and conducted the "quiet" test. The results are shown in the table. Perform a Mann-Whitney U test.

Car	Company Y ranks	Car	Company Z ranks
B	1	H	3
I	2	A	5
F	4	G	7
C	6	E	8
		D	9

THE WILCOXON MATCHED-PAIRS SIGNED-RANKS *T* TEST

The **Wilcoxon matched-pairs signed-ranks *T* test** (1945) is appropriate for testing the difference between two correlated samples. In Chapter 8, you learned of three kinds of correlated-samples designs: natural pairs, matched pairs, and repeated measures (before and after). In each of these designs, a score in one group is logically paired with a score in the other group. If you are not sure of your understanding of the difference between a correlated-samples and an independent-samples design, you should review that section in Chapter 8. Being sure of the difference is necessary if you are to choose correctly between a Mann-Whitney U test and a Wilcoxon matched-pairs signed-ranks T test.

The result of a Wilcoxon matched-pairs signed-ranks test is a **T value**,[6] which is interpreted using Table J. Finding T involves some steps that are different from finding U, so be alert.

Table 13.4 provides you with a few numbers from a correlated-samples design. Using **Table 13.4**, work through the following steps, which lead to a T value for a Wilcoxon matched-pairs signed-ranks T:

1. Find a difference, D, for every pair of scores. The order of subtraction doesn't matter, but it *must* be the same for every pair.

T A B L E   13.4   Illustration of how to calculate *T*

Pair	Variable 1	Variable 2	D	Rank	Signed rank
A	16	24	−8	3	−3
B	14	17	−3	1	−1
C	23	18	5	2	2
D	23	9	14	4	4

$$\Sigma(\text{positive ranks}) = 66$$
$$\Sigma(\text{negative ranks}) = -4$$
$$T = 4$$

[6] Be alert when you see a capital T in your outside readings; it has uses other than to symbolize the Wilcoxon matched-pairs signed-ranks test. Also note that this T is capitalized, whereas the t in the t test and t distribution is not capitalized except by some computer printers. Some writers avoid these problems by designating the Wilcoxon matched-pairs signed-ranks test with a W or a W_s.

2. Using the *absolute value* for each difference, rank the differences. The rank of 1 is given to the *smallest* difference, 2 goes to the next smallest, and so on.
3. Attach to each rank the sign of its difference. Thus, if a difference produces a negative value, the rank for that pair will be negative.
4. Sum the positive ranks and sum the negative ranks.
5. *T* is the *smaller of the absolute values* of the two sums.[7]

For the Wilcoxon matched-pairs signed-ranks *T* test, it is the differences that are ranked, and not the scores themselves. For this test, a rank of 1 always goes to the smallest difference.

The rationale of the Wilcoxon matched-pairs signed-ranks *T* test is that, *if* there is no true difference between the two populations, the absolute value of the negative sum should be equal to the positive sum, with any deviations being due to sampling fluctuations.

Table J shows the critical values for both one- and two-tailed tests for several α levels. To enter the table, use *N*, the number of *pairs* of subjects. Reject H_0 when the obtained *T* is *equal to* or *smaller than* the critical value in the table.[8]

To illustrate the calculation and interpretation of a Wilcoxon matched-pairs signed-ranks *T* test, here is an experiment based on some early work of Muzafer Sherif (1935). Sherif was interested in whether a person's basic perception could be influenced by others. The basic perception he used was a judgment of the size of the *autokinetic effect*. The autokinetic effect is obtained when a person views a stationary point of light in an otherwise dark room. After a few moments, the light appears to move erratically. Sherif asked his participants to judge how many inches the light moved. Under such conditions, judgments differ widely between individuals but they are fairly consistent for each individual.

In Sherif's experiment, participants began by working alone. They estimated the distance the light moved and, after a few judgments, their estimates stabilized. These are the before measurements. Next, additional observers were brought into the dark room and everyone announced aloud their perceived movement. These additional observers were confederates of Sherif, and they always judged the movement to be somewhat less than that of the participant. Finally, the confederates left and the participant made another series of judgments (the after measurements). The perceived movement (in inches) of the light is shown in Table 13.5 for 12 participants.

Table 13.5 also shows the calculation of a Wilcoxon matched-pairs signed-ranks *T* test. The *D* column is simply the after measurements minus the before measurements. These *D* scores are then ranked by absolute size and the sign of the difference attached in the Signed Ranks column. Notice that when $D = 0$, that pair of scores is dropped from further analysis and *N* is reduced by 1. The negative ranks have the smaller sum, so $T = 4$.

This *T* is smaller than the *T* value of 5 shown in Table J under $\alpha = .01$ (two-tailed test) for $N = 11$. Thus, the null hypothesis is rejected. The after scores represent a

---

[7] The Wilcoxon test is like the Mann-Whitney test in that you have a choice of two values for your test statistic. For both tests, choose the smaller value.
[8] Like the Mann-Whitney test, your obtained statistic (*T*) must be equal to or smaller than the tabled critical value if you are to reject H_0.

TABLE 13.5 Wilcoxon matched-pairs signed-ranks analysis of the effects of others' judgments on perception

Subject	Mean movement (inches) Before	After	D	Signed ranks
1	3.7	2.1	1.6	4.5
2	12.0	7.3	4.7	10
3	6.9	5.0	1.9	6
4	2.0	2.6	−0.6	−3
5	17.6	16.0	1.6	4.5
6	9.4	6.3	3.1	8
7	1.1	1.1	0.0	Eliminated
8	15.5	11.4	4.1	9
9	9.7	9.3	0.4	2
10	20.3	11.2	9.1	11
11	7.1	5.0	2.1	7
12	2.2	2.3	−0.1	−1

$$\Sigma \text{ positive} = 62$$
$$\Sigma \text{ negative} = -4$$
$$T = 4$$
$$N = 11$$

Check: $62 + 4 = 66$ and $\dfrac{11(12)}{2} = 66$

distribution that is different from the before scores. Now let's interpret this in terms of the experiment.

By examining the D column, you can see that all scores except two are positive. This means that, after hearing others give judgments smaller than their own, the participants saw the amount of movement as less. Thus, you may conclude (as did Sherif) that even basic perceptions tend to conform to perceptions expressed by others.

■ Tied Scores and $D = 0$

Ties among the D scores are handled in the usual way—that is, each tied score is assigned the mean of the ranks that would have been assigned if there had been no ties. Ties do not affect the probability of the rank sum unless they are numerous (10 percent or more of the ranks are tied). In the case of numerous ties, the probabilities in Table J associated with a given critical T value may be too large. In a situation with numerous ties, the test is described as too conservative because it may fail to ascribe significance to differences that are in fact significant (Wilcoxon and Wilcox, 1964).

As you already know, when *one* of the D scores is zero, it is not assigned a rank and N is reduced by 1. When *two* of the D scores are tied at zero, each is given the average rank of 1.5. Each is kept in the computation, with one being assigned a plus sign and the other a minus sign. If *three D* scores are zero, one is dropped, N is reduced by 1, and the remaining two are given signed ranks of $+1.5$ and -1.5. You can generalize from these three cases to situations with four, five, or more zeros.

■ Wilcoxon Matched-Pairs Signed-Ranks T Test for Large Samples

When the number of pairs exceeds 50, the T statistic may be evaluated using the normal curve. The test statistic is

$$z = \frac{(T + c) - \mu_T}{\sigma_T}$$

where T = smaller sum of the signed ranks
$c = 0.5$

$$\mu_T = \frac{N(N + 1)}{4}$$

$$\sigma_T = \sqrt{\frac{N(N + 1)(2N + 1)}{24}}$$

N = number of pairs

PROBLEMS

5. A private consultant was asked to evaluate a government job-retraining program. As part of the evaluation, she gathered information on the income of 112 individuals before and after retraining. She found a T value of 4077. Complete the analysis and draw a conclusion. Be especially careful in wording your conclusion.

6. Six industrial workers were chosen for a study of the effects of rest periods on production. Output was measured for one week before the new rest periods were instituted and again during the first week of the new schedule. Perform an appropriate nonparametric statistical test on the results shown in the table.

Worker	Without rests	With rests
1	2240	2421
2	2069	2260
3	2132	2333
4	2095	2314
5	2162	2297
6	2203	2389

7. A political science student was interested in the differences between two sets of Americans in attitudes toward government regulation of business. One set of Americans was made up of Canadians and the other of people from the United

States. At a model United Nations session, the student managed to get 30 participants to fill out his questionnaire. High scores on this questionnaire indicated a favorable attitude toward government regulation of business. Analyze the data in the table with the appropriate nonparametric test and write a conclusion.

Canadians	United States	Canadians	United States
12	16	27	26
39	19	33	15
34	6	34	14
29	14	18	25
7	20	31	21
10	13	17	30
17	28	8	3
5	9		

8. On the first day of class, one professor always gave his General Psychology class a beginning-point test. The purpose was to find out the students' beliefs about punishment, reward, mental breakdowns, and so forth—topics that would be covered during the course. Many of the items were phrased so that they represented a common misconception. For example, "mental breakdowns run in families and are usually caused by defective genes" was one item on the test. At the end of the course, the same test was given again. High scores mean lots of misconceptions. Analyze the before and after data and write a conclusion about the effects the General Psychology course had on misconceptions.

Student	Before	After
1	18	4
2	14	14
3	20	10
4	6	9
5	15	10
6	17	5
7	29	16
8	5	4
9	8	8
10	10	4
11	26	15
12	17	9
13	14	10
14	12	12

9. A health specialist conducted an eight-week workshop on weight control during which all 17 of the people who completed the course lost weight. In order to assess the long-term effects of the workshop, she weighed the participants again 10

months later. The weight lost or gained is shown here with a positive sign for those who continued to lose weight and a negative sign for those who gained some back. What can you conclude about the long-term effects of the workshop?

−5	24	0	13	9	6	−7	−3	2
−10	−16	7	12	−19	−4	8	15	

THE WILCOXON-WILCOX MULTIPLE-COMPARISONS TEST

So far in this chapter on the analysis of ranked data, you have covered both designs for the two-group case (independent and correlated samples). The next step is to analyze results from three or more independent groups. The method presented here is one that allows you to compare all possible *pairs* of groups, regardless of the number of groups in the experiment. This is the nonparametric equivalent of a one-way ANOVA followed by Tukey HSD tests. A direct analogue of the overall F test is the Kruskal-Wallis one-way ANOVA on ranks, which is explained in many elementary statistics texts.

The **Wilcoxon-Wilcox multiple-comparisons test** (1964) is a method that allows you to compare all possible pairs of treatments. This is like running several Mann-Whitney tests, with a test for each pair of treatments. However, the Wilcoxon-Wilcox multiple-comparisons test keeps your α level at .05 or .01 regardless of how many pairs you have. The test is an extension of the procedures in the Mann-Whitney U test and, like it, requires independent samples. (Remember that Wilcoxon devised a test identical to the Mann-Whitney U test.)

The Wilcoxon-Wilcox method requires you to order the scores from the K samples into one overall ranking. Then, for each sample, add the ranks of its individual members together. The rationale is that if the null hypothesis is true, these sums should be about equal. A large *difference* between two sums makes you suspicious that the two samples came from different populations. Of course, the larger K is, the greater the likelihood of large differences by chance alone, and this is taken into account in the table of critical values, Table K.

The Wilcoxon-Wilcox test can be used only when the N's for all groups are equal. A common solution to the problem of unequal N's is to reduce the too-large group(s) by throwing out one or more randomly chosen scores. A better solution is to design the experiment so that you have equal N's.

The data in **Table 13.6** represents the results of an experiment conducted on a solar collector by two designer-entrepreneurs. These two had designed and built a 4-foot by 8-foot solar collector they planned to market, and they wanted to know the optimal rate at which to pump water through the collector. The rule of thumb for pumping is one-half gallon per hour per square foot of collector, so they chose values of 14, 15, 16, and 17 gallons per hour for their experiment. Starting with the reservoir full of ice water, the water was pumped for one hour through the collector and back to the reservoir. At the end of the hour, the temperature of the water in the reservoir was measured in degrees Celsius. Then the water was replaced with ice water, the flow rate was changed, and the

TABLE   13.6   Temperature increases (°C) for four flow rates in a solar collector

			Flow rate (gallons/hour)				
14	Rank	15	Rank	16	Rank	17	Rank
28.1	7	28.9	3	25.1	14	24.7	15
28.8	4	27.7	8.5	25.3	13	23.5	18
29.4	1	27.7	8.5	23.7	17	22.6	19
29.0	2	28.6	5	25.9	11	21.7	20
28.3	6	26.0	10	24.2	16	25.8	12
Σ (ranks)	20		35		71		84

Check: $20 + 35 + 71 + 84 = 210$

$$\frac{N(N + 1)}{2} = \frac{20(21)}{2} = 210$$

process was repeated. The numbers in the body of Table 13.6 are the temperature measurements (to the nearest tenth of a degree).

In Table 13.6, ranks are given to each temperature, ignoring the group the temperature is in. The ranks of all those in a group are summed, producing values that range from 20 to 84 for flow rates of 14 to 17 gallons per hour. Note that you can confirm your arithmetic for the Wilcoxon-Wilcox test using the same checks you used with the Mann-Whitney U test. That is, the sum of the four group sums, 210, is equal to $N(N + 1)/2$, where N is the total number of observations. Also, the largest rank, 20, is equal to the total number of observations.

The next step is to make pairwise comparisons. With four groups, six pairwise comparisons are possible. The rate of 14 gallons per hour can be paired with 15, 16, and 17; the rate of 15 can be paired with 16 and 17; and the rate of 16 can be paired with 17. For each pair, a difference in the sum of ranks is found and the absolute value of that difference is compared with the critical value in Table K to see if it is significant.

Table K appears in two parts—one for the .05 level and one for the .01 level. In both cases, critical values are given for a two-tailed test. In the case of the data in Table 13.6, where $K = 4, N = 5$, you will find in Table K that rank-sum differences of 48.1 and 58.2 are required to reject H_0 at the .05 and .01 levels, respectively.

A convenient summary table for the Wilcoxon-Wilcox multiple-comparisons test is shown in **Table 13.7**. The table displays, for each pair of independent-variable values, the difference in the sum of the ranks. The next task is simply to see if any of these obtained differences are *equal to* or *larger than* the critical values of 48.1 and 58.2. (Note that for the multiple-comparisons test, unlike U and T, the data-based value that you calculate must *exceed* the tabled value.) At the .05 level, rates of 14 and 16 are significantly different from each other, as are rates of 15 and 17. In addition, a rate of 14 is significantly different from a rate of 17 at the .01 level. What does all this mean for our two designer-entrepreneurs? Let's listen to their explanation to their old statistics professor.

TABLE 13.7 Summary table for differences in the sums
of ranks in the flow-rate experiment

		Flow rates		
	ΣR	14 (20)	15 (35)	16 (71)
Flow rates	15 (35)	15		
	16 (71)	51*	36	
	17 (84)	64**	49*	13

* $p < .05$; for $\alpha = .05$, the critical value is 48.1.
** $p < .01$; for $\alpha = .01$, the critical value is 58.2.

"How did the flow-rate experiment come out, fellows?" inquired the kindly old
gentleman.

"OK, but we are going to have to do a follow-up experiment using different flow
rates. We know that 16 and 17 gallons per hour are not as good as 14, but we don't
know if 14 is optimal for our design. Fourteen was the best of the rates we tested,
though. On our next experiment, we are going to test rates of 12, 13, 14, and 15."

The professor stroked his beard and nodded thoughtfully. "Typical experiment.
You know more after it than you did before, . . . but not yet enough."

PROBLEMS

10. Given the summary data in the table, test all possible comparisons. Each group had
a sample of eight subjects.

Sum of ranks for six groups						
Groups	1	2	3	4	5	6
ΣR	196	281	227	214	93	165

11. The effect of three types of leadership on group productivity and satisfaction was
investigated.[9] Groups of five children were randomly constituted and assigned an
authoritarian, a democratic, or a laissez-faire leader. Nine groups were formed—
three with each type of leader. The groups worked for a week on various projects.
During this time, four different measures of each child's personal satisfaction were
taken. These were pooled to give one score for each child. The data are presented in
the table. Test all possible comparisons with a Wilcoxon-Wilcox test.

[9] For a summary of a similar investigation, see Lewin (1958).

	Leaders	
Authoritarian	Democratic	Laissez-faire
120	108	100
102	156	69
141	92	103
90	132	76
130	161	99
153	90	126
77	105	79
97	125	114
135	146	141
121	131	82
100	107	84
147	118	120
137	110	101
128	132	62
86	100	50

12. List the tests presented so far in this chapter and the design for which each is appropriate. Be sure you can do this from memory. ■

CORRELATION OF RANKED DATA

Here is a short review of what you learned about correlation in Chapter 4.

1. Correlation requires a bivariate distribution (a logical pairing of scores).
2. Correlation is a method of describing the degree of relationship between two variables—that is, the degree to which high scores on one variable are associated with low or high scores on the other variable.
3. Correlation coefficients range in value from +1.00 (perfect positive) to −1.00 (perfect negative). A value of .00 indicates that there is no relationship between the two variables.
4. Statements about cause and effect may not be made on the basis of a correlation coefficient alone.

In 1901, Charles Spearman (1863–1945) was "inspired by a book by Galton" and began experimenting at a little village school. He wanted to see if there was a relationship between intellect (school grades) and sensory ability (detecting musical discord). Spearman thought there was and he wanted to find a mathematical way to express the *degree* of this relationship. He came up with a coefficient that did this, though he later found that others were ahead of him in developing such a coefficient (Spearman, 1930).

Spearman's name is attached to the coefficient that is used to show the degree of correlation between two sets of *ranked* data. He used the Greek letter ρ (rho) as the symbol for his coefficient. Later statisticians began to reserve Greek letters to indicate parameters, so the modern symbol for Spearman's statistic has become r_s, the s honoring Spearman. **Spearman r_s** is a special case of the Pearson product-moment correlation coefficient and is most often used when the number of pairs of scores is small (less than 20).

Actually, r_s is a *descriptive statistic* that could have been introduced in the first part of this book. I waited until now to introduce it because r_s is a rank-order statistic, and this is a chapter about ranks. The next section shows you how to calculate this descriptive statistic; determining the statistical significance of r_s follows.

■ Calculation of r_s

The formula for r_s is

$$r_s = 1 - \frac{6\Sigma D^2}{N(N^2 - 1)}$$

where D = difference in ranks of a pair of scores

N = number of pairs of scores

I started this chapter with speculation about male tennis players; I will end it with data about female tennis players. Suppose you were interested in the relationship between age and rank among professional women tennis players. The statistic r_s will give you a numerical index of the degree of the relationship. If a rank of 1 were assigned to the oldest player, a positive r_s would mean that, the older the player, the higher her rank. A negative r_s would mean that, the older the player, the lower her rank. A zero or near zero r_s would indicate that there is no relationship between age and rank.

Table 13.8 shows the ten top-ranked women tennis players for 1995, their age as a rank score among the ten, and an $r_s = .13$. The next issue is interpretation. Can you say with confidence to a friend, "There is distinct tendency for older women to rank higher in tennis"? Or, perhaps an $r_s = .13$ is not trustworthy and reliable. Phrasing the question in statistical language, Is it likely that such an r_s would come from a population in which the true correlation is zero? You will recall that you answered this kind of question for a Pearson r at the end of Chapter 7.

ERROR DETECTION

For the Wilcoxon matched-pairs signed-ranks T test, find the difference between the scores and then rank the differences. For the Spearman r_s, assign ranks to the scores first and then find the difference.

■ Testing the Significance of r_s

Table L in Appendix B gives critical values of r_s for the .05 and .01 levels of significance (when the number of pairs is 16 or fewer). In using Table L, use the same decision rule you used in Table A, the critical values for a Pearson r. Reject the

TABLE 13.8 The top ten women tennis players in 1995: their rank in age, and the calculation of Spearman r_s

Player	Rank in tennis	Rank in age	D	D^2
Graf (Germany)	1	1	0	0
Seles (USA)	2	7	−5	25
Martinez (Spain)	3	6	−3	9
Sanchez Vicario (Spain)	4	5	−1	1
Maleeva (Bulgaria)	5	9	−5	16
Huber (Germany)	6	8	−2	4
Date (Japan)	7	3	4	16
Sabatini (Argentina)	8	2	6	36
Majoli (Croatia)	9	10	−1	1
Fernandez (USA)	10	4	6	36
				$\Sigma = 144$

$$r_s = 1 - \frac{6\Sigma D^2}{N(N^2 - 1)} = 1 - \frac{6(144)}{10(99)} = 1 - .87 = .13$$

hypothesis that the population correlation coefficient is zero if the obtained r_s is *equal to* or *larger than* the tabled value.

The tennis data in Table 13.8 produced $r_s = .13$ based on ten pairs. **Table L** shows that a correlation of .648 (either positive or negative) is required for significance at the .05 level. Thus, a correlation of .13 is not statistically significant.

Notice in Table L that rather large correlations are required for significance. As with r, not much confidence can be placed in low or moderate correlation coefficients that are based on only a few pairs of scores.

For samples larger than 16, you may test the significance of r_s by comparing it to the tabled values in Table A, just as you did for a Pearson r. Note, however, that **Table A** requires df, which, for r_s, is $N - 2$ (the number of pairs minus 2), which is the same formula used for r.

■ When $D = 0$ and Tied Ranks

In calculating r_s, you may get a D value of zero. Should a zero be dropped from further analysis as it is in a Wilcoxon matched-pairs signed-ranks T test? You can probably answer this by taking a moment to decide what a $D = 0$ means in an r_s correlation problem.

A zero means there is a perfect correspondence between the two ranks. If all differences were zero, the r_s would be 1.00. Thus, differences of zero should *not* be dropped when calculating an r_s.

The formula for r_s that you are working with is not designed to handle ranks that are tied. With r_s, ties are troublesome. A tedious procedure has been devised to overcome ties, but your best solution is to arrange your data collection so there are no ties. Sometimes, however, you are stuck with tied ranks, perhaps as a result of working with someone else's data. Kirk (1990) recommends assigning average ranks to ties, as you did for the three other procedures in this chapter, and then computing a Pearson r on the data.

MY FINAL WORD

In the first chapter, I said that the essence of applied statistics is to let numbers stand for things of interest, manipulate the numbers according to the rules of statistics, translate the numbers back into the things, and finally describe the relationship between the things of interest. The question, "What things are better understood when translated into numbers?" was not raised in this book, but it is important nonetheless. Here is an anecdote by E. F. Schumacher (1979), an English economist, that helps emphasize that importance.

I will tell you a moment in my life when I almost missed learning something. It was during the war and I was a farm laborer and my task was before breakfast to go to yonder hill and to a field there and count the cattle. I went and I counted the cattle—there were always thirty-two—and then I went back to the bailiff, touched my cap, and said "Thirty-two, sir," and went and had my breakfast. One day when I arrived at the field an old farmer was standing at the gate, and he said, "Young man, what do you do here every morning?" I said, "Nothing much. I just count the cattle." He shook his head and said, "If you count them every day they won't flourish." I went back, I reported thirty-two, and on the way back I thought, Well, after all, I am a professional statistician, this is only a country yokel, how stupid can he get. One day I went back, I counted and counted again, there were only thirty-one. Well, I didn't want to spend all day there so I went back and reported thirty-one. The bailiff was very angry. He said, "Have your breakfast and then we'll go up there together." And we went together and we searched the place and indeed, under a bush, was a dead beast. I thought to myself, Why have I been counting them all the time? I haven't prevented this beast dying. Perhaps that's what the farmer meant. They won't flourish if you don't look and watch the quality of each individual beast. Look him in the eye. Study the sheen on his coat. Then I might have gone back and said, "Well, I don't know how many I saw but one looks mimsey."

PROBLEMS

13. For each situation described, tell whether you would use a Mann-Whitney test, a Wilcoxon matched-pairs signed-ranks test, a Wilcoxon-Wilcox multiple-comparisons test, or an r_s.

   a. A limnologist (a scientist who studies freshwater streams and lakes) measured algae growth in a lake before and after the construction of a copper smelting plant to see what effect the plant had.

   b. An educational psychologist compared the sociability scores of firstborn children with scores of their next-born brother or sister to see if the two groups differed in sociability.

   c. A child psychologist wanted to determine the degree of relationship between IQ scores obtained at age 3 and scores obtained from the same individuals at age 12.

   d. A nutritionist randomly and evenly divided boxes of cornflakes cereal into three groups. One group was stored at 40°F, one group at 80°F, and one group alternated from day to day between the two temperatures. After 30 days, the vitamin content of each box was assayed.

e. The effect of STP gas treatment on gasoline mileage was assessed by driving six cars over a 10-mile course, adding STP, and then again driving over the same 10-mile course.

14. Fill in the descriptions for the following table.

	Symbol of statistic (if any)	Appropriate for what design?	Calculated statistic must be (larger, smaller) than the tabled statistic to reject H_0
Mann-Whitney test			
Wilcoxon matched-pairs signed-ranks test			
Wilcoxon-Wilcox multiple-comparisons test			

15. Calculate an appropriate correlation coefficient for the hypothetical data in the table. Determine whether the correlation coefficient is significant at the .001 level of significance.

Years	Number of marriages ($\times 10,000$)	Total grain crop ($ billions)
1918	131	1.7
1919	142	2.5
1920	125	1.4
1921	129	1.7
1922	145	2.8
1923	151	3.3
1924	142	2.6
1925	160	3.7
1926	157	3.4
1927	163	3.5
1928	141	2.6
1929	138	2.6
1930	173	3.6
1931	166	3.9

16. With $N = 16$ and $\Sigma D^2 = 308$, calculate r_s and test its significance at the .05 level.

17. Two members of the department of philosophy (Locke and Kant) were responsible for hiring a new professor. Each privately ranked from 1 to 10 the ten candidates who met the objective requirements (degree, specialty, and so forth). The rankings are shown in the table. Begin by writing an interpretation of a low correlation and a high correlation. Calculate an r_s.

Candidates	Professors	
	Locke	Kant
A	7	8
B	10	10
C	3	5
D	9	9
E	1	1
F	8	7
G	5	3
H	2	4
I	6	6
J	4	2

18. You are once again asked to give advice to a friend who comes to you for criticism of an experimental design. This friend has four pairs of scores obtained randomly from a population. Her intention is to calculate a correlation coefficient r_s and decide whether there is any significant correlation in the population. Give advice.

19. One last time. Review the objectives at the beginning of the chapter. ∎

WHAT WOULD YOU RECOMMEND? Chapters 9–13

Here is the last set of *What would you recommend?* problems. Your task is to choose an appropriate statistical technique from among the several you have learned in the previous five chapters. No calculations are necessary. For each problem that follows, (1) recommend a statistic that will either answer the question or make the comparison, and (2) explain why you recommend that statistic.

A. One behavior that is anxiety arousing is strapping on a parachute and jumping out of an airplane. Flooding and systematic desensitization are the names of two different techniques that help people overcome anxiety about specific behaviors. To compare the two techniques, military trainees experienced one technique or the other. Later, at the time of their first five actual jumps, each trainee's "delay time" was measured. (The assumption was that longer delay means more anxiety.) On the basis of the severely skewed distribution of delay times, each trainee was given a rank. What statistical test should be used to compare the two techniques?

[handwritten margin note: Mann-Whitney u variable measured is dependent]

Chi-square goodness of fit against chance **B.** Stanley Milgram did a series of studies on the question, "When do people submit to demands for compliance?" In one experiment, participants were told to administer a shock to another participant, who was in another room. (No one was actually shocked in this experiment.) The experimenter gave directions to half the participants over a telephone; the other half were given directions in person. The dependent variable in this experiment was whether or not a participant complied with the directions.

2x2 factorial ANOVA b/w groups **C.** Forensic psychologists study factors that influence people as they judge defendents in criminal cases. Berry and McArthur (1986) supplied participants in their study with a fictitious "pretrial report" and asked them to estimate the likelihood, on a scale of .00 to 1.00, that the defendant would be convicted. The pretrial reports contained a photograph showing either a person who was baby-faced or mature-faced. In addition, half the reports indicated that the defendant was simply negligent in the alleged incident and half indicated that the defendant had been intentionally deceptive in the incident.

Spearman's rs **D.** As part of a class project, two students administered a creativity test to a group of undergraduate seniors. In addition, they obtained each person's rank in class. In what way could the two students determine the degree of <u>relationship</u> between the two variables?

Chi-square goodness of fit against a model **E.** Suppose <u>a theory predicts</u> that the proportion of people who participate in a rather difficult exercise will drop by one-half each time the exercise is repeated. When the data were gathered, 100 people participated on the first day. On each of the next three occasions, fewer and fewer people participated. What statistical test could be used to arrive at a judgment about the theory?

F. Subjects in an experiment were told to "remember what you hear." A series of sentences followed. (An example was "The plumber slips $50 into his wife's purse.") Later, when the subjects were asked to recall the sentences, one group was given disposition cues (such as "generous"), one was given semantic (associative) cues (such as "pipes"), and one was given no cues at all. Each subject received a recall score between 0 and 100. <u>The recall scores for the "no cues" condition were much more variable than they were for the other two conditions.</u> The researcher's question was, "Which of the three conditions produces the best recall?" (See Winter, Uleman, and Cunniff, 1985.)

G. To answer a question about the effect of an empty stomach on eating, investigators in the 1930s studied dogs that ate a normal-sized meal (measured in ounces). Because of a surgical operation, however, no food reached the stomach. Thirty minutes later, the dogs were allowed to eat again. The data showed that the dogs, on the average, ate less the second time (even though their stomachs were just as empty). At both time periods, the data were positively skewed.

H. College cafeteria personnel are often dismayed by the amount of food that is left on plates. One way to reduce waste is to prepare the food using a tastier recipe. Suppose a particular vegetable is served on three different occasions, and each time a different recipe is used. As the plates are cleaned in the kitchen, the amount of the vegetable left on the plate is recorded. How might each pair of recipes be compared to determine if one recipe has resulted in significantly less waste?

I. Many programs seek to change behavior. One way to assess the effect of a program is to give a survey to the participants about their behavior, administer the program, and give the survey again. To demonstrate the lasting value of the program, however, follow-up data some months later are required. If the data consist of each participant's three scores on the same 100-point survey (pretest, posttest, and follow-up), what statistical test is needed to determine the effect of the program?

Choosing Tests and Writing Interpretations

OBJECTIVES FOR CHAPTER 14

After studying the text and working the problems in this chapter, you should be able to:

1. List the descriptive and inferential statistical techniques that you studied in this book

2. Describe, using a table of characteristics or decision trees or both, the descriptive and inferential statistics that you studied in this book

3. Read the description of a study and identify a statistical technique that is appropriate for such data

4. Read the description of a study and recognize that you have not studied a statistical technique that is appropriate for such data

5. Given the statistical analysis of a study, write an interpretation of what the study reveals

This chapter is designed to help you consolidate the many bits and pieces that you learned in your statistics course. It will help you consolidate, regardless of the number of chapters your course covered (as long as you completed through Chapter 8, "Hypothesis Testing and Effect Size: Two-Sample Tests"). First, there is a review of the major concepts of each chapter. The second section describes statistical techniques that you *did not* read about in this book. These two sections should help prepare you to do the exercises in the last section of this chapter.

A REVIEW

If you have been reading for integration, the kind of reading in which you are continually asking the question, "How does this stuff relate to what I learned before?," you may have clearly in your head a major thread that ties this book together from

Chapter 2 onward. The name of that thread is distributions. It is the importance of distributions that justifies the subtitle of this book, *Tales of Distributions*. I want to summarize explicitly some of what you have learned about distributions. I hope this will help you integrate the material and facilitate your understanding of other statistical techniques that were not covered in this course (but that you might encounter in future courses or in future encounters with statistics).

In Chapter 2, you were confronted with a disorganized array of scores of one variable. You learned to organize those scores into a frequency distribution, graph the distribution, and find central tendency values. In Chapter 3, you learned about the variability of distributions—especially the standard deviation—and how to express scores as z scores. In Chapter 4, where you were introduced to bivariate distributions, you learned to express the relationship between variables with a correlation coefficient and, using a linear regression equation, predict scores on one variable from scores on another. In Chapter 5, you learned about the normal distribution and how to use means and standard deviations to find probabilities associated with distributions. In Chapter 6, you were introduced to the important concept of the sampling distribution. You used two sampling distributions, the normal distribution and the t distribution to determine probabilities about sample means and to establish confidence intervals about a sample mean. Chapter 7 presented you with the concept of hypothesis testing. You used the t distribution to test the hypothesis that a sample mean came from a population with a hypothesized mean, and you calculated the size of the effect. In Chapter 8, you studied the design of a simple two-group experiment. You then applied the hypothesis-testing procedure to both independent-samples designs and correlated-samples designs, again using the t distribution as a source of probabilities. Again, you calculated effect sizes. In Chapter 9, you learned about still another sampling distribution (F distribution) that is used when there are two or more groups in an experiment. Chapter 10 discussed an even more complex experiment, one with *two* independent variables for which probability figures could be obtained with the F distribution. The last of the three chapters on ANOVA (Chapter 11), covered the analysis of two or more correlated groups. Chapter 12 taught you how to use the chi square distribution for data that consist of frequencies. In Chapter 13, you analyzed data using nonparametric statistics—techniques that use distributions that do not require the assumptions that the populations of dependent-variable scores be normally distributed and have equal variances.

Look back at the preceding paragraph. You have learned a lot, even if your course did not cover *every* chapter. Allow me to offer my congratulations!

FUTURE STEPS

Several paths might be taken from here. One more or less obvious next step is the analysis of an experiment with three independent variables. The three main effects and the several interactions are tested with an F distribution.

A second possibility is to study techniques for the analysis of experiments that have more than one *dependent* variable. Such techniques are called multivariate statistics. Many of these techniques are analogous to those you have studied in this book, except there are two or more dependent variables instead of just one. For example, Hotelling's T^2 is analogous to the t test; one-way multivariate analysis of variance (MANOVA) is analogous to one-way ANOVA; and higher-order MANOVA is analogous to factorial ANOVA. [For a good nontechnical introduction to many of the techniques of multivariate statistics, see Harris (1985, chap. 1).]

Finally, there is the task of combining the results of many *separate* studies that all deal with the same topic. It is not unusual to have conflicting results among 10 or 15 or 50 studies, and researchers have been at the task of synthesizing since the beginning. Recently the task has been facilitated by a set of statistical procedures called **meta-analysis**. For an introduction and two descriptive examples, see Rosenthal and Rosnow (1991, pp. 140–141). A more complete description (along with a history of meta-analysis) is in Bangert-Drowns (1986).

CHOOSING TESTS AND WRITING INTERPRETATIONS

In Chapter 1 (page 16), I described the fourfold task of a statistics student:

1. reading a problem,
2. deciding what statistical procedure to use,
3. performing that procedure, and
4. writing an interpretation of the results

My students tell me that the parts of this they still need work on at the end of the course are (2) deciding what statistical procedure to use, and (4) writing an interpretation of the results.

Even though you may have some uncertainties, you are well on your way to being able to do all four tasks. What you may need at this point is exercises that help you "put it all together." Pick and choose from the four exercises that follow; the emphasis is heavily on the side of deciding what statistical test to use.

Exercise A: Reread all the chapters you have studied and all the tests and homework that you have. If you choose to do this exercise (or part of it), spend most of your time on concepts that you recognize as especially important (such as hypothesis testing). It will take a considerable amount of time for you to do this—probably 8 to 16 hours. However, the benefits are great, according to students who have invested the time. A typical quote is, "Well, during the course I would read a chapter and work the problems without too much trouble, but I didn't relate it to the earlier stuff. When I went back over the whole book, though, I got the big picture. All the pieces do fit together."

Exercise B: Construct a table with all the statistical procedures and their characteristics. List the procedures down the left side. I would suggest column heads such as *Description* (brief description of procedure), *Purpose* (descriptive or inferential), *Null Hypothesis* (parametric or nonparametric), *Number of Independent Variables* (if applicable), and so forth.

I've included my version of this table on page 336 (Table 14.1), but please don't look at what I've done until you have a copy of your own. However, after you

TABLE 14.1 Summary Table for Elementary Statistics

Procedure (text page numbers)	Purpose	Description	Null hypothesis	No. of IVs	No. of IV levels	Design
Graphs (Chapter 2)	D	Gives picture of data	NA	NA	NA	NA
Mean (39, 42)	D	Central tendency	NA	NA	NA	NA
Median (41, 43)	D	Central tendency	NA	NA	NA	NA
Mode (42, 44)	D	Central tendency	NA	NA	NA	NA
Range (56)	D	Variability	NA	NA	NA	NA
Standard deviation (60, 66)	D	Variability	NA	NA	NA	NA
Variance (68)	D	Variability	NA	NA	NA	NA
z Score (71)	D	Relative position of one score	NA	NA	NA	NA
Regression (102)	D/I	Linear equation; 2 variables	P	NA	NA	Corr
Pearson r (86)	D/I	Degree of relationship; 2 variables	P	NA	NA	Corr
Spearman r_s (326)	D/I	Degree of relationship; 2 variables	NP	NA	NA	Corr
Confidence interval (143, 153)	D/I	Limits around population mean	P	1	NA	Indep
t test—one sample (170)	I	Sample $\bar{X}$ from pop. w/mean μ?	P	NA	NA	Indep
one-sample—effect size (177)	D	Assesses size of IV effect	P	1	NA	Indep
t test—independent samples (193)	I	2 sample $\bar{X}$'s w/same μ?	P	1	2	Indep
independent samples—effect size (202)	D	Assesses size of IV effect	P	1	2	Indep
t test—correlated samples (197)	I	2 sample $\bar{X}$'s w/same μ?	P	1	2	Corr
correlated samples—effect size (203)	D	Assesses size of IV effect	P	1	2	Corr
F test—two variances (229)	I	2 sample variances w/same σ^2?	P	1	NA	Indep
F test—One-way ANOVA (228)	I	$\geq$ 2 sample $\bar{X}$'s w/same μ?	P	1	2+	Indep
one-way ANOVA—effect size (235)	D	Assesses size of IV effect	P	1	2+	Indep
F test—factorial ANOVA (257)	I	2 or more IVs plus interactions	P	2+	2+	Indep
F test—correlated measures (277)	I	$\geq$ 2 sample $\bar{X}$'s w/same μ?	P	1	2+	Corr
Tukey HSD (233, 267, 278)	I	Tests all pairwise differences	P	1	2	Indep
Chi-square—goodness of fit (290)	I	Do data fit theory?	NP	NA	NA	Indep
Chi-square—independence (293)	I	Are variables independent?	NP	NA	NA	Indep
Mann-Whitney U (311)	I	2 samples from same population?	NP	1	2	Indep
Wilcoxon matched-pairs signed-ranks T (317)	I	2 samples from same population?	NP	1	2	Corr
Wilcoxon-Wilcox multiple comparisons (322)	I	Tests all pairwise differences	NP	1	2+	Indep

Key: Corr = Correlated   Indep = Independent   NP = Nonparametric
D = Descriptive   IV = Independent Variable   P = Parametric
I = Inferential   NA = Not Applicable

have worked for a while on your own, it wouldn't hurt to peek at the column heads I used.

Exercise C: Construct two decision trees. One should show all the descriptive statistics you have learned. To begin, you might divide these statistics into categories of univariate and bivariate distributions. Under univariate, you might list the different categories of descriptive statistics you have learned and then, under those categories, identify specific descriptive statistics. Under bivariate, choose categories under which specific descriptive statistics belong.

The second decision tree is for inferential statistics. You might begin with the categories of hypothesis testing and confidence intervals, or you might decide to divide inferential statistics into the type of statistics analyzed (means, frequency counts, and ranks). Work in issues such as the number of independent variables, whether the design is correlated samples or independent samples, and other issues that will occur to you.

There are several good ways you might organize your decision trees. My versions are Figure 14.1 and Figure 14.2, but please don't look at them now. Looking at the figures now would prevent you from going back over your text and digging out the information you learned earlier, which you now need for your decision trees. My experience has been that students who create their own decision trees learn more than those who are handed someone else's. After you have *your* trees in hand, *then* look at mine.

Exercise D: Work the two sets of problems that follow. The first set requires you to decide what statistical test or descriptive statistic will answer the question the investigator is asking. The second set requires you to interpret the results of an analyzed experiment. You will probably be glad that no number crunching is required for these problems. You won't need your calculator or computer.

However, here is a word of warning. For a few of the problems, the statistical techniques necessary for analysis are not covered in this book. (In addition, perhaps your course did not cover all the chapters.) Thus, a part of your task is to recognize what problems you are prepared to solve and what problems would require digging into a statistics textbook.

PROBLEMS

Set A. Determine what descriptive statistic or inferential test is appropriate for each problem in this set.

1. A company had three separate divisions. Based on the capital invested in 1997, Division A made 10 cents per dollar, Division B made 20 cents per dollar, and Division C made 30 cents per dollar. How can the overall profit for the company be found?

2. Reaction time scores tend to be severely skewed. The majority of the responses are quick and there are diminishing numbers of scores in the slower categories. A student wanted to find out the effects of alcohol on reaction time, so he found the reaction time of each subject under both conditions—alcohol and no alcohol.

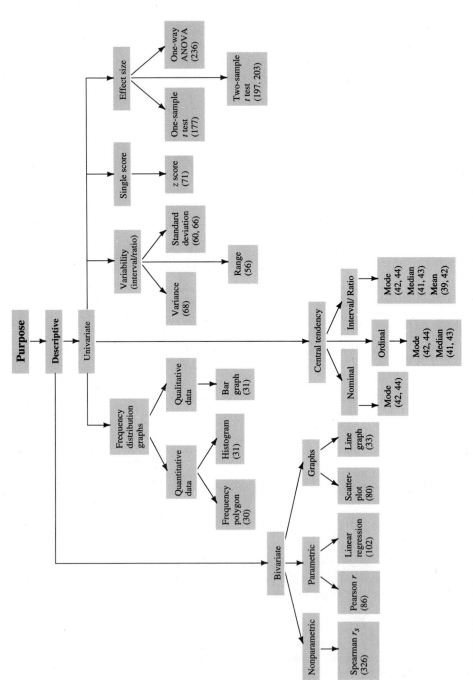

F I G U R E 14.1 Decision tree for descriptive statistics. (Text page numbers are in parentheses.)

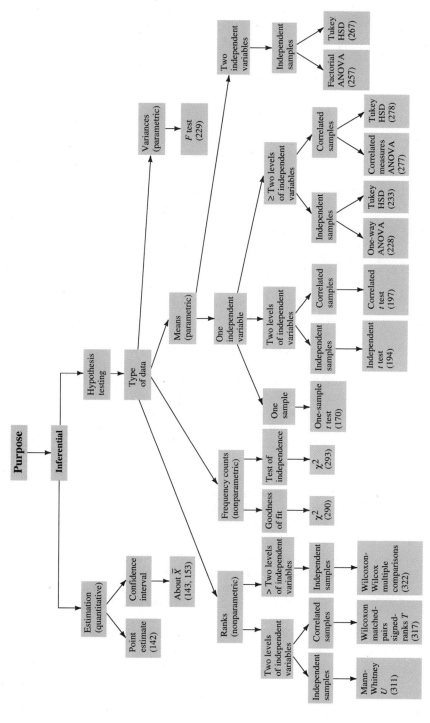

FIGURE 14.2 Decision tree for inferential statistics. (Text page numbers are in parentheses.)

3. As part of a two-week treatment for phobias, a therapist measured the general anxiety level of each client four times: before, after one week of treatment, at the end of treatment, and four months later.

4. "I want a number that will describe the typical score on this test. Most did very well, some scored in the middle, and a very few did quite poorly."

5. The designer for a city park had benches built that were 16.7 inches high. The mean distance from the bottom of the shoe to the back of the knee is 18.1 inches for women. The standard deviation is 0.70 inch. What proportion of the women who use the bench will have their feet dangle (unless they sit forward on the bench)?

6. A promoter of low-cost, do-it-yourself buildings found the insulation scores (R values) for 12 samples of concrete mixed with clay and for 12 samples of concrete mixed with hay. Each sample was 6 inches thick. He wanted to know if clay or straw was superior as an insulating additive.

7. A large school district wanted to know the range of scores within which they could expect the mean reading achievement of their sixth-grade students. A random sample of 50 of these students and their scores on a standardized reading achievement test was available.

8. Based on several large studies, a researcher knew that 40 percent of the public in the United States favored capital punishment, 35 percent were against it, and 25 percent expressed no strong feelings. The researcher polled 150 Canadians on this question with the intention of comparing the two countries in attitudes toward capital punishment.

9. What descriptive statistic could be used to identify the most typical choice of major at your college?

10. A researcher wanted to express with a correlation coefficient the strength of the relationship between stimulus intensity and pleasantness. This researcher worked with music and she knew that both very low and very high intensities are not pleasant and that the middle range of intensity produces the highest pleasantness ratings. Would you recommend a Pearson r or an r_s?

11. A consumer testing group compared Boraxo and Tide to determine which got laundry whiter. White towels that had been subjected to a variety of filthy treatments were identified on each end and were cut in half. Each was then washed in either Boraxo or Tide. After washing, each half was tested with a photometer for the amount of light reflected.

12. An investigator wanted to predict a male adult's height from his length at birth. He obtained records of both measures from a sample of male military personnel.

13. An experimenter was interested in the effect of expectations and drugs on alertness. Each subject was given an amphetamine (stimulant) or a placebo (an inert substance). In addition, half the subjects in each group were told that the drug taken was a stimulant, and half that the drug was a depressant. An alertness score for each subject was obtained from a composite of measures that included a questionnaire and observation of subjects.

14. A professor told a class that she found a correlation coefficient of .20 between the alphabetical order of a person's last name and that person's cumulative point total for 50 students in a General Psychology course. One student in the class wondered

whether the correlation coefficient was reliable; that is, would a nonzero correlation be likely to be found in other classes? How can a probability figure be found without calculating an r on another class?

15. As part of a test for some advertising copy, a company assembled 120 people who rated four toothpastes. They then read several ads (including the ad being tested). Then they rated all four toothpastes again. The data to be analyzed consisted of the number of people who rated Crest highest before reading the ad and the number who rated it highest after reading the ad.

16. Most light-bulb manufacturing companies claim that their 60-watt bulbs have an average life of 1000 hours. A skeptic with some skill as an electrician wired up some sockets and timers and burned 40 bulbs until they failed. The life of each bulb was recorded automatically. The skeptic wanted to know if the companies' claim was justified.

17. An investigator wanted to know the degree to which a person's education was related to his or her satisfaction in life. The investigator had a way to measure both of these variables on a quantitative scale.

18. In an effort to find out the effect of a "terrible drug" on reaction time, an investigator administered 0, 25, 50, or 75 milligrams to four groups of volunteer rats. The quickness of their response to a tipping platform was measured in seconds.

19. The experimenters (a male and a female) staged 140 "shoplifting" incidents in a grocery store. In each case, the "shoplifter" (one experimenter) picked up a carton of cigarettes in full view of a customer and then walked out. Half the time the experimenter was well dressed, and half the time sloppily dressed. Half the incidents involved a male shoplifter and half involved a female. For each incident, the experimenters (with the cooperation of the checkout person) simply noted whether the shoplifter was reported or not.

20. A psychologist wanted to illustrate for a group of seventh graders the fact that items in the middle of a series are more difficult to learn than those at either end. The students learned an eight-item series. The psychologist found the mean number of repetitions necessary to learn each of the items.

21. A nutritionist, with the help of an anthropologist, gathered data on the incidence of cancer among 21 different cultures. The incidence data were quite skewed, with many cultures having low incidences and the rest scattered among the higher incidences. In eight of the cultures, red meat was a significant portion of the diet. For the other 13 groups, the diet consisted primarily of grains.

22. A military science teacher wanted to find if there was any relationship between the rank in class of graduates of West Point and their military rank in the service at age 45.

23. The boat dock operators at a lake sponsored a fishing contest. Prizes were to be awarded to the persons who caught the largest bass, the largest crappie, and the largest bream, and to the overall winner. The problem in deciding the overall winner is that the bass are by nature larger than the other two species of fish. Describe an objective way to find the overall winner that will be fair to all three kinds of contestants.

24. "Yes" is the answer to the question, "Is psychotherapy effective?" Studies

comparing a quantitative measure of groups who did or did not receive psychotherapy show that the therapy group improves significantly more than the nontherapy group. What statistical technique will answer the question, "*How effective is psychotherapy?*"

Set B.   Read each problem, look at the statistical analysis, and write an appropriate conclusion. For each conclusion, use the terms used in the problem; that is, rather than saying "The null hypothesis was rejected at the .05 level," say "Those with medium anxiety solved problems more quickly than those with high anxiety."

For some of the problems, the design of the study has flaws. There are uncontrolled extraneous variables that would make it impossible to draw precise conclusions about the experiment. It is good for you to be able to recognize such design flaws, but the task here is to interpret the statistics. Thus, don't let design flaws keep you from drawing a conclusion based on the statistic.

For all problems, use a two-tailed test with $\alpha = .05$. If the results are significant at the .01 or .001 level, report that, but treat any difference that has a probability greater than .05 as not significant.

25. Four kinds of herbicides were compared for their weed-killing characteristics. Eighty plots were randomly but evenly divided and one herbicide was applied to each. Because the effects of two of the herbicides were quite variable, the dependent variable was reduced to ranks and a Wilcoxon-Wilcox multiple-comparisons test was run. The plot with the fewest weeds remaining was given a rank of 1. The summary table is shown here.

	A (936.5)	B (316.5)	C (1186)
B (316.5)	620		
C (1186)	249.5	869.5	
D (801)	135.5	482.5	385

26. A company that offered instruction in taking college entrance examinations correctly claimed that their methods, "produced improvement that was statistically significant." A parent, with some skill as a statistician, obtained the information the claim was based on (a one-sample *t* test) and calculated an effect size. The effect size was .10.

27. In a particular recycling process, the breakeven point for each batch occurred when 54 kilograms of raw material was unusable foreign matter. An engineer developed a screening process that left a mean of 38 kilograms of foreign material. The upper and lower limits of a 95 percent interval were 30 and 46 kilograms.

28. The interference theory of the serial position effect predicts that subjects who are given extended practice will perform more poorly on the initial items of a series

than will subjects who are not given extended practice. An experiment was done and the mean number of errors on initial items is shown in the table. A t test with 54 df produced a value of 3.88.

	Extended practice	No extended practice
Mean number of errors	13.7	18.4

29. A developmental psychologist developed a theory that predicted the proportion of children who would, during a period of stress, cling to their mother, attack the mother, or attack a younger sibling. The stress situation was set up and the responses of 50 children recorded. A χ^2 was calculated and found to be 5.30.

30. An experimental psychologist at a Veterans Administration hospital obtained approval from the Institutional Review Board to conduct a study of the efficacy of Zoloft for the treatment of depression. Patients whose principal problem was depression were chosen for the study. The experiment lasted 30 days, during which one group was given placebos, one was given a low dose of Zoloft, and one was given a high dose of Zoloft. At the end of the experiment, the patients were observed by two outside psychologists who rated several behaviors (eye contact, posture, verbal output, activity, and such) for degree of depression. A composite score was obtained on each patient, with low scores indicating depression. The mean scores were: placebo, 9.86; low dose, 23.00; and high dose, 9.21. An ANOVA summary table and two Tukey HSD tests are shown in the table.

Source	df	F
Between treatments	2	21.60
Error	18	

HSD (placebo vs. low) $= 13.14$
HSD (placebo vs. high) $= 0.65$

31. One sample of 16 third graders had been taught to read by the "look-say" method. A second sample of 18 had been taught by the phonics method. Both were given a reading achievement test, and a Mann-Whitney U test was performed on the results. A U value of 51 was found.

32. Sociologists sometimes make a profile of the attitudes of a group by asking each individual to rate his or her tolerance of people with particular characteristics. For example, members of a group might be asked to assess, on a scale of 1–7, their tolerance of prostitutes, atheists, former mental patients, intellectuals, college graduates, and so on. The mean score for each category is then calculated and the categories are ranked from low to high. When 18 categories of people were ranked from low to high by both a group of people who owned businesses and a group of college students, an r_s of .753 between the rankings was found.

33. In a large high school, one group of juniors took an English course that included a 9-week unit on poetry. Another group of juniors studied plays during that 9-week period. Afterward, both groups completed a questionnaire on their attitudes toward poetry. High scores mean favorable attitudes. The means and variances are presented in the table. To compare the means, a t test was run: t (78 df) $= .48$. To compare the variances, an F test was performed: F (37, 41) $= 3.62$. (See footnote 6 in Chapter 9.)

	Attitudes toward poetry	
	Studied poetry	Studied plays
Mean	23.7	21.7
Variance	51.3	16.1

34. Here is an example of the results of experiments that asked the question: Can you get more attitude change from an audience by presenting just one side of an argument and claim that it is correct, or should you present both sides and then claim that one side is correct? In the following experiment, a second independent variable is also examined: level of education. The first table contains the mean change in attitude and the second is an ANOVA summary table.

	Presentation	
Level of education	One side	Both sides
Less than a high school diploma	4.3	2.7
At least 1 year of college	2.1	4.5

Source	df	F
Between presentations	1	2.01
Between education	1	1.83
Presentations × education	1	7.93
Error	44	

35. A college student was interested in the relationship between handedness and verbal ability. She gave three classes of third-grade students a test of verbal ability that she had devised and then classified each child as right-handed, left-handed, or mixed. The obtained F value was 2.63.

	Handedness		
	Left	Right	Mixed
Mean verbal ability score	21.6	25.9	31.3
N	16	36	12

36. As part of a large-scale study on alcoholism, the alcoholics and nonalcoholics were classified according to when they were toilet-trained as children. The three categories were early, normal, or late. A χ^2 of 7.90 was found. ■

Arithmetic and Algebra Review

OBJECTIVES FOR APPENDIX A

After studying the text and working the problems, you should be able to:

1. Estimate answers

2. Round numbers

3. Find answers to problems with decimals, fractions, negative numbers, proportions, percents, absolute values, ± signs, exponents, square roots, complex expressions, and simple algebraic expressions

This appendix reviews the basic mathematical skills you will need to work the problems in this course. This math is not complex. The highest level of mathematical sophistication is simple algebra.

However, although the mathematical reasoning is not very complex, a good bit of arithmetic is required. To be good in statistics, you need to be good at arithmetic. A wrong answer is wrong whether the error is arithmetical or logical. Most people are better at arithmetic when they use a calculator with a square root key and a memory. With a calculator, you can do your work in less time and with fewer computational errors.

Of course, calculators can do more than simple arithmetic. Answers to many of the computational problems in this book can be found using a calculator with buttons for means, standard deviations, and correlation coefficients. By using a computer program you can find numerical answers to all the computational problems. The question for you is, At what stage of my statistical education should I use these aids?

Remember that the major goal of this course is for you to understand the logic of statistical reasoning so well that you can tell the story of what the answers mean. My advice is not to use your calculator or computer to produce *final statistics* like the mean, standard deviation, correlation coefficient, and *t*-test values *until* you have gone through the step-by-step, part-by-part calculations several times.

Having given you my advice about when to use your calculator to find final answers, let me encourage you to use the special features of calculators for intermediate values that go into formulas. For example, suppose a problem had just three scores: 1, 2, and 3. For many problems, you will need to know the sum of the numbers (6) and the sum when each number is squared (14, which is $1 + 4 + 9$). If your calculator has a $\Sigma+$ key or a *stat* function, it will give you both the sum of the numbers and the sum of the squares with just one entry of each of the three numbers. This feature will save you time. (It takes less than 15 seconds to calculate and write down the values 6 and 14 using a $\Sigma+$ key.) Your work is also less likely to have errors. Now, early in the course, is an excellent time to learn to exploit the capability of your calculator.

This appendix is divided into two parts.

- Part 1 is a self-administered, self-scored test of your skills in arithmetic and algebra. If you do very well on this, I suggest you skip Part 2.
- Part 2 summarizes the rules of arithmetic and algebra and gives you problems and worked-out answers.

The pretest gives you problems that are similar to those you will have to work later in the course. Take the pretest and then check your answers against the answers in the back of the book. If you find that you made any mistakes, work through those sections of Part 2 that explain the problems you missed.

As stated earlier, to be good at statistics, you need to be good at arithmetic. To be good at arithmetic, you need to know the rules and to be careful in your computations. Part 2 is designed to refresh your memory of the rules. It is up to you to be careful in your computations.

PART 1: PRETEST

Estimating answers (Estimate whole-number answers to the problems.)
1. $(4.02)^2$ **2.** 1.935×7.89 **3.** $31.219 \div 2.0593$

Rounding numbers (Round numbers to the nearest tenth.)
4. 6.06   **5.** 0.35   **6.** 10.348

Decimals

Add:
7. 3.12, 6.3, 2.004   **8.** 12, 8.625, 2.316, 4.2

Subtract:
9. $28.76 - 8.91$ **10.** $3.2 - 1.135$

Multiply:
11. 6.2×8.06 **12.** 0.35×0.162

Divide:
13. $64.1 \div 21.25$ **14.** $0.065 \div 0.0038$

Fractions

Add:

15. $\dfrac{1}{8} + \dfrac{3}{4} + \dfrac{1}{2}$

16. $\dfrac{1}{3} + \dfrac{1}{2}$

Subtract:

17. $\dfrac{11}{16} - \dfrac{1}{2}$

18. $\dfrac{5}{9} - \dfrac{1}{2}$

Multiply:

19. $\dfrac{4}{5} \times \dfrac{3}{4}$

20. $\dfrac{2}{3} \times \dfrac{1}{4} \times \dfrac{3}{5}$

Divide:

21. $\dfrac{3}{8} \div \dfrac{1}{4}$

22. $\dfrac{7}{9} \div \dfrac{1}{4}$

Negative numbers

Add:

23. $-5, 16, -1, -4$

24. $-11, -2, -12, 3$

Subtract:

25. $(-10) - (-3)$

26. $(-8) - (-2)$

Multiply:

27. $(-5) \times (-5)$

28. $(-8) \times (3)$

Divide:

29. $(-10) \div (-3)$

30. $(-21) \div 4$

Percents and proportions

31. 12 is what percent of 36?

32. Find 27 percent of 84.

33. What proportion of 112 is 21?

34. A proportion .40 of the tagged birds was recovered. 150 were tagged in all. How many were recovered?

Absolute value

35. $|-5|$

36. $|8 - 12|$

± Problems

37. $8 \pm 2 =$

38. $13 \pm 9 =$

Exponents

39. 4^2

40. 2.5^2

41. 0.35^2

Square roots

42. $\sqrt{9.30}$

43. $\sqrt{0.93}$

44. $\sqrt{0.093}$

Complex problems (Round all answers to two decimal places.)

45. $\dfrac{3 + 4 + 7 + 2 + 5}{5}$

46. $\dfrac{(5 - 3)^2 + (3 - 3)^2 + (1 - 3)^2}{3 - 1}$

47. $\dfrac{(8 - 6.5)^2 + (5 - 6.5)^2}{2 - 1}$

48. $\left(\dfrac{8 + 4}{8 + 4 - 2}\right)\left(\dfrac{1}{8} + \dfrac{1}{4}\right)$

49. $\left(\dfrac{3.6}{1.2}\right)^2 + \left(\dfrac{6.0}{2.4}\right)^2$

50. $\dfrac{(3)(4) + (6)(8) + (5)(6) + (2)(3)}{(4)(4 - 1)}$

51. $\dfrac{190 - (25^2/5)}{5 - 1}$

52. $\dfrac{12(50 - 20)^2}{(8)(10)(12)(11)}$

53. $\dfrac{6[(3 + 5)(4 - 1) + 2]}{5^2 - (6)(2)}$

54. For the numbers: 1, 2, 3, and 4: Find the sum and the sum of the squared numbers.
55. For the numbers: 2, 4, and 6: Find the sum and the sum of the squared numbers.

Simple algebra (Solve for x.)

56. $\dfrac{x - 3}{4} = 2.5$

57. $\dfrac{14 - 8.5}{x} = 0.5$

58. $\dfrac{20 - 6}{2} = 4x - 3$

59. $\dfrac{6 - 2}{3} = \dfrac{x + 9}{5}$

■ Statistical Symbols

Although, as far as I know, there has never been a clinical case of neoiconophobia,[1] some students show a mild form of this behavior. Symbols like $\bar{X}$, σ, μ, and Σ may cause a grimace, a frown, or a droopy eyelid. In more severe cases, the behavior involves avoiding a statistics course entirely. I'm sure that you don't have such a severe case because you have read this far. Even so, if you are a typical beginning student in statistics, symbols like σ, μ, Σ, and $\bar{X}$ are not very meaningful to you, and they may even elicit feelings of uneasiness. However, by the end of the course, you will know what these symbols mean and be able to approach them with an unruffled psyche—and perhaps even approach them joyously.

Some of the symbols will stand for concepts you are already quite familiar with. For example, the capital letter X with a bar over it, $\bar{X}$, stands for the mean or average. ($\bar{X}$ is pronounced "mean" or sometimes "ex-bar.") You already know about means. You just add up the scores and divide by the number of scores. This verbal instruction can be put into symbols: $\bar{X} = \Sigma X/N$. The Greek uppercase sigma (Σ) is the instruction to add and X is the symbol for scores. $\bar{X} = \Sigma X/N$ is something you already know about, even if you have not been using these symbols.

[1] An extreme and unreasonable fear of new symbols.

Pay careful attention to symbols. They serve as shorthand notations for the ideas and concepts you are learning. Each time a new symbol is introduced, concentrate on it, learn it, memorize its definition and pronunciation. The more meaning a symbol has for you, the better you understand the concepts it represents and, of course, the easier the course will be.

Sometimes I will need to distinguish between two different σ's or two X's. I will use subscripts, and the results will look like σ_1 and σ_2, or X_1 and X_2. Later, subscripts other than numbers will be used to identify a symbol. You will see σ_X and $\sigma_{\bar{X}}$. The point to learn here is that subscripts are for identification purposes only; they never indicate multiplication. Thus $\sigma_{\bar{X}}$ does not mean $(\sigma)(\bar{X})$.

Two additional comments—to encourage and to caution you. I encourage you to do more in this course than just read the text, work the problems, and pass the tests, however exciting that may be. I encourage you to occasionally get beyond this elementary text and read journal articles or short portions of other statistics textbooks. I will indicate recommendations with footnotes at appropriate places. The word of caution that goes with this encouragement is that reading statistics texts is like reading a Russian novel—the same characters have different names in different places. For example, the mean of a sample in some texts is symbolized M rather than $\bar{X}$. There is even more variety when it comes to symbolizing the standard deviation. If you expect such differences, it will be less difficult for you to fit the new symbol into your established scheme of understanding.

PART 2: REVIEW OF FUNDAMENTALS

This section gives you a quick review of the pencil and paper rules of arithmetic and simple algebra. I assume that you once knew all these rules but that refresher exercises will be helpful. Thus, there isn't much explanation. To obtain basic explanations, ask a teacher in your school's mathematics department to recommend one of the many "refresher" books that are available.

If your concern about this course comes from your anxiety about math, then I may have another recommendation for you. If you are a good reader, I recommend Sheila Tobias's book, *Overcoming Math Anxiety* (1980). This well-written, 288-page book will give you some insight into your problem and provide you with exercises that should help you reduce your math anxiety.

▨ Definitions

1. *Sum*. The answer to an addition problem is called a sum. In Chapter 9, you will calculate a *sum of squares*, a quantity that is obtained by adding together some squared numbers.
2. *Difference*. The answer to a subtraction problem is called a difference. Much of

what you will learn in statistics deals with differences and whether or not they should be attributed to chance.

3. *Product.* The answer to a multiplication problem is called a product. Chapter 4 is about the *product-moment correlation coefficient*, which requires multiplication. Multiplication problems are indicated either by a × or by parentheses. Thus, 6 × 4 and (6)(4) call for the same operation.

4. *Quotient.* The answer to a division problem is called a quotient. The IQ, or *intelligence quotient*, is based on the division of two numbers. The three ways to indicate a division problem are ÷, / (slash mark), and – (division line). Thus, 9 ÷ 4, 9/4, and $\frac{9}{4}$ call for the same operation. It is a good idea to think of any common fraction as a division problem. The numerator is to be divided by the denominator.

■ Estimating Answers

It is a **very good idea** to just look at a problem and make an estimate of the answer before you do any calculating. This is referred to as "eyeballing the data" and Edward Minium (1993) has captured its importance with Minium's First Law of Statistics: "The eyeball is the statistician's most powerful instrument."

Estimating answers should keep you from making gross errors, such as misplacing a decimal point. For example, $\frac{31.5}{5}$ can be estimated as a little more than 6. If you make this estimate before you divide, you are likely to recognize that an answer of 63 or 0.63 is incorrect.

The estimated answer to the problem (21)(108) is 2000, since (20)(100) = 2000. The problem (0.47)(0.20) suggests an estimated answer of 0.10 because $(\frac{1}{2})(0.20) = 0.10$. With 0.10 in mind, you are not likely to write 0.94 for the answer, which is 0.094. Estimating answers is also important if you are finding a square root. You can estimate that $\sqrt{95}$ is about 10 because $\sqrt{100} = 10$; $\sqrt{1.034}$ is about 1; $\sqrt{0.093}$ is about 0.1.

To calculate a mean, eyeball the numbers and estimate the mean. If you estimate a mean of 30 for a group of numbers that are primarily in the 20s, 30s, and 40s, a calculated mean of 60 will arouse your suspicion that you have made an error.

■ Rounding Numbers

There are two parts to the rule for rounding a number. If the first digit of the part that is to be dropped is less than 5, simply drop it. If the first digit of the part to be dropped is 5 or greater, increase the number to the left of it by one. These rules are built into most electronic calculators. Here are some illustrations of these rules.

Rounding to the nearest whole number:

6.2 = 6	4.5 = 5
12.7 = 13	163.5 = 164
6.49 = 6	9.5 = 10

Rounding to the nearest hundredth:

$$13.614 = 13.61 \qquad 12.065 = 12.07$$
$$0.049 = 0.05 \qquad 4.005 = 4.01$$
$$1.097 = 1.10 \qquad 0.675 = 0.68$$
$$3.6248 = 3.62 \qquad 1.995 = 2.00$$

A reasonable question is, How many decimal places should an answer in statistics have? A good rule of thumb in statistics is to carry all operations to three decimal places and then, for the final answer, round back to two decimal places.

Sometimes this rule of thumb can get you into trouble, though. For example, if halfway through some work you had a division problem of $0.0016 \div 0.0074$, and you dutifully rounded those four decimals to three ($0.002 \div 0.007$), you would get an answer of 0.2857, which becomes 0.29. However, division without rounding gives you an answer of 0.2162 or 0.22. The difference between 0.22 and 0.29 may be quite substantial. I will often give you cues if more than two decimal places are necessary but you will always need to be alert to the problems of rounding.

Most calculators carry more decimal places in memory than they show in the display. If you have such a calculator, it will protect you from the problem of rounding too much or too soon.

PROBLEMS

1. Define (a) sum, (b) quotient, (c) product, and (d) difference.
2. Estimate answers to the following expressions.
   a. $\sqrt{103.48}$ **b.** $74.16 \div 9.87$ **c.** $(11.4)^2$ **d.** $\sqrt{0.0459}$
   e. 0.41^2 **f.** 11.92×4.60 **g.** $\sqrt{0.888}$ **h.** $\sqrt{0.0098}$
3. Round the following numbers to the nearest whole number.
   a. 13.9   **b.** 126.4   **c.** 9.0
   d. 0.4   **e.** 127.5   **f.** 12.51
   g. 12.49   **h.** 12.50   **i.** 9.46
4. Round the following numbers to the nearest hundredth.
   a. 6.3348   **b.** 12.997   **c.** 0.050
   d. 0.965   **e.** 2.605   **f.** 0.3445
   g. 0.003   **h.** 0.015   **i.** 0.9949

■ Decimals

1. *Addition and subtraction of decimals.* There is only one rule about the addition and subtraction of numbers that have decimals: *Keep the decimal points in a vertical line.* The decimal point in the answer goes directly below those in the problem. This rule is illustrated in the five problems here.

Add			*Subtract*	
	0.004	6.0		
1.26	1.310	18.0	14.032	16.00
10.00	4.039	0.5	8.26	4.32
11.26	5.353	24.5	5.772	11.68

2. *Multiplication of decimals.* The basic rule for multiplying decimals is that the number of decimal places in the answer is found by adding up the number of decimal places in the two numbers that are being multiplied. To place the decimal point in the product, count from the right.

1.3	0.21	1.47
$\times 4.2$	$\times 0.4$	$\times 3.12$
26	0.084	294
52		147
5.46		441
		4.5864

3. *Division of decimals.* Two methods have been used to teach division of decimals. The older method required the student to move the decimal in the divisor (the number you are dividing by) enough places to the right to make the divisor a whole number. The decimal in the dividend is then moved to the right the same number of places, and division is carried out in the usual way. The new decimal places are identified with carets, and the decimal place in the quotient is just above the caret in the dividend. For example,

$$20. \atop 0.016_\wedge \overline{)0.320_\wedge}$$

Decimal moved three places in both the divisor and the dividend

$$2.072 \atop 6\overline{)12.432}$$

Divisor is already a whole number

$$38.46 \atop 0.39_\wedge \overline{)15.00_\wedge}$$

Decimal moved two places in both the divisor and the dividend

$$.004 \atop 9.1_\wedge \overline{)0.0_\wedge 369}$$

Decimal moved one place in both the divisor and the dividend

The newer method of teaching the division of decimals is to multiply both the divisor and the dividend by the number that will make both of them whole numbers. (Actually, this is the way the caret method works also.) For example,

$$\frac{0.32}{0.016} \times \frac{1000}{1000} = \frac{320}{16} = 20.00$$

$$\frac{0.0369}{9.1} \times \frac{10,000}{10,000} = \frac{369}{91,000} = 0.004$$

$$\frac{15}{0.39} \times \frac{100}{100} = \frac{1500}{39} = 38.46$$

$$\frac{12.432}{6} \times \frac{1000}{1000} = \frac{12,432}{6000} = 2.072$$

Both of these methods work. Use the one you are more familiar with.

PROBLEMS

Perform the operations indicated.

5. $0.001 + 10 + 3.652 + 2.5$

6. $14.2 - 7.31$

7. 0.04×1.26

8. $143.3 + 16.92 + 2.307 + 8.1$

9. $3.06 \div 0.04$

10. $\dfrac{24}{11.75}$

11. $152.12 - 127.4$

12. $(0.5)(0.07)$

■ Fractions

In general, there are two ways to deal with fractions.

1. Convert the fraction to a decimal and perform the operations on the decimals.
2. Work directly with the fractions, using a set of rules for each operation. The rule for addition and subtraction is: Convert the fractions to ones with common denominators, add or subtract the numerators, and place the result over the common denominator. The rule for multiplication is: Multiply the numerators together to get the numerator of the answer, and multiply the denominators together for the denominator of the answer. The rule for division is: Invert the divisor and multiply the fractions.

For statistics problems, it is usually easier to convert the fractions to decimals and then work with the decimals. Therefore, this is the method that is illustrated. However, if you are a whiz at working directly with fractions, by all means continue with your method. To convert a fraction to a decimal, divide the lower number into the upper one. Thus, $3/4 = 0.75$, and $13/17 = 0.765$.

Addition of fractions

$$\frac{1}{2} + \frac{1}{4} = 0.50 + 0.25 = 0.75$$

$$\frac{13}{17} + \frac{21}{37} = 0.765 + 0.568 = 1.33$$

$$\frac{2}{3} + \frac{3}{4} = 0.667 + 0.75 = 1.42$$

Subtraction of fractions

$$\frac{1}{2} - \frac{1}{4} = 0.50 - 0.25 = 0.25$$

$$\frac{11}{12} - \frac{2}{3} = 0.917 - 0.667 = 0.25$$

$$\frac{41}{53} - \frac{17}{61} = 0.774 - 0.279 = 0.50$$

Multiplication of fractions

$$\frac{1}{2} \times \frac{3}{4} = (0.5)(0.75) = 0.38$$

$$\frac{10}{19} \times \frac{61}{90} = 0.526 \times 0.678 = 0.36$$

$$\frac{1}{11} \times \frac{2}{3} = (0.09)(0.67) = 0.06$$

Division of fractions

$$\frac{9}{21} \div \frac{13}{19} = 0.429 \div 0.684 = 0.63$$

$$14 \div \frac{1}{3} = 14 \div 0.33 = 42$$

$$\frac{7}{8} \div \frac{3}{4} = 0.875 \div 0.75 = 1.17$$

PROBLEMS

Perform the operations indicated.

13. $\dfrac{9}{10} + \dfrac{1}{2} + \dfrac{2}{5}$

14. $\dfrac{9}{20} \div \dfrac{19}{20}$

15. $\left(\dfrac{1}{3}\right)\left(\dfrac{5}{6}\right)$

16. $\dfrac{4}{5} - \dfrac{1}{6}$

17. $\dfrac{1}{3} \div \dfrac{5}{6}$

18. $\dfrac{3}{4} \times \dfrac{5}{6}$

19. $18 \div \dfrac{1}{3}$ ■

■ Negative Numbers

1. *Addition of negative numbers.* (Any number without a sign is understood to be positive.)

   a. To add a series of negative numbers, add the numbers in the usual way and attach a negative sign to the total.

$$\begin{array}{r} -3 \\ -8 \\ -12 \\ \underline{-5} \\ -28 \end{array} \qquad (-1) + (-6) + (-3) = -10$$

   b. To add two numbers, one positive and one negative, subtract the smaller number from the larger number and attach the sign of the larger number to the result.

$$\begin{array}{r} 140 \\ \underline{-55} \\ 85 \end{array} \qquad \begin{array}{r} -14 \\ \underline{8} \\ -6 \end{array} \qquad \begin{array}{l} (28) + (-9) = 19 \\[4pt] 74 + (-96) = -22 \end{array}$$

   c. To add a series of numbers, of which some are positive and some negative, add all the positive numbers together, add all the negative numbers together (see 1a), and then combine the two sums (see 1b).

$$(-4) + (-6) + (12) + (-5) + (2) + (-9) = 14 + (-24) = -10$$
$$(-7) + (10) + (4) + (-5) = 14 + (-12) = 2$$

2. *Subtraction of negative numbers.* To subtract a negative number, change it to positive and add it.

$$\begin{array}{r} (-14) \\ \underline{-(-2)} \\ -12 \end{array} \qquad \begin{array}{r} 5 \\ \underline{-(-7)} \\ 12 \end{array} \qquad \begin{array}{r} 8 \\ \underline{-(-3)} \\ 11 \end{array} \qquad \begin{array}{r} (-7) \\ \underline{-(-5)} \\ -2 \end{array}$$

3. *Multiplication of negative numbers.* When the two numbers to be multiplied are both negative, the product is positive.

$$(-3)(-3) = 9 \qquad (-6)(-8) = 48$$

When one of the numbers is negative and the other is positive, the product is negative.

$$(-8)(3) = -24 \qquad 14 \times -2 = -28$$

4. *Division of negative numbers.* The rule in division is the same as the rule in multiplication. If the two numbers are both negative, the quotient is positive.

$$(-10) \div (-2) = 5 \qquad (-4) \div (-20) = 0.20$$

If one number is negative and the other positive, the quotient is negative.

$$(-10) \div 2 = -5 \qquad 6 \div (-18) = -0.33$$

$$14 \div (-7) = -2 \qquad (-12) \div 3 = -4$$

PROBLEMS

20. Add the following numbers.
  a. $-3, 19, -14, 5, -11$ **b.** $-8, -12, -3$
  c. $-8, 11$ **d.** $3, -6, -2, 5, -7$
21. $(-8)(5)$ **22.** $(-4)(-6)$ **23.** $(4)(12)$
24. $(11)(-3)$ **25.** $(-18) - (-9)$ **26.** $14 \div (-6)$
27. $12 - (-3)$ **28.** $(-6) - (-7)$ **29.** $(-9) \div (-3)$
30. $(-10) \div 5$ **31.** $4 \div (-12)$ **32.** $(-7) - 5$ ∎

■ Proportions and Percents

A **proportion** is a part of a whole and can be expressed as a fraction or as a decimal. Usually proportions are expressed as decimals. If eight students in a class of 44 received A's, we may express 8 as a proportion of the whole (44). Thus, we have 8/44, or 0.18. The proportion of the class that received A's is 0.18.

To convert a proportion to a percent (per one hundred), multiply by 100. Thus: $0.18 \times 100 = 18$; 18 percent of the students received A's. As you can see, proportions and percents are two ways to express the same idea.

If you know a proportion (or percent) and the size of the original whole, you can find the number that the proportion represents. If 0.28 of the students were absent due to illness and there are 50 students in all, then 0.28 of the 50 were absent. $(0.28)(50) = 14$ students who were absent. Here are some more examples.

1. 26 out of 31 completed the course. What proportion completed the course?

$$\frac{26}{31} = 0.83$$

2. What percent completed the course?

$$0.83 \times 100 = 83 \text{ percent}$$

3. What percent of 19 is 5?

$$\frac{5}{19} = 0.26 \qquad 0.26 \times 100 = 26 \text{ percent}$$

4. If 90 percent of the population agreed and the population consisted of 210 members, how many members agreed?

$$0.90 \times 210 = 189 \text{ members}$$

■ Absolute Value

The **absolute value** of a number ignores the sign of the number. Thus, the absolute value of -6 is 6. This is expressed with symbols as $|-6| = 6$. It is expressed verbally as "the absolute value of negative six is six." In a similar way, the absolute value of $4 - 7$ is 3; that is, $|4 - 7| = |-3| = 3$.

■ ± Signs

A $\pm$ sign ("plus or minus" sign) means to both add *and* subtract. A $\pm$ problem *always* has two answers.

$$10 \pm 4 = 6, \ 14$$
$$8 \pm (3)(2) = 8 \pm 6 = 2, \ 14$$
$$\pm (4)(3) + 21 = \pm 12 + 21 = 9, \ 33$$
$$\pm 4 - 6 = -10, \ -2$$

■ Exponents

In the expression 5^2 ("five squared"), 2 is the exponent. The 2 means that 5 is to be multiplied by itself. Thus, $5^2 = 5 \times 5 = 25$.

In elementary statistics, the only exponent used is 2, but it will be used frequently. When a number has an exponent of 2, the number is said to be squared. The expression 4^2 ("four squared") means 4×4, and the product is 16.

$$8^2 = 8 \times 8 = 64 \qquad\qquad \left(\frac{3}{4}\right)^2 = (0.75)^2 = 0.5625$$

$$1.2^2 = (1.2)(1.2) = 1.44 \qquad\qquad 12^2 = 12 \times 12 = 144$$

▨ Square Roots

Statistics problems often require that a square root be found. There are three possible ways to find the square root:

1. A calculator with a square root key
2. A table of square roots
3. The paper and pencil method

Use a calculator with a square root key. If this is not possible, find a book with a table of square roots. The third solution, the paper and pencil method, is so tedious and error-prone that I do not recommend it.

PROBLEMS

33. Six is what proportion of 13?　　**34.** Six is what percent of 13?
35. What proportion of 25 is 18?　　**36.** What percent of 115 is 85?
37. Find 72 percent of 36.　　**38.** 22 is what proportion of 72?
39. If 0.40 of the 320 members voted against the proposal, how many voted no?
40. 22 percent of the 850 coupons were redeemed. How many were redeemed?
41. $|-31|$　　　　　　　　　　　**42.** $|21-25|$
43. $12 \pm (2)(5)$　　　　　　　**44.** $\pm(5)(6) + 10$
45. $\pm(2)(2) - 6$　　　　　　　**46.** $(2.5)^2$
47. 9^2　　　　　　　　　　　　**48.** $(0.3)^2$
49. $\left(\dfrac{1}{4}\right)^2$

50. Find the square root of the following numbers to two decimal places (or more if appropriate).

a. 625	**b.** 6.25	**c.** 0.625
d. 0.0625	**e.** 16.85	**f.** 0.003
g. 181,476	**h.** 0.25	**i.** 22.51

▨ Complex Expressions

Two rules will suffice for the kinds of complex expressions encountered in statistics.

1. Perform the operations within the parentheses. If there are brackets in the expression, perform the operations within the parentheses and then the operations within the brackets.
2. Perform the operations in the numerator separately from those in the

denominator and, finally, carry out the division.

$$\frac{(8-6)^2 + (7-6)^2 + (3-6)^2}{3-1} = \frac{2^2 + 1^2 + (-3)^2}{2} = \frac{4+1+9}{2}$$

$$= \frac{14}{2} = 7.00$$

$$\left(\frac{10+12}{4+3-2}\right)\left(\frac{1}{4}+\frac{1}{3}\right) = \left(\frac{22}{5}\right)(0.25 + 0.333) = (4.40)(0.583)$$

$$= 2.57$$

$$\left(\frac{8.2}{4.1}\right)^2 + \left(\frac{4.2}{1.2}\right)^2 = (2)^2 + (3.5)^2 = 4 + 12.25 = 16.25$$

$$\frac{6[(13-10)^2 - 5]}{6(6-1)} = \frac{6(3^2-5)}{6(5)} = \frac{6(9-5)}{30} = \frac{6(4)}{30}$$

$$= \frac{24}{30} = 0.80$$

$$\frac{18 - \frac{6^2}{4}}{4-1} = \frac{18 - \frac{36}{4}}{3} = \frac{18-9}{3} = \frac{9}{3} = 3.00$$

■ Simple Algebra

To solve a simple algebra problem, isolate the unknown (x) on one side of the equal sign and combine the numbers on the other side. To do this, remember that you can multiply or divide both sides of the equation by the same number without affecting the value of the unknown. For example,

$$\frac{x}{6} = 9 \qquad\qquad 11x = 30 \qquad\qquad \frac{3}{x} = 14$$

$$(6)\left(\frac{x}{6}\right) = (6)(9) \qquad \frac{11x}{11} = \frac{30}{11} \qquad (x)\left(\frac{3}{x}\right) = (14)(x)$$

$$x = 54 \qquad\qquad x = 2.73 \qquad\qquad 3 = 14x$$

$$\frac{3}{14} = \frac{14x}{14}$$

$$0.21 = x$$

In a similar way, the same number can be *added to* or *subtracted from* both sides of the equation without affecting the value of the unknown. For example,

$$x - 5 = 12 \qquad\qquad x + 7 = 9$$

$$x - 5 + 5 = 12 + 5 \qquad x + 7 - 7 = 9 - 7$$

$$x = 17 \qquad\qquad x = 2$$

I will combine some of these steps in the problems I will work for you. Be sure you see what operation is being performed on both sides in each step.

$$\frac{x - 2.5}{1.3} = 1.96 \qquad\qquad \frac{21.6 - 15}{x} = 0.04$$

$$x - 2.5 = (1.3)(1.96) \qquad\qquad 6.6 = 0.04x$$

$$x = 2.548 + 2.5 \qquad\qquad \frac{6.6}{0.04} = x$$

$$x = 5.048 \qquad\qquad\qquad 165 = x$$

$$4x - 9 = 5^2 \qquad\qquad \frac{14 - x}{6} = 1.9$$

$$4x = 25 + 9 \qquad\qquad 14 - x = (1.9)(6)$$

$$x = \frac{34}{4} \qquad\qquad\qquad 14 = 11.4 + x$$

$$x = 8.50 \qquad\qquad\qquad 2.6 = x$$

PROBLEMS

Reduce these complex expressions to a single number rounded to two decimal places.

51. $\dfrac{(4 - 2)^2 + (0 - 2)^2}{6}$

52. $\dfrac{(12 - 8)^2 + (8 - 8)^2 + (5 - 8)^2 + (7 - 8)^2}{4 - 1}$

53. $\left(\dfrac{5 + 6}{3 + 2 - 2}\right)\left(\dfrac{1}{3} + \dfrac{1}{2}\right)$

54. $\left(\dfrac{13 + 18}{6 + 8 - 2}\right)\left(\dfrac{1}{6} + \dfrac{1}{8}\right)$

55. $\dfrac{8[(6 - 2)^2 - 5]}{(3)(2)(4)}$

56. $\dfrac{[(8 - 2)(5 - 1)]^2}{5(10 - 7)}$

57. $\dfrac{6}{1/2} + \dfrac{8}{1/3}$

58. $\left(\dfrac{9}{2/3}\right)^2 + \left(\dfrac{8}{3/4}\right)^2$

59. $\dfrac{10 - (6^2/9)}{8}$

60. $\dfrac{104 - (12^2/6)}{5}$

Find the value of x in the problems.

61. $\dfrac{x - 4}{2} = 2.58$

62. $\dfrac{x - 21}{6.1} = 1.04$

63. $x = \dfrac{14 - 11}{2.5}$

64. $x = \dfrac{36 - 41}{8.2}$

I will combine some of these steps in the problems I will work for you. Be sure you see what operation is being performed on both sides in each step.

$$\frac{x - 2.5}{1.3} = 1.96$$

$$x - 2.5 = (1.3)(1.96)$$

$$x = 2.548 + 2.5$$

$$x = 5.048$$

$$\frac{21.6 - 15}{x} = 0.04$$

$$6.6 = 0.04x$$

$$\frac{6.6}{0.04} = x$$

$$165 = x$$

$$4x - 9 = 5^2$$

$$4x = 25 + 9$$

$$x = \frac{34}{4}$$

$$x = 8.50$$

$$\frac{14 - x}{6} = 1.9$$

$$14 - x = (1.9)(6)$$

$$14 = 11.4 + x$$

$$2.6 = x$$

PROBLEMS

Reduce these complex expressions to a single number rounded to two decimal places.

51. $\dfrac{(4 - 2)^2 + (0 - 2)^2}{6}$

52. $\dfrac{(12 - 8)^2 + (8 - 8)^2 + (5 - 8)^2 + (7 - 8)^2}{4 - 1}$

53. $\left(\dfrac{5 + 6}{3 + 2 - 2}\right)\left(\dfrac{1}{3} + \dfrac{1}{2}\right)$

54. $\left(\dfrac{13 + 18}{6 + 8 - 2}\right)\left(\dfrac{1}{6} + \dfrac{1}{8}\right)$

55. $\dfrac{8[(6 - 2)^2 - 5]}{(3)(2)(4)}$

56. $\dfrac{[(8 - 2)(5 - 1)]^2}{5(10 - 7)}$

57. $\dfrac{6}{1/2} + \dfrac{8}{1/3}$

58. $\left(\dfrac{9}{2/3}\right)^2 + \left(\dfrac{8}{3/4}\right)^2$

59. $\dfrac{10 - (6^2/9)}{8}$

60. $\dfrac{104 - (12^2/6)}{5}$

Find the value of x in the problems.

61. $\dfrac{x - 4}{2} = 2.58$

62. $\dfrac{x - 21}{6.1} = 1.04$

63. $x = \dfrac{14 - 11}{2.5}$

64. $x = \dfrac{36 - 41}{8.2}$

Tables

TABLE A Critical values for Pearson product-moment correlation coefficients, r*

	α Levels (two-tailed test)				
df	.10	.05	.02	.01	.001
	α Levels (one-tailed test)				
$(df = N - 2)$	.05	.025	.01	.005	.0005
1	.98769	.99692	.999507	.999877	.9999988
2	.90000	.95000	.98000	.990000	.99900
3	.8054	.8783	.93433	.95873	.99116
4	.7293	.8114	.8822	.91720	.97406
5	.6694	.7545	.8329	.8745	.95074
6	.6215	.7067	.7887	.8343	.92493
7	.5822	.6664	.7498	.7977	.8982
8	.5494	.6319	.7155	.7646	.8721
9	.5214	.6021	.6851	.7348	.8371
10	.4973	.5760	.6581	.7079	.8233
11	.4762	.5529	.6339	.6835	.8010
12	.4575	.5324	.6120	.6614	.7800
13	.4409	.5139	.5923	.6411	.7603
14	.4259	.4973	.5742	.6226	.7420
15	.4124	.4821	.5577	.6055	.7246
16	.4000	.4683	.5425	.5897	.7084
17	.3887	.4555	.5285	.5751	.6932
18	.3783	.4438	.5155	.5614	.6787
19	.3687	.4329	.5034	.5487	.6652
20	.3598	.4227	.4921	.5368	.6524
25	.3233	.3809	.4451	.4869	.5974
30	.2960	.3494	.4093	.4487	.5541
35	.2746	.3246	.3810	.4182	.5189
40	.2573	.3044	.3578	.3932	.4896
45	.2428	.2875	.3384	.3721	.4648
50	.2306	.2732	.3218	.3541	.4433
60	.2108	.2500	.2948	.3248	.4078
70	.1954	.2319	.2737	.3017	.3799
80	.1829	.2172	.2565	.2830	.3568
90	.1726	.2050	.2422	.2673	.3375
100	.1638	.1946	.2301	.2540	.3211

* To be significant the r obtained from the data must be equal to or **larger than** the value shown in the table.

SOURCE: From Table VII in R. A. Fisher and F. Yates, *Statistical Tables for Biological, Agricultural, and Medical Research*, Sixth Edition, published by Addison Wesley Longman Ltd., (1963). Reprinted with permission.

T A B L E B   Random digits

	00–04	05–09	10–14	15–19	20–24	25–29	30–34	35–39	40–44	45–49
00	54463	22662	65905	70639	79365	67382	29085	69831	47058	08186
01	15389	85205	18850	39226	42249	90669	96325	23248	60933	26927
02	85941	40756	82414	02015	13858	78030	16269	65978	01385	15345
03	61149	69440	11286	88218	58925	03638	52862	62733	33451	77455
04	05219	81619	10651	67079	92511	59888	84502	72095	83463	75577
05	41417	98326	87719	92294	46614	50948	64886	20002	97365	30976
06	28357	94070	20652	35774	16249	75019	21145	05217	47286	76305
07	17783	00015	10806	83091	91530	36466	39981	62481	49177	75779
08	40950	84820	29881	85966	62800	70326	84740	62660	77379	90279
09	82995	64157	66164	41180	10089	41757	78258	96488	88629	37231
10	96754	17676	55659	44105	47361	34833	86679	23930	53249	27083
11	34357	88040	53364	71726	45690	66334	60332	22554	90600	71113
12	06318	37403	49927	57715	50423	67372	63116	48888	21505	80182
13	62111	52820	07243	79931	89292	84767	85693	73947	22278	11551
14	47534	09243	67879	00544	23410	12740	02540	54440	32949	13491
15	98614	75993	84460	62846	59844	14922	48730	73443	48167	34770
16	24856	03648	44898	09351	98795	18644	39765	71058	90368	44104
17	96887	12479	80621	66223	86085	78285	02432	53342	42846	94771
18	90801	21472	42815	77408	37390	76766	52615	32141	30268	18106
19	55165	77312	83666	36028	28420	70219	81369	41943	47366	41067
20	75884	12952	84318	95108	72305	64620	91318	89872	45375	85436
21	16777	37116	58550	42958	21460	43910	01175	87894	81378	10620
22	46230	43877	80207	88877	89380	32992	91380	03164	98656	59337
23	42902	66892	46134	01432	94710	23474	20423	60137	60609	13119
24	81007	00333	39693	28039	10154	95425	39220	19774	31782	49037
25	68089	01122	51111	72373	06902	74373	96199	97017	41273	21546
26	20411	67081	89950	16944	93054	87687	96693	87236	77054	33848
27	58212	13160	06468	15718	82627	76999	05999	58680	96739	63700
28	70577	42866	24969	61210	76046	67699	42054	12696	93758	03283
29	94522	74358	71659	62038	79643	79169	44741	05437	39038	13163
30	42626	86819	85651	88678	17401	03252	99547	32404	17918	62880
31	16051	33763	57194	16752	54450	19031	58580	47629	54132	60631
32	08244	27647	33851	44705	94211	46716	11738	55784	95374	72655
33	59497	04392	09419	89964	51211	04894	72882	17805	21896	83864
34	97155	13428	40293	09985	58434	91412	69124	82171	59058	82859
35	98409	66162	95763	47420	20792	61527	29441	39435	11859	41567
36	45476	84882	65109	96597	25930	66790	65706	61203	53634	22557
37	89300	69700	50741	30329	11658	23166	05400	66669	48708	02306
38	50051	95137	91631	66315	91428	12275	24816	68091	71710	33258
39	31753	85178	31310	89642	98364	92396	24617	09609	83942	22716
40	79152	53829	77250	20190	56535	18760	69942	77448	33278	48805
41	44560	38750	83635	56540	64900	42912	13953	79149	18710	68618
42	68328	83378	63369	71381	39564	95615	42451	64559	97501	65747
43	46939	38689	58625	08342	30459	85863	20781	09284	26333	91777
44	83544	86141	15707	96256	23068	13782	08467	89469	93842	55349
45	91621	00881	04900	54224	46177	55309	17852	27491	89415	23466
46	91896	67126	04151	03795	59077	11848	12630	98375	52068	60142
47	55751	62515	21108	80830	02263	29303	37204	96926	30506	09808
48	85156	87689	95493	88842	00664	55017	55539	17771	69448	87530
49	07521	56898	12236	60277	39102	62315	12239	07105	11844	01117

(continued)

SOURCE: From *Statistical Methods*, by G. W. Snedecor and W. G. Cochran, Seventh Edition. Copyright © 1980 Iowa State University Press. Reprinted by permission.

T A B L E B (*continued*)

	50–54	55–59	60–64	65–69	70–74	75–79	80–84	85–89	90–94	95–99
00	59391	58030	52098	82718	87024	82848	04190	96574	90464	29065
01	99567	76364	77204	04615	27062	96621	43918	01896	83991	51141
02	10363	97518	51400	25670	98342	61891	27101	37855	06235	33316
03	86859	19558	64432	16706	99612	59798	32803	67708	15297	28612
04	11258	24591	36863	55368	31721	94335	34936	02566	80972	08188
05	95068	88628	35911	14530	33020	80428	39936	31855	34334	64865
06	54463	47237	73800	91017	36239	71824	83671	39892	60518	37092
07	16874	62677	57412	13215	31389	62233	80827	73917	82802	84420
08	92494	63157	76593	91316	03505	72389	96363	52887	01087	66091
09	15669	56689	35682	40844	53256	81872	35213	09840	34471	74441
10	99116	75486	84989	23476	52967	67104	39495	39100	17217	74073
11	15696	10703	65178	90637	63110	17622	53988	71087	84148	11670
12	97720	15369	51269	69620	03388	13699	33423	67453	43269	56720
13	11666	13841	71681	98000	35979	39719	81899	07449	47985	46967
14	71628	73130	78783	75691	41632	09847	61547	18707	65489	69944
15	40501	51089	99943	91843	41995	88931	73631	69361	05375	15417
16	22518	55576	98215	82068	10798	86211	36584	67466	69373	40054
17	75112	30485	62173	02132	14878	92879	22281	16783	86352	00077
18	80327	02671	98191	84342	90813	49268	95441	15496	20168	09271
19	60251	45548	02146	05597	48228	81366	34598	72856	66762	17002
20	57430	82270	10421	05540	43648	75888	66049	21511	47676	33444
21	73528	39559	34434	88596	54086	71693	43132	14414	79949	85193
22	25991	65959	70769	64721	86413	33475	42740	06175	82758	66248
23	78388	16638	09134	59880	63806	48472	39318	35434	24057	74739
24	12477	09965	96657	57994	59439	76330	24596	77515	09577	91871
25	83266	32883	42451	15579	38155	29793	40914	65990	16255	17777
26	76970	80876	10237	39515	79152	74798	39357	09054	73579	92359
27	37074	65198	44785	68624	98336	84481	97610	78735	46703	98265
28	83712	06514	30101	78295	54656	85417	43189	60048	72781	72606
29	20287	56862	69727	94443	64936	08366	27227	05158	50326	59566
30	74261	32592	86538	27041	65172	85532	07571	80609	39285	65340
31	64081	49863	08478	96001	18888	14810	70545	89755	59064	07210
32	05617	75818	47750	67814	29575	10526	66192	44464	27058	40467
33	26793	74951	95466	74307	13330	42664	85515	20632	05497	33625
34	65988	72850	48737	54719	52056	01596	03845	35067	03134	70322
35	27366	42271	44300	73399	21105	03280	73457	43093	05192	48657
36	56760	10909	98147	34736	33863	95256	12731	66598	50771	83665
37	72880	43338	93643	58904	59543	23943	11231	83268	65938	81581
38	77888	38100	03062	58103	47961	83841	25878	23746	55903	44115
39	28440	07819	21580	51459	47971	29882	13990	29226	23608	15873
40	63525	94441	77033	12147	51054	49955	58312	76923	96071	05813
41	47606	93410	16359	89033	89696	47231	64498	31776	05383	39902
42	52669	45030	96279	14709	52372	87832	02735	50803	72744	88208
43	16738	60159	07425	62369	07515	82721	37875	71153	21315	00132
44	59348	11695	45751	15865	74739	05572	32688	20271	65128	14551
45	12900	71775	29845	60774	94924	21810	38636	33717	67598	82521
46	75086	23537	49939	33595	13484	97588	28617	17979	70749	35234
47	99495	51434	29181	09993	38190	42553	68922	52125	91077	40197
48	26075	31671	45386	36583	93159	48599	52022	41330	60651	91321
49	13636	93596	23377	51133	95126	61496	42474	45141	46660	42338

(*continued*)

T A B L E B (*continued*)

	00–04	05–09	10–14	15–19	20–24	25–29	30–34	35–39	40–44	45–49
50	64249	63664	39652	40646	97306	31741	07294	84149	46797	82487
51	26538	44249	04050	48174	65570	44072	40192	51153	11397	58212
52	05845	00512	78630	55328	18116	69296	91705	86224	29503	57071
53	74897	68373	67359	51014	33510	83048	17056	72506	82949	54600
54	20872	54570	35017	88132	25730	22626	86723	91691	13191	77212
55	31432	96156	89177	75541	81355	24480	77243	76690	42507	84362
56	66890	61505	01240	00660	05873	13568	76082	79172	57913	93448
57	48194	57790	79970	33106	86904	48119	52503	24130	72824	21627
58	11303	87118	81471	52936	08555	28420	49416	44448	04269	27029
59	54374	57325	16947	45356	78371	10563	97191	53798	12693	27928
60	64852	34421	61046	90849	13966	39810	42699	21753	76192	10508
61	16309	20384	09491	91588	97720	89846	30376	76970	23063	35894
62	42587	37065	24526	72602	57589	98131	37292	05967	26002	51945
63	40177	98590	97161	41682	84533	67588	62036	49967	01990	72308
64	82309	76128	93965	26743	24141	04838	40254	26065	07938	76236
65	79788	68243	59732	04257	27084	14743	17520	95401	55811	76099
66	40538	79000	89559	25026	42274	23489	34502	75508	06059	86682
67	64016	73598	18609	73150	62463	33102	45205	87440	96767	67042
68	49767	12691	17903	93871	99721	79109	09425	26904	07419	76013
69	76974	55108	29795	08404	82684	00497	51126	79935	57450	55671
70	23854	08480	85983	96025	50117	64610	99425	62291	86943	21541
71	68973	70551	25098	78033	98573	79848	31778	29555	61446	23037
72	36444	93600	65350	14971	25325	00427	52073	64280	18847	24768
73	03003	87800	07391	11594	21196	00781	32550	57158	58887	73041
74	17540	26188	36647	78386	04558	61463	57842	90382	77019	24210
75	38916	55809	47982	41968	69760	79422	80154	91486	19180	15100
76	64288	19843	69122	42502	48508	28820	59933	72998	99942	10515
77	86809	51564	38040	39418	49915	19000	58050	16899	79952	57849
78	99800	99566	14742	05028	30033	94889	53381	23656	75787	79223
79	92345	31890	95712	08279	91794	94068	49037	88674	35355	12267
80	90363	65162	32245	82279	79256	80834	06088	99462	56705	06118
81	64437	32242	48431	04835	39070	59702	31508	60935	22390	52246
82	91714	53662	28373	34333	55791	74758	51144	18827	10704	76803
83	20902	17646	31391	31459	33315	03444	55743	74701	58851	27427
84	12217	86007	70371	52281	14510	76094	96579	54863	78339	20839
85	45177	02863	42307	53571	22532	74921	17735	42201	80540	54721
86	28325	90814	08804	52746	47913	54577	47525	77705	95330	21866
87	29019	28776	56116	54791	64604	08815	46049	71186	34650	14994
88	84979	81353	56219	67062	26146	82567	33122	14124	46240	92973
89	50371	26347	48513	63915	11158	25563	91915	18431	92978	11591
90	53422	06825	69711	67950	64716	18003	49581	45378	99878	61130
91	67453	35651	89316	41620	32048	70225	47597	33137	31443	51445
92	07294	85353	74819	23445	68237	07202	99515	62282	53809	26685
93	79544	00302	45338	16015	66613	88968	14595	63836	77716	79596
94	64144	85442	82060	46471	24162	39500	87351	36637	42833	71875
95	90919	11883	58318	00042	52402	28210	34075	33272	00840	73268
96	06670	57353	86275	92276	77591	46924	60839	55437	03183	13191
97	36634	93976	52062	83678	41256	60948	18685	48992	19462	96062
98	75101	72891	85745	67106	26010	62107	60885	37503	55461	71213
99	05112	71222	72654	51583	05228	62056	57390	42746	39272	96659

(*continued*)

LE B (*continued*)

	50–54	55–59	60–64	65–69	70–74	75–79	80–84	85–89	90–94	95–99
	32847	31282	03345	89593	69214	70381	78285	20054	91018	16742
	16916	00041	30236	55023	14253	76582	12092	86533	92426	37655
	66176	34047	21005	27137	03191	48970	64625	22394	39622	79085
	46299	13335	12180	16861	38043	59292	62675	63631	37020	78195
54	22847	47839	45385	23289	47526	54098	45683	55849	51575	64689
55	41851	54160	92320	69936	34803	92479	33399	71160	64777	83378
56	28444	59497	91586	95917	68553	28639	06455	34174	11130	91994
57	47520	62378	98855	83174	13088	16561	68559	26679	06238	51254
58	34978	63271	13142	82681	05271	08822	06490	44984	49307	62717
59	37404	80416	69035	92980	49486	74378	75610	74976	70056	15478
60	32400	65482	52099	53676	74648	94148	65095	69597	52771	71551
61	89262	86332	51718	70663	11623	29834	79820	73002	84886	03591
62	86866	09127	98021	03871	27789	58444	44832	36505	40672	30180
63	90814	14833	08759	74645	05046	94056	99094	65091	32663	73040
64	19192	82756	20553	58446	55376	88914	75096	26119	83898	43816
65	77585	52593	56612	95766	10019	29531	73064	20953	53523	58136
66	23757	16364	05096	03192	62386	45389	85332	18877	55710	96459
67	45989	96257	23850	26216	23309	21526	07425	50254	19455	29315
68	92970	94243	07316	41467	64837	52406	25225	51553	31220	14032
69	74346	59596	40088	98176	17896	86900	20249	77753	19099	48885
70	87646	41309	27636	45153	29988	94770	07255	70908	05340	99751
71	50099	71038	45146	06146	55211	99429	43169	66259	97786	59180
72	10127	46900	64984	75348	04115	33624	68774	60013	35515	62556
73	67995	81977	18984	64091	02785	27762	42529	97144	80407	64524
74	26304	80217	84934	82657	69291	35397	98714	35104	08187	48109
75	81994	41070	56642	64091	31229	02595	13513	45148	78722	30144
76	59537	34662	79631	89403	65212	09975	06118	86197	58208	16162
77	51228	10937	62396	81460	47331	91403	95007	06047	16846	64809
78	31089	37995	29577	07828	42272	54016	21950	86192	99046	84864
79	38207	97938	93459	75174	79460	55436	57206	87644	21296	43395
80	88666	31142	09474	89712	63153	62333	42212	06140	42594	43671
81	53365	56134	67582	92557	89520	33452	05134	70628	27612	33738
82	89807	74530	38004	90102	11693	90257	05500	79920	62700	43325
83	18682	81038	85662	90915	91631	22223	91588	80774	07716	12548
84	63571	32579	63942	25371	09234	94592	98475	76884	37635	33608
85	68927	56492	67799	95398	77642	54613	91853	08424	81450	76229
86	56401	63186	39389	99798	31356	89235	97036	32341	33292	73757
87	24333	95603	02359	72942	46287	95382	08452	62862	97869	71775
88	17025	84202	95199	62272	06366	16175	97577	99304	41587	03686
89	02804	08253	52133	20224	68034	50865	57868	22343	55111	03607
90	08298	03879	20995	19850	73090	13191	18963	82244	78479	99121
91	59883	01785	82403	96062	03785	03488	12970	64896	38336	30030
92	46982	06682	62864	91837	74021	89094	39952	64158	79614	78235
93	31121	47266	07661	02051	67599	24471	69843	83696	71402	76287
94	97867	56641	63416	17577	30161	87320	37752	73276	48969	41915
95	57364	86746	08415	14621	49430	22311	15836	72492	49372	44103
96	09559	26263	69511	28064	75999	44540	13337	10918	79846	54809
97	53873	55571	00608	42661	91332	63956	74087	59008	47493	99581
98	35531	19162	86406	05299	77511	24311	57257	22826	77555	05941
99	28229	88629	25695	94932	30721	16197	78742	34974	97528	45447

TABLE C Area under the normal curve between μ and z and beyo

A	B	C	A	B	C	A	B	C
	Area between mean and	Area beyond		Area between mean and	Area beyond		Area b mean	
z	z	z	z	z	z	z	z	
0.00	.0000	.5000	0.55	.2088	.2912	1.10	.3643	.1357
0.01	.0040	.4960	0.56	.2123	.2877	1.11	.3665	.1335
0.02	.0080	.4920	0.57	.2157	.2843	1.12	.3686	.1314
0.03	.0120	.4880	0.58	.2190	.2810	1.13	.3708	.1292
0.04	.0160	.4840	0.59	.2224	.2776	1.14	.3729	.1271
0.05	.0199	.4801	0.60	.2257	.2743	1.15	.3749	.1251
0.06	.0239	.4761	0.61	.2291	.2709	1.16	.3770	.1230
0.07	.0279	.4721	0.62	.2324	.2676	1.17	.3790	.1210
0.08	.0319	.4681	0.63	.2357	.2643	1.18	.3810	.1190
0.09	.0359	.4641	0.64	.2389	.2611	1.19	.3830	.1170
0.10	.0398	.4602	0.65	.2422	.2578	1.20	.3849	.1151
0.11	.0438	.4562	0.66	.2454	.2546	1.21	.3869	.1131
0.12	.0478	.4522	0.67	.2486	.2514	1.22	.3888	.1112
0.13	.0517	.4483	0.68	.2517	.2483	1.23	.3907	.1093
0.14	.0557	.4443	0.69	.2549	.2451	1.24	.3925	.1075
0.15	.0596	.4404	0.70	.2580	.2420	1.25	.3944	.1056
0.16	.0636	.4364	0.71	.2611	.2389	1.26	.3962	.1038
0.17	.0675	.4325	0.72	.2642	.2358	1.27	.3980	.1020
0.18	.0714	.4286	0.73	.2673	.2327	1.28	.3997	.1003
0.19	.0753	.4247	0.74	.2704	.2296	1.29	.4015	.0985
0.20	.0793	.4207	0.75	.2734	.2266	1.30	.4032	.0968
0.21	.0832	.4168	0.76	.2764	.2236	1.31	.4049	.0951
0.22	.0871	.4129	0.77	.2794	.2206	1.32	.4066	.0934
0.23	.0910	.4090	0.78	.2823	.2177	1.33	.4082	.0918
0.24	.0948	.4052	0.79	.2852	.2148	1.34	.4099	.0901
0.25	.0987	.4013	0.80	.2881	.2119	1.35	.4115	.0885
0.26	.1026	.3974	0.81	.2910	.2090	1.36	.4131	.0869
0.27	.1064	.3936	0.82	.2939	.2061	1.37	.4147	.0853
0.28	.1103	.3897	0.83	.2967	.2033	1.38	.4162	.0838
0.29	.1141	.3859	0.84	.2995	.2005	1.39	.4177	.0823
0.30	.1179	.3821	0.85	.3023	.1977	1.40	.4192	.0808
0.31	.1217	.3783	0.86	.3051	.1949	1.41	.4207	.0793
0.32	.1255	.3745	0.87	.3078	.1922	1.42	.4222	.0778
0.33	.1293	.3707	0.88	.3106	.1894	1.43	.4236	.0764
0.34	.1331	.3669	0.89	.3133	.1867	1.44	.4251	.0749
0.35	.1368	.3632	0.90	.3159	.1841	1.45	.4265	.0735
0.36	.1406	.3594	0.91	.3186	.1814	1.46	.4279	.0721
0.37	.1443	.3557	0.92	.3212	.1788	1.47	.4292	.0708
0.38	.1480	.3520	0.93	.3238	.1762	1.48	.4306	.0694
0.39	.1517	.3483	0.94	.3264	.1736	1.49	.4319	.0681
0.40	.1554	.3446	0.95	.3289	.1711	1.50	.4332	.0668
0.41	.1591	.3409	0.96	.3315	.1685	1.51	.4345	.0655
0.42	.1628	.3372	0.97	.3340	.1660	1.52	.4357	.0643
0.43	.1664	.3336	0.98	.3365	.1635	1.53	.4370	.0630
0.44	.1700	.3300	0.99	.3389	.1611	1.54	.4382	.0618
0.45	.1736	.3264	1.00	.3413	.1587	1.55	.4394	.0606
0.46	.1772	.3228	1.01	.3438	.1562	1.56	.4406	.0594
0.47	.1808	.3192	1.02	.3461	.1539	1.57	.4418	.0582
0.48	.1844	.3156	1.03	.3485	.1515	1.58	.4429	.0571
0.49	.1879	.3121	1.04	.3508	.1492	1.59	.4441	.0559
0.50	.1915	.3085	1.05	.3531	.1469	1.60	.4452	.0548
0.51	.1950	.3050	1.06	.3554	.1446	1.61	.4463	.0537
0.52	.1985	.3015	1.07	.3577	.1423	1.62	.4474	.0526
0.53	.2019	.2981	1.08	.3599	.1401	1.63	.4484	.0516
0.54	.2054	.2946	1.09	.3621	.1379	1.64	.4495	.0505

(continued)

TABLE C (*continued*)

A	B	C	A	B	C	A	B	C
	Area between mean and	Area beyond		Area between mean and	Area beyond		Area between mean and	Area beyond
z	z	z	z	z	z	z	z	z
1.65	.4505	.0495	2.22	.4868	.0132	2.79	.4974	.0026
1.66	.4515	.0485	2.23	.4871	.0129	2.80	.4974	.0026
1.67	.4525	.0475	2.24	.4875	.0125	2.81	.4975	.0025
1.68	.4535	.0465	2.25	.4878	.0122	2.82	.4976	.0024
1.69	.4545	.0455	2.26	.4881	.0119	2.83	.4977	.0023
1.70	.4554	.0446	2.27	.4884	.0116	2.84	.4977	.0023
1.71	.4564	.0436	2.28	.4887	.0113	2.85	.4978	.0022
1.72	.4573	.0427	2.29	.4890	.0110	2.86	.4979	.0021
1.73	.4582	.0418	2.30	.4893	.0107	2.87	.4979	.0021
1.74	.4591	.0409	2.31	.4896	.0104	2.88	.4980	.0020
1.75	.4599	.0401	2.32	.4898	.0102	2.89	.4981	.0019
1.76	.4608	.0392	2.33	.4901	.0099	2.90	.4981	.0019
1.77	.4616	.0384	2.34	.4904	.0096	2.91	.4982	.0018
1.78	.4625	.0375	2.35	.4906	.0094	2.92	.4982	.0018
1.79	.4633	.0367	2.36	.4909	.0091	2.93	.4983	.0017
1.80	.4641	.0359	2.37	.4911	.0089	2.94	.4984	.0016
1.81	.4649	.0351	2.38	.4913	.0087	2.95	.4984	.0016
1.82	.4656	.0344	2.39	.4916	.0084	2.96	.4985	.0015
1.83	.4664	.0336	2.40	.4918	.0082	2.97	.4985	.0015
1.84	.4671	.0329	2.41	.4920	.0080	2.98	.4986	.0014
1.85	.4678	.0322	2.42	.4922	.0078	2.99	.4986	.0014
1.86	.4686	.0314	2.43	.4925	.0075	3.00	.4987	.0013
1.87	.4693	.0307	2.44	.4927	.0073	3.01	.4987	.0013
1.88	.4699	.0301	2.45	.4929	.0071	3.02	.4987	.0013
1.89	.4706	.0294	2.46	.4931	.0069	3.03	.4988	.0012
1.90	.4713	.0287	2.47	.4932	.0068	3.04	.4988	.0012
1.91	.4719	.0281	2.48	.4934	.0066	3.05	.4989	.0011
1.92	.4726	.0274	2.49	.4936	.0064	3.06	.4989	.0011
1.93	.4732	.0268	2.50	.4938	.0062	3.07	.4989	.0011
1.94	.4738	.0262	2.51	.4940	.0060	3.08	.4990	.0010
1.95	.4744	.0256	2.52	.4941	.0059	3.09	.4990	.0010
1.96	.4750	.0250	2.53	.4943	.0057	3.10	.4990	.0010
1.97	.4756	.0244	2.54	.4945	.0055	3.11	.4991	.0009
1.98	.4761	.0239	2.55	.4946	.0054	3.12	.4991	.0009
1.99	.4767	.0233	2.56	.4948	.0052	3.13	.4991	.0009
2.00	.4772	.0228	2.57	.4949	.0051	3.14	.4992	.0008
2.01	.4778	.0222	2.58	.4951	.0049	3.15	.4992	.0008
2.02	.4783	.0217	2.59	.4952	.0048	3.16	.4992	.0008
2.03	.4788	.0212	2.60	.4953	.0047	3.17	.4992	.0008
2.04	.4793	.0207	2.61	.4955	.0045	3.18	.4993	.0007
2.05	.4798	.0202	2.62	.4956	.0044	3.19	.4993	.0007
2.06	.4803	.0197	2.63	.4957	.0043	3.20	.4993	.0007
2.07	.4808	.0192	2.64	.4959	.0041	3.21	.4993	.0007
2.08	.4812	.0188	2.65	.4960	.0040	3.22	.4994	.0006
2.09	.4817	.0183	2.66	.4961	.0039	3.23	.4994	.0006
2.10	.4821	.0179	2.67	.4962	.0038	3.24	.4994	.0006
2.11	.4826	.0174	2.68	.4963	.0037	3.25	.4994	.0006
2.12	.4830	.0170	2.69	.4964	.0036	3.30	.4995	.0005
2.13	.4834	.0166	2.70	.4965	.0035	3.35	.4996	.0004
2.14	.4838	.0162	2.71	.4966	.0034	3.40	.4997	.0003
2.15	.4842	.0158	2.72	.4967	.0033	3.45	.4997	.0003
2.16	.4846	.0154	2.73	.4968	.0032	3.50	.4998	.0002
2.17	.4850	.0150	2.74	.4969	.0031	3.60	.4998	.0002
2.18	.4854	.0146	2.75	.4970	.0030	3.70	.4999	.0001
2.19	.4857	.0143	2.76	.4971	.0029	3.80	.4999	.0001
2.20	.4861	.0139	2.77	.4972	.0028	3.90	.49995	.00005
2.21	.4864	.0136	2.78	.4973	.0027	4.00	.49997	.00003

T A B L E   D   The *t* distribution*

df	Confidence interval percents (two-tailed)					
	80%	90%	95%	98%	99%	99.9%
	α level for two-tailed test					
	.20	.10	.05	.02	.01	.001
	α level for one-tailed test					
	.10	.05	.025	.01	.005	.0005
1	3.078	6.314	12.706	31.821	63.657	636.619
2	1.886	2.920	4.303	6.965	9.925	31.598
3	1.638	2.353	3.182	4.541	5.841	12.924
4	1.533	2.132	2.776	3.747	4.604	8.610
5	1.476	2.015	2.571	3.365	4.032	6.869
6	1.440	1.943	2.447	3.143	3.707	5.959
7	1.415	1.895	2.365	2.998	3.499	5.408
8	1.397	1.860	2.306	2.896	3.355	5.041
9	1.383	1.833	2.262	2.821	3.250	4.781
10	1.372	1.812	2.228	2.764	3.169	4.587
11	1.363	1.796	2.201	2.718	3.106	4.437
12	1.356	1.782	2.179	2.681	3.055	4.318
13	1.350	1.771	2.160	2.650	3.012	4.221
14	1.345	1.761	2.145	2.624	2.977	4.140
15	1.341	1.753	2.131	2.602	2.947	4.073
16	1.337	1.746	2.120	2.583	2.921	4.015
17	1.333	1.740	2.110	2.567	2.898	3.965
18	1.330	1.734	2.101	2.552	2.878	3.922
19	1.328	1.729	2.093	2.539	2.861	3.883
20	1.325	1.725	2.086	2.528	2.845	3.850
21	1.323	1.721	2.080	2.518	2.831	3.819
22	1.321	1.717	2.074	2.508	2.819	3.792
23	1.319	1.714	2.069	2.500	2.807	3.767
24	1.318	1.711	2.064	2.492	2.797	3.745
25	1.316	1.708	2.060	2.485	2.787	3.725
26	1.315	1.706	2.056	2.479	2.779	3.707
27	1.314	1.703	2.052	2.474	2.771	3.690
28	1.313	1.701	2.048	2.467	2.763	3.674
29	1.311	1.699	2.045	2.462	2.756	3.659
30	1.310	1.697	2.042	2.457	2.750	3.646
40	1.303	1.684	2.021	2.423	2.704	3.551
60	1.296	1.671	2.000	2.390	2.660	3.460
120	1.289	1.658	1.980	2.358	2.617	3.373
∞	1.282	1.645	1.960	2.326	2.576	3.291

* To be significant the *t* obtained from the data must be equal to or **larger than** the value shown in the table.

SOURCE: From Table III in R. A. Fisher and F. Yates, *Statistical Tables for Biological, Agricultural, and Medical Research*, Sixth Edition, published by Addison Wesley Longman Ltd., (1974). Reprinted with permission.

TABLE E   Chi square distribution*

df	.10	.05	.02	.01	.001
			α levels		
1	2.71	3.84	5.41	6.64	10.83
2	4.60	5.99	7.82	9.21	13.82
3	6.25	7.82	9.84	11.34	16.27
4	7.78	9.49	11.67	13.28	18.46
5	9.24	11.07	13.39	15.09	20.52
6	10.64	12.59	15.03	16.81	22.46
7	12.02	14.07	16.62	18.48	24.32
8	13.36	15.51	18.17	20.09	26.12
9	14.68	16.92	19.68	21.67	27.88
10	15.99	18.31	21.16	23.21	29.59
11	17.28	19.68	22.62	24.72	31.26
12	18.55	21.03	24.05	26.22	32.91
13	19.81	22.36	25.47	27.69	34.53
14	21.06	23.68	26.87	29.14	36.12
15	22.31	25.00	28.26	30.58	37.70
16	23.54	26.30	29.63	32.00	39.25
17	24.77	27.59	31.00	33.41	40.79
18	25.99	28.87	32.35	34.80	42.31
19	27.20	30.14	33.69	36.19	43.82
20	28.41	31.41	35.02	37.57	45.32
21	29.62	32.67	36.34	38.93	46.80
22	30.81	33.92	37.66	40.29	48.27
23	32.01	35.17	38.97	41.64	49.73
24	33.20	36.42	40.27	42.98	51.18
25	34.38	37.65	41.57	44.31	52.62
26	35.56	38.88	42.86	45.64	54.05
27	36.74	40.11	44.14	46.96	55.48
28	37.92	41.34	45.42	48.28	56.89
29	39.09	42.56	46.69	49.59	58.30
30	40.26	43.77	47.96	50.89	59.70

* To be significant the χ^2 obtained from the data must be equal to or **larger than** the value shown in the table.
SOURCE: From Table IV in R. A. Fisher and F. Yates, *Statistical Tables for Biological, Agricultural, and Medical Research*, Sixth Edition, published by Addison Wesley Longman Ltd., (1974). Reprinted with permission.

TABLE F The F distribution*

α levels of .05 (lightface) and .01 (**boldface**) for the distribution of F

Degrees of freedom (for the numerator of F ratio) — b/w group

Degrees of freedom (for the denominator of F ratio) — w/i group

df	1	2	3	4	5	6	7	8	9	10	11	12	14	16	20	24	30	40	50	75	100	200	500	∞
1	161	200	216	225	230	234	237	239	241	242	243	244	245	246	248	249	250	251	252	253	253	254	254	254
	4,052	**4,999**	**5,403**	**5,625**	**5,764**	**5,859**	**5,928**	**5,981**	**6,022**	**6,056**	**6,082**	**6,106**	**6,142**	**6,169**	**6,208**	**6,234**	**6,258**	**6,286**	**6,302**	**6,323**	**6,334**	**6,352**	**6,361**	**6,366**
2	18.51	19.00	19.16	19.25	19.30	19.33	19.36	19.37	19.38	19.39	19.40	19.41	19.42	19.43	19.44	19.45	19.46	19.47	19.47	19.48	19.49	19.49	19.50	19.50
	98.49	**99.00**	**99.17**	**99.25**	**99.30**	**99.33**	**99.34**	**99.36**	**99.38**	**99.40**	**99.41**	**99.42**	**99.43**	**99.44**	**99.45**	**99.46**	**99.47**	**99.48**	**99.48**	**99.49**	**99.49**	**99.49**	**99.50**	**99.50**
3	10.13	9.55	9.28	9.12	9.01	8.94	8.88	8.84	8.81	8.78	8.76	8.74	8.71	8.69	8.66	8.64	8.62	8.60	8.58	8.57	8.56	8.54	8.54	8.53
	34.12	**30.82**	**29.46**	**28.71**	**28.24**	**27.91**	**27.67**	**27.49**	**27.34**	**27.23**	**27.13**	**27.05**	**26.92**	**26.83**	**26.69**	**26.60**	**26.50**	**26.41**	**26.35**	**26.27**	**26.23**	**26.18**	**26.14**	**26.12**
4	7.71	6.94	6.59	6.39	6.26	6.16	6.09	6.04	6.00	5.96	5.93	5.91	5.87	5.84	5.80	5.77	5.74	5.71	5.70	5.68	5.66	5.66	5.64	5.63
	21.20	**18.00**	**16.69**	**15.98**	**15.52**	**15.21**	**14.98**	**14.80**	**14.66**	**14.54**	**14.45**	**14.37**	**14.24**	**14.15**	**14.02**	**13.93**	**13.83**	**13.74**	**13.69**	**13.61**	**13.57**	**13.52**	**13.48**	**13.46**
5	6.61	5.79	5.41	5.19	5.05	4.95	4.88	4.82	4.78	4.74	4.70	4.68	4.64	4.60	4.56	4.53	4.50	4.46	4.44	4.42	4.40	4.38	4.37	4.36
	16.26	**13.27**	**12.06**	**11.39**	**10.97**	**10.67**	**10.45**	**10.27**	**10.15**	**10.05**	**9.96**	**9.89**	**9.77**	**9.68**	**9.55**	**9.47**	**9.38**	**9.29**	**9.24**	**9.17**	**9.13**	**9.07**	**9.04**	**9.02**
6	5.99	5.14	4.76	4.53	4.39	4.28	4.21	4.15	4.10	4.06	4.03	4.00	3.96	3.92	3.87	3.84	3.81	3.77	3.75	3.72	3.71	3.69	3.68	3.67
	13.74	**10.92**	**9.78**	**9.15**	**8.75**	**8.47**	**8.26**	**8.10**	**7.98**	**7.87**	**7.79**	**7.72**	**7.60**	**7.52**	**7.39**	**7.31**	**7.23**	**7.14**	**7.09**	**7.02**	**6.99**	**6.94**	**6.90**	**6.88**
7	5.59	4.74	4.35	4.12	3.97	3.87	3.79	3.73	3.68	3.63	3.60	3.57	3.52	3.49	3.44	3.41	3.38	3.34	3.32	3.29	3.28	3.25	3.24	3.23
	12.25	**9.55**	**8.45**	**7.85**	**7.46**	**7.19**	**7.00**	**6.84**	**6.71**	**6.62**	**6.54**	**6.47**	**6.35**	**6.27**	**6.15**	**6.07**	**5.98**	**5.90**	**5.85**	**5.78**	**5.75**	**5.70**	**5.67**	**5.65**
8	5.32	4.46	4.07	3.84	3.69	3.58	3.50	3.44	3.39	3.34	3.31	3.28	3.23	3.20	3.15	3.12	3.08	3.05	3.03	3.00	2.98	2.96	2.94	2.93
	11.26	**8.65**	**7.59**	**7.01**	**6.63**	**6.37**	**6.19**	**6.03**	**5.91**	**5.82**	**5.74**	**5.67**	**5.56**	**5.48**	**5.36**	**5.28**	**5.20**	**5.11**	**5.06**	**5.00**	**4.96**	**4.91**	**4.88**	**4.86**
9	5.12	4.26	3.86	3.63	3.48	3.37	3.29	3.23	3.18	3.13	3.10	3.07	3.02	2.98	2.93	2.90	2.86	2.82	2.80	2.77	2.76	2.73	2.72	2.71
	10.56	**8.02**	**6.99**	**6.42**	**6.06**	**5.80**	**5.62**	**5.47**	**5.35**	**5.26**	**5.18**	**5.11**	**5.00**	**4.92**	**4.80**	**4.73**	**4.64**	**4.56**	**4.51**	**4.45**	**4.41**	**4.36**	**4.33**	**4.31**
10	4.96	4.10	3.71	3.48	3.33	3.22	3.14	3.07	3.02	2.97	2.94	2.91	2.86	2.82	2.77	2.74	2.70	2.67	2.64	2.61	2.59	2.56	2.55	2.54
	10.04	**7.56**	**6.55**	**5.99**	**5.64**	**5.39**	**5.21**	**5.06**	**4.95**	**4.85**	**4.78**	**4.71**	**4.60**	**4.52**	**4.41**	**4.33**	**4.25**	**4.17**	**4.12**	**4.05**	**4.01**	**3.96**	**3.93**	**3.91**
11	4.84	3.98	3.59	3.36	3.20	3.09	3.01	2.95	2.90	2.86	2.82	2.79	2.74	2.70	2.65	2.61	2.57	2.53	2.50	2.47	2.45	2.42	2.41	2.40
	9.65	**7.20**	**6.22**	**5.67**	**5.32**	**5.07**	**4.88**	**4.74**	**4.63**	**4.54**	**4.46**	**4.40**	**4.29**	**4.21**	**4.10**	**4.02**	**3.94**	**3.86**	**3.80**	**3.74**	**3.70**	**3.66**	**3.62**	**3.60**
12	4.75	3.88	3.49	3.26	3.11	3.00	2.92	2.85	2.80	2.76	2.72	2.69	2.64	2.60	2.54	2.50	2.46	2.42	2.40	2.36	2.35	2.32	2.31	2.30
	9.33	**6.93**	**5.95**	**5.41**	**5.06**	**4.82**	**4.65**	**4.50**	**4.39**	**4.30**	**4.22**	**4.16**	**4.05**	**3.98**	**3.86**	**3.78**	**3.70**	**3.61**	**3.56**	**3.49**	**3.46**	**3.41**	**3.38**	**3.36**
13	4.67	3.80	3.41	3.18	3.02	2.92	2.84	2.77	2.72	2.67	2.63	2.60	2.55	2.51	2.46	2.42	2.38	2.34	2.32	2.28	2.26	2.24	2.22	2.21
	9.07	**6.70**	**5.74**	**5.20**	**4.86**	**4.62**	**4.44**	**4.30**	**4.19**	**4.10**	**4.02**	**3.96**	**3.85**	**3.78**	**3.67**	**3.59**	**3.51**	**3.42**	**3.37**	**3.30**	**3.27**	**3.21**	**3.18**	**3.16**

(continued)

* To be significant the F obtained from the data must be equal to or **larger than** the value shown in the table.
SOURCE: From *Statistical Methods*, by G. W. Snedecor and W. G. Cochran, Seventh Edition. Copyright © 1980 Iowa State University Press. Reprinted with permission.

TABLE F (continued)

Degrees of freedom (for the denominator of F ratio) — rows; Degrees of freedom (for the numerator of F ratio) — columns. In each cell the upper (light) value is the .05 level and the lower (**bold**) value is the .01 level.

	1	2	3	4	5	6	7	8	9	10	11	12	14	16	20	24	30	40	50	75	100	200	500	∞
14	4.60 **8.86**	3.74 **6.51**	3.34 **5.56**	3.11 **5.03**	2.96 **4.69**	2.85 **4.46**	2.77 **4.28**	2.70 **4.14**	2.65 **4.03**	2.60 **3.94**	2.56 **3.86**	2.53 **3.80**	2.48 **3.70**	2.44 **3.62**	2.39 **3.51**	2.35 **3.43**	2.31 **3.34**	2.27 **3.26**	2.24 **3.21**	2.21 **3.14**	2.19 **3.11**	2.16 **3.06**	2.14 **3.02**	2.13 **3.00**
15	4.54 **8.68**	3.68 **6.36**	3.29 **5.42**	3.06 **4.89**	2.90 **4.56**	2.79 **4.32**	2.70 **4.14**	2.64 **4.00**	2.59 **3.89**	2.55 **3.80**	2.51 **3.73**	2.48 **3.67**	2.43 **3.56**	2.39 **3.48**	2.33 **3.36**	2.29 **3.29**	2.25 **3.20**	2.21 **3.12**	2.18 **3.07**	2.15 **3.00**	2.12 **2.97**	2.10 **2.92**	2.08 **2.89**	2.07 **2.87**
16	4.49 **8.53**	3.63 **6.23**	3.24 **5.29**	3.01 **4.77**	2.85 **4.44**	2.74 **4.20**	2.66 **4.03**	2.59 **3.89**	2.54 **3.78**	2.49 **3.69**	2.45 **3.61**	2.42 **3.55**	2.37 **3.45**	2.33 **3.37**	2.28 **3.25**	2.24 **3.18**	2.20 **3.10**	2.16 **3.01**	2.13 **2.96**	2.09 **2.89**	2.07 **2.86**	2.04 **2.80**	2.02 **2.77**	2.01 **2.75**
17	4.45 **8.40**	3.59 **6.11**	3.20 **5.18**	2.96 **4.67**	2.81 **4.34**	2.70 **4.10**	2.62 **3.93**	2.55 **3.79**	2.50 **3.68**	2.45 **3.59**	2.41 **3.52**	2.38 **3.45**	2.33 **3.35**	2.29 **3.27**	2.23 **3.16**	2.19 **3.08**	2.15 **3.00**	2.11 **2.92**	2.08 **2.86**	2.04 **2.79**	2.02 **2.76**	1.99 **2.70**	1.97 **2.67**	1.96 **2.65**
18	4.41 **8.28**	3.55 **6.01**	3.16 **5.09**	2.93 **4.58**	2.77 **4.25**	2.66 **4.01**	2.58 **3.85**	2.51 **3.71**	2.46 **3.60**	2.41 **3.51**	2.37 **3.44**	2.34 **3.37**	2.29 **3.27**	2.25 **3.19**	2.19 **3.07**	2.15 **3.00**	2.11 **2.91**	2.07 **2.83**	2.04 **2.78**	2.00 **2.71**	1.98 **2.68**	1.95 **2.62**	1.93 **2.59**	1.92 **2.57**
19	4.38 **8.18**	3.52 **5.93**	3.13 **5.01**	2.90 **4.50**	2.74 **4.17**	2.63 **3.94**	2.55 **3.77**	2.48 **3.63**	2.43 **3.52**	2.38 **3.43**	2.34 **3.36**	2.31 **3.30**	2.26 **3.19**	2.21 **3.12**	2.15 **3.00**	2.11 **2.92**	2.07 **2.84**	2.02 **2.76**	2.00 **2.70**	1.96 **2.63**	1.94 **2.60**	1.91 **2.54**	1.90 **2.51**	1.88 **2.49**
20	4.35 **8.10**	3.49 **5.85**	3.10 **4.94**	2.87 **4.43**	2.71 **4.10**	2.60 **3.87**	2.52 **3.71**	2.45 **3.56**	2.40 **3.45**	2.35 **3.37**	2.31 **3.30**	2.28 **3.23**	2.23 **3.13**	2.18 **3.05**	2.12 **2.94**	2.08 **2.86**	2.04 **2.77**	1.99 **2.69**	1.96 **2.63**	1.92 **2.56**	1.90 **2.53**	1.87 **2.47**	1.85 **2.44**	1.84 **2.42**
21	4.32 **8.02**	3.47 **5.78**	3.07 **4.87**	2.84 **4.37**	2.68 **4.04**	2.57 **3.81**	2.49 **3.65**	2.42 **3.51**	2.37 **3.40**	2.32 **3.31**	2.28 **3.24**	2.25 **3.17**	2.20 **3.07**	2.15 **2.99**	2.09 **2.88**	2.05 **2.80**	2.00 **2.72**	1.96 **2.63**	1.93 **2.58**	1.89 **2.51**	1.87 **2.47**	1.84 **2.42**	1.82 **2.38**	1.81 **2.36**
22	4.30 **7.94**	3.44 **5.72**	3.05 **4.82**	2.82 **4.31**	2.66 **3.99**	2.55 **3.76**	2.47 **3.59**	2.40 **3.45**	2.35 **3.35**	2.30 **3.26**	2.26 **3.18**	2.23 **3.12**	2.18 **3.02**	2.13 **2.94**	2.07 **2.83**	2.03 **2.75**	1.98 **2.67**	1.93 **2.58**	1.91 **2.53**	1.87 **2.46**	1.84 **2.42**	1.81 **2.37**	1.80 **2.33**	1.78 **2.31**
23	4.28 **7.88**	3.42 **5.66**	3.03 **4.76**	2.80 **4.26**	2.64 **3.94**	2.53 **3.71**	2.45 **3.54**	2.38 **3.41**	2.32 **3.30**	2.28 **3.21**	2.24 **3.14**	2.20 **3.07**	2.14 **2.97**	2.10 **2.89**	2.04 **2.78**	2.00 **2.70**	1.96 **2.62**	1.91 **2.53**	1.88 **2.48**	1.84 **2.41**	1.82 **2.37**	1.79 **2.32**	1.77 **2.28**	1.76 **2.26**
24	4.26 **7.82**	3.40 **5.61**	3.01 **4.72**	2.78 **4.22**	2.62 **3.90**	2.51 **3.67**	2.43 **3.50**	2.36 **3.36**	2.30 **3.25**	2.26 **3.17**	2.22 **3.09**	2.18 **3.03**	2.13 **2.93**	2.09 **2.85**	2.02 **2.74**	1.98 **2.66**	1.94 **2.58**	1.89 **2.49**	1.86 **2.44**	1.82 **2.36**	1.80 **2.33**	1.76 **2.27**	1.74 **2.23**	1.73 **2.21**
25	4.24 **7.77**	3.38 **5.57**	2.99 **4.68**	2.76 **4.18**	2.60 **3.86**	2.49 **3.63**	2.41 **3.46**	2.34 **3.32**	2.28 **3.21**	2.24 **3.13**	2.20 **3.05**	2.16 **2.99**	2.11 **2.89**	2.06 **2.81**	2.00 **2.70**	1.96 **2.62**	1.92 **2.54**	1.87 **2.45**	1.84 **2.40**	1.80 **2.32**	1.77 **2.29**	1.74 **2.23**	1.72 **2.19**	1.71 **2.17**
26	4.22 **7.72**	3.37 **5.53**	2.98 **4.64**	2.74 **4.14**	2.59 **3.82**	2.47 **3.59**	2.39 **3.42**	2.32 **3.29**	2.27 **3.17**	2.22 **3.09**	2.18 **3.02**	2.15 **2.96**	2.10 **2.86**	2.05 **2.77**	1.99 **2.66**	1.95 **2.58**	1.90 **2.50**	1.85 **2.41**	1.82 **2.36**	1.78 **2.28**	1.76 **2.25**	1.72 **2.19**	1.70 **2.15**	1.69 **2.13**

(continued)

T A B L E F *(continued)*

Degrees of freedom (for the numerator of *F* ratio)

	1	2	3	4	5	6	7	8	9	10	11	12	14	16	20	24	30	40	50	75	100	200	500	∞	
27	4.21 **7.68**	3.35 **5.49**	2.96 **4.60**	2.73 **4.11**	2.57 **3.79**	2.46 **3.56**	2.37 **3.39**	2.30 **3.26**	2.25 **3.14**	2.20 **3.06**	2.16 **2.98**	2.13 **2.93**	2.08 **2.83**	2.03 **2.74**	1.97 **2.63**	1.93 **2.55**	1.88 **2.47**	1.84 **2.38**	1.80 **2.33**	1.76 **2.25**	1.74 **2.21**	1.71 **2.16**	1.68 **2.12**	1.67 **2.10**	27
28	4.20 **7.64**	3.34 **5.45**	2.95 **4.57**	2.71 **4.07**	2.56 **3.76**	2.44 **3.53**	2.36 **3.36**	2.29 **3.23**	2.24 **3.11**	2.19 **3.03**	2.15 **2.95**	2.12 **2.90**	2.06 **2.80**	2.02 **2.71**	1.96 **2.60**	1.91 **2.52**	1.87 **2.44**	1.81 **2.35**	1.78 **2.30**	1.75 **2.22**	1.72 **2.18**	1.69 **2.13**	1.67 **2.09**	1.65 **2.06**	28
29	4.18 **7.60**	3.33 **5.42**	2.93 **4.54**	2.70 **4.04**	2.54 **3.73**	2.43 **3.50**	2.35 **3.33**	2.28 **3.20**	2.22 **3.08**	2.18 **3.00**	2.14 **2.92**	2.10 **2.87**	2.05 **2.77**	2.00 **2.68**	1.94 **2.57**	1.90 **2.49**	1.85 **2.41**	1.80 **2.32**	1.77 **2.27**	1.73 **2.19**	1.71 **2.15**	1.68 **2.10**	1.65 **2.06**	1.64 **2.03**	29
30	4.17 **7.56**	3.32 **5.39**	2.92 **4.51**	2.69 **4.02**	2.53 **3.70**	2.42 **3.47**	2.34 **3.30**	2.27 **3.17**	2.21 **3.06**	2.16 **2.98**	2.12 **2.90**	2.09 **2.84**	2.04 **2.74**	1.99 **2.66**	1.93 **2.55**	1.89 **2.47**	1.84 **2.38**	1.79 **2.29**	1.76 **2.24**	1.72 **2.16**	1.69 **2.13**	1.66 **2.07**	1.64 **2.03**	1.62 **2.01**	30
32	4.15 **7.50**	3.30 **5.34**	2.90 **4.46**	2.67 **3.97**	2.51 **3.66**	2.40 **3.42**	2.32 **3.25**	2.25 **3.12**	2.19 **3.01**	2.14 **2.94**	2.10 **2.86**	2.07 **2.80**	2.02 **2.70**	1.97 **2.62**	1.91 **2.51**	1.86 **2.42**	1.82 **2.34**	1.76 **2.25**	1.74 **2.20**	1.69 **2.12**	1.67 **2.08**	1.64 **2.02**	1.61 **1.98**	1.59 **1.96**	32
34	4.13 **7.44**	3.28 **5.29**	2.88 **4.42**	2.65 **3.93**	2.49 **3.61**	2.38 **3.38**	2.30 **3.21**	2.23 **3.08**	2.17 **2.97**	2.12 **2.89**	2.08 **2.82**	2.05 **2.76**	2.00 **2.66**	1.95 **2.58**	1.89 **2.47**	1.84 **2.38**	1.80 **2.30**	1.74 **2.21**	1.71 **2.15**	1.67 **2.08**	1.64 **2.04**	1.61 **1.98**	1.59 **1.94**	1.57 **1.91**	34
36	4.11 **7.39**	3.26 **5.25**	2.86 **4.38**	2.63 **3.89**	2.48 **3.58**	2.36 **3.35**	2.28 **3.18**	2.21 **3.04**	2.15 **2.94**	2.10 **2.86**	2.06 **2.78**	2.03 **2.72**	1.98 **2.62**	1.93 **2.54**	1.87 **2.43**	1.82 **2.35**	1.78 **2.26**	1.72 **2.17**	1.69 **2.12**	1.65 **2.04**	1.62 **2.00**	1.59 **1.94**	1.56 **1.90**	1.55 **1.87**	36
38	4.10 **7.35**	3.25 **5.21**	2.85 **4.34**	2.62 **3.86**	2.46 **3.54**	2.35 **3.32**	2.26 **3.15**	2.19 **3.02**	2.14 **2.91**	2.09 **2.82**	2.05 **2.75**	2.02 **2.69**	1.96 **2.59**	1.92 **2.51**	1.85 **2.40**	1.80 **2.32**	1.76 **2.22**	1.71 **2.14**	1.67 **2.08**	1.63 **2.00**	1.60 **1.97**	1.57 **1.90**	1.54 **1.86**	1.53 **1.84**	38
40	4.08 **7.31**	3.23 **5.18**	2.84 **4.31**	2.61 **3.83**	2.45 **3.51**	2.34 **3.29**	2.25 **3.12**	2.18 **2.99**	2.12 **2.88**	2.07 **2.80**	2.04 **2.73**	2.00 **2.66**	1.95 **2.56**	1.90 **2.49**	1.84 **2.37**	1.79 **2.29**	1.74 **2.20**	1.69 **2.11**	1.66 **2.05**	1.61 **1.97**	1.59 **1.94**	1.55 **1.88**	1.53 **1.84**	1.51 **1.81**	40
42	4.07 **7.27**	3.22 **5.15**	2.83 **4.29**	2.59 **3.80**	2.44 **3.49**	2.32 **3.26**	2.24 **3.10**	2.17 **2.96**	2.11 **2.86**	2.06 **2.77**	2.02 **2.70**	1.99 **2.64**	1.94 **2.54**	1.89 **2.46**	1.82 **2.35**	1.78 **2.26**	1.73 **2.17**	1.68 **2.08**	1.64 **2.02**	1.60 **1.94**	1.57 **1.91**	1.54 **1.85**	1.51 **1.80**	1.49 **1.78**	42
44	4.06 **7.24**	3.21 **5.12**	2.82 **4.26**	2.58 **3.78**	2.43 **3.46**	2.31 **3.24**	2.23 **3.07**	2.16 **2.94**	2.10 **2.84**	2.05 **2.75**	2.01 **2.68**	1.98 **2.62**	1.92 **2.52**	1.88 **2.44**	1.81 **2.32**	1.76 **2.24**	1.72 **2.15**	1.66 **2.06**	1.63 **2.00**	1.58 **1.92**	1.56 **1.88**	1.52 **1.82**	1.50 **1.78**	1.48 **1.75**	44
46	4.05 **7.21**	3.20 **5.10**	2.81 **4.24**	2.57 **3.76**	2.42 **3.44**	2.30 **3.22**	2.22 **3.05**	2.14 **2.92**	2.09 **2.82**	2.04 **2.73**	2.00 **2.66**	1.97 **2.60**	1.91 **2.50**	1.87 **2.42**	1.80 **2.30**	1.75 **2.22**	1.71 **2.13**	1.65 **2.04**	1.62 **1.98**	1.57 **1.90**	1.54 **1.86**	1.51 **1.80**	1.48 **1.76**	1.46 **1.72**	46
48	4.04 **7.19**	3.19 **5.08**	2.80 **4.22**	2.56 **3.74**	2.41 **3.42**	2.30 **3.20**	2.21 **3.04**	2.14 **2.90**	2.08 **2.80**	2.03 **2.71**	1.99 **2.64**	1.96 **2.58**	1.90 **2.48**	1.86 **2.40**	1.79 **2.28**	1.74 **2.20**	1.70 **2.11**	1.64 **2.02**	1.61 **1.96**	1.56 **1.88**	1.53 **1.84**	1.50 **1.78**	1.47 **1.73**	1.45 **1.70**	48

Degrees of freedom (for the denominator of *F* ratio)

(continued)

TABLE F (continued)

Degrees of freedom (for the numerator of F ratio)

df₂	1	2	3	4	5	6	7	8	9	10	11	12	14	16	20	24	30	40	50	75	100	200	500	∞
50	4.03 / **7.17**	3.18 / **5.06**	2.79 / **4.20**	2.56 / **3.72**	2.40 / **3.41**	2.29 / **3.18**	2.20 / **3.02**	2.13 / **2.88**	2.07 / **2.78**	2.02 / **2.70**	1.98 / **2.62**	1.95 / **2.56**	1.90 / **2.46**	1.85 / **2.39**	1.78 / **2.26**	1.74 / **2.18**	1.69 / **2.10**	1.63 / **2.00**	1.60 / **1.94**	1.55 / **1.86**	1.52 / **1.82**	1.48 / **1.76**	1.46 / **1.71**	1.44 / **1.68**
55	4.02 / **7.12**	3.17 / **5.01**	2.78 / **4.16**	2.54 / **3.68**	2.38 / **3.37**	2.27 / **3.15**	2.18 / **2.98**	2.11 / **2.85**	2.05 / **2.75**	2.00 / **2.66**	1.97 / **2.59**	1.93 / **2.53**	1.88 / **2.43**	1.83 / **2.35**	1.76 / **2.23**	1.72 / **2.15**	1.67 / **2.06**	1.61 / **1.96**	1.58 / **1.90**	1.52 / **1.82**	1.50 / **1.78**	1.46 / **1.71**	1.43 / **1.66**	1.41 / **1.64**
60	4.00 / **7.08**	3.15 / **4.98**	2.76 / **4.13**	2.52 / **3.65**	2.37 / **3.34**	2.25 / **3.12**	2.17 / **2.95**	2.10 / **2.82**	2.04 / **2.72**	1.99 / **2.63**	1.95 / **2.56**	1.92 / **2.50**	1.86 / **2.40**	1.81 / **2.32**	1.75 / **2.20**	1.70 / **2.12**	1.65 / **2.03**	1.59 / **1.93**	1.56 / **1.87**	1.50 / **1.79**	1.48 / **1.74**	1.44 / **1.68**	1.41 / **1.63**	1.39 / **1.60**
65	3.99 / **7.04**	3.14 / **4.95**	2.75 / **4.10**	2.51 / **3.62**	2.36 / **3.31**	2.24 / **3.09**	2.15 / **2.93**	2.08 / **2.79**	2.02 / **2.70**	1.98 / **2.61**	1.94 / **2.54**	1.90 / **2.47**	1.85 / **2.37**	1.80 / **2.30**	1.73 / **2.18**	1.68 / **2.09**	1.63 / **2.00**	1.57 / **1.90**	1.54 / **1.84**	1.49 / **1.76**	1.46 / **1.71**	1.42 / **1.64**	1.39 / **1.60**	1.37 / **1.56**
70	3.98 / **7.01**	3.13 / **4.92**	2.74 / **4.08**	2.50 / **3.60**	2.35 / **3.29**	2.23 / **3.07**	2.14 / **2.91**	2.07 / **2.77**	2.01 / **2.67**	1.97 / **2.59**	1.93 / **2.51**	1.89 / **2.45**	1.84 / **2.35**	1.79 / **2.28**	1.72 / **2.15**	1.67 / **2.07**	1.62 / **1.98**	1.56 / **1.88**	1.53 / **1.82**	1.47 / **1.74**	1.45 / **1.69**	1.40 / **1.62**	1.37 / **1.56**	1.35 / **1.53**
80	3.96 / **6.96**	3.11 / **4.88**	2.72 / **4.04**	2.48 / **3.56**	2.33 / **3.25**	2.21 / **3.04**	2.12 / **2.87**	2.05 / **2.74**	1.99 / **2.64**	1.95 / **2.55**	1.91 / **2.48**	1.88 / **2.41**	1.82 / **2.32**	1.77 / **2.24**	1.70 / **2.11**	1.65 / **2.03**	1.60 / **1.94**	1.54 / **1.84**	1.51 / **1.78**	1.45 / **1.70**	1.42 / **1.65**	1.38 / **1.57**	1.35 / **1.52**	1.32 / **1.49**
100	3.94 / **6.90**	3.09 / **4.82**	2.70 / **3.98**	2.46 / **3.51**	2.30 / **3.20**	2.19 / **2.99**	2.10 / **2.82**	2.03 / **2.69**	1.97 / **2.59**	1.92 / **2.51**	1.88 / **2.43**	1.85 / **2.36**	1.79 / **2.26**	1.75 / **2.19**	1.68 / **2.06**	1.63 / **1.98**	1.57 / **1.89**	1.51 / **1.79**	1.48 / **1.73**	1.42 / **1.64**	1.39 / **1.59**	1.34 / **1.51**	1.30 / **1.46**	1.28 / **1.43**
125	3.92 / **6.84**	3.07 / **4.78**	2.68 / **3.94**	2.44 / **3.47**	2.29 / **3.17**	2.17 / **2.95**	2.08 / **2.79**	2.01 / **2.65**	1.95 / **2.56**	1.90 / **2.47**	1.86 / **2.40**	1.83 / **2.33**	1.77 / **2.23**	1.72 / **2.15**	1.65 / **2.03**	1.60 / **1.94**	1.55 / **1.85**	1.49 / **1.75**	1.45 / **1.68**	1.39 / **1.59**	1.36 / **1.54**	1.31 / **1.46**	1.27 / **1.40**	1.25 / **1.37**
150	3.91 / **6.81**	3.06 / **4.75**	2.67 / **3.91**	2.43 / **3.44**	2.27 / **3.14**	2.16 / **2.92**	2.07 / **2.76**	2.00 / **2.62**	1.94 / **2.53**	1.89 / **2.44**	1.85 / **2.37**	1.82 / **2.30**	1.76 / **2.20**	1.71 / **2.12**	1.64 / **2.00**	1.59 / **1.91**	1.54 / **1.83**	1.47 / **1.72**	1.44 / **1.66**	1.37 / **1.56**	1.34 / **1.51**	1.29 / **1.43**	1.25 / **1.37**	1.22 / **1.33**
200	3.89 / **6.76**	3.04 / **4.71**	2.65 / **3.88**	2.41 / **3.41**	2.26 / **3.11**	2.14 / **2.90**	2.05 / **2.73**	1.98 / **2.60**	1.92 / **2.50**	1.87 / **2.41**	1.83 / **2.34**	1.80 / **2.28**	1.74 / **2.17**	1.69 / **2.09**	1.62 / **1.97**	1.57 / **1.88**	1.52 / **1.79**	1.45 / **1.69**	1.42 / **1.62**	1.35 / **1.53**	1.32 / **1.48**	1.26 / **1.39**	1.22 / **1.33**	1.19 / **1.28**
400	3.86 / **6.70**	3.02 / **4.66**	2.62 / **3.83**	2.39 / **3.36**	2.23 / **3.06**	2.12 / **2.85**	2.03 / **2.69**	1.96 / **2.55**	1.90 / **2.46**	1.85 / **2.37**	1.81 / **2.29**	1.78 / **2.23**	1.72 / **2.12**	1.67 / **2.04**	1.60 / **1.92**	1.54 / **1.84**	1.49 / **1.74**	1.42 / **1.64**	1.38 / **1.57**	1.32 / **1.47**	1.28 / **1.42**	1.22 / **1.32**	1.16 / **1.24**	1.13 / **1.19**
1000	3.85 / **6.66**	3.00 / **4.62**	2.61 / **3.80**	2.38 / **3.34**	2.22 / **3.04**	2.10 / **2.82**	2.02 / **2.66**	1.95 / **2.53**	1.89 / **2.43**	1.84 / **2.34**	1.80 / **2.26**	1.76 / **2.20**	1.70 / **2.09**	1.65 / **2.01**	1.58 / **1.89**	1.53 / **1.81**	1.47 / **1.71**	1.41 / **1.61**	1.36 / **1.54**	1.30 / **1.44**	1.26 / **1.38**	1.19 / **1.28**	1.13 / **1.19**	1.08 / **1.11**
∞	3.84 / **6.64**	2.99 / **4.60**	2.60 / **3.78**	2.37 / **3.32**	2.21 / **3.02**	2.09 / **2.80**	2.01 / **2.64**	1.94 / **2.51**	1.88 / **2.41**	1.83 / **2.32**	1.79 / **2.24**	1.75 / **2.18**	1.69 / **2.07**	1.64 / **1.99**	1.57 / **1.87**	1.52 / **1.79**	1.46 / **1.69**	1.40 / **1.59**	1.35 / **1.52**	1.28 / **1.41**	1.24 / **1.36**	1.17 / **1.25**	1.11 / **1.15**	1.00 / **1.00**

Degrees of freedom (for the denominator of F ratio)

T A B L E G   Critical values of the studentized range statistic (for Tukey HSD tests)

$\alpha = .05$

df_{wg}	\multicolumn{14}{c}{Number of levels of the independent variable}

df_{wg}	2	3	4	5	6	7	8	9	10	11	12	13	14	15
1	17.97	26.98	32.82	37.08	40.41	43.12	45.40	47.36	49.07	50.59	51.96	53.20	54.33	55.36
2	6.08	8.33	9.80	10.88	11.74	12.44	13.03	13.54	13.99	14.39	14.75	15.08	15.38	15.65
3	4.50	5.91	6.82	7.50	8.04	8.48	8.85	9.18	9.46	9.72	9.95	10.15	10.35	10.53
4	3.93	5.04	5.76	6.29	6.71	7.05	7.35	7.60	7.83	8.03	8.21	8.37	8.52	8.66
5	3.64	4.60	5.22	5.67	6.03	6.33	6.58	6.80	7.00	7.17	7.32	7.47	7.60	7.72
6	3.46	4.34	4.90	5.31	5.63	5.90	6.12	6.32	6.49	6.65	6.79	6.92	7.03	7.14
7	3.34	4.16	4.68	5.06	5.36	5.61	5.82	6.00	6.16	6.30	6.43	6.55	6.66	6.76
8	3.26	4.04	4.53	4.89	5.17	5.40	5.60	5.77	5.92	6.05	6.18	6.29	6.39	6.48
9	3.20	3.95	4.42	4.76	5.02	5.24	5.43	5.60	5.74	5.87	5.98	6.09	6.19	6.28
10	3.15	3.88	4.33	4.65	4.91	5.12	5.30	5.46	5.60	5.72	5.83	5.94	6.03	6.11
11	3.11	3.82	4.26	4.57	4.82	5.03	5.20	5.35	5.49	5.60	5.71	5.81	5.90	5.98
12	3.08	3.77	4.20	4.51	4.75	4.95	5.12	5.26	5.40	5.51	5.62	5.71	5.79	5.88
13	3.06	3.74	4.15	4.45	4.69	4.88	5.05	5.19	5.32	5.43	5.53	5.63	5.71	5.79
14	3.03	3.70	4.11	4.41	4.64	4.83	4.99	5.13	5.25	5.36	5.46	5.55	5.64	5.71
15	3.01	3.67	4.08	4.37	4.60	4.78	4.94	5.08	5.20	5.31	5.40	5.49	5.57	5.65
16	3.00	3.65	4.05	4.33	4.56	4.74	4.90	5.03	5.15	5.26	5.35	5.44	5.52	5.59
17	2.98	3.63	4.02	4.30	4.52	4.70	4.86	4.99	5.11	5.21	5.31	5.39	5.47	5.54
18	2.97	3.61	4.00	4.28	4.50	4.67	4.82	4.96	5.07	5.17	5.27	5.35	5.43	5.50
19	2.96	3.59	3.98	4.25	4.47	4.64	4.79	4.92	5.04	5.14	5.23	5.32	5.39	5.46
20	2.95	3.58	3.96	4.23	4.44	4.62	4.77	4.90	5.01	5.11	5.20	5.28	5.36	5.43
24	2.92	3.53	3.90	4.17	4.37	4.54	4.68	4.81	4.92	5.01	5.10	5.18	5.25	5.32
30	2.89	3.49	3.84	4.10	4.30	4.46	4.60	4.72	4.82	4.92	5.00	5.08	5.15	5.21
40	2.86	3.44	3.79	4.04	4.23	4.39	4.52	4.64	4.74	4.82	4.90	4.98	5.04	5.11
60	2.83	3.40	3.74	3.98	4.16	4.31	4.44	4.55	4.65	4.73	4.81	4.88	4.94	5.00
120	2.80	3.36	3.69	3.92	4.10	4.24	4.36	4.47	4.56	4.64	4.71	4.78	4.84	4.90
∞	2.77	3.31	3.63	3.86	4.03	4.17	4.29	4.39	4.47	4.55	4.62	4.68	4.74	4.80

(*continued*)

SOURCE: From "Tables of Range and Studentized Range," by M. L. Harter, 1960, *Annals of Mathematical Statistics, 31*, 1122–1147. Copyright © 1960 The Institute of Mathematical Statistics. Reprinted with permission.

TABLE G (continued)

$\alpha = .01$

df_{wg}	\multicolumn{14}{c}{Number of levels of the independent variable}
	2	3	4	5	6	7	8	9	10	11	12	13	14	15
1	90.03	135.00	164.30	185.60	202.20	215.80	227.20	237.00	245.60	253.20	260.00	266.20	271.80	277.00
2	14.04	19.02	22.29	24.72	26.63	28.20	29.53	30.68	31.69	32.59	33.40	34.13	34.81	35.43
3	8.26	10.62	12.17	13.33	14.24	15.00	15.64	16.20	16.69	17.13	17.53	17.89	18.22	18.52
4	6.51	8.12	9.17	9.96	10.58	11.10	11.55	11.93	12.27	12.57	12.84	13.09	13.32	13.53
5	5.70	6.98	7.80	8.42	8.91	9.32	9.67	9.97	10.24	10.48	10.70	10.89	11.08	11.24
6	5.24	6.33	7.03	7.56	7.97	8.32	8.62	8.87	9.10	9.30	9.48	9.65	9.81	9.95
7	4.95	5.92	6.54	7.00	7.37	7.68	7.94	8.17	8.37	8.55	8.71	8.86	9.00	9.12
8	4.75	5.64	6.20	6.62	6.96	7.24	7.47	7.68	7.86	8.03	8.18	8.31	8.44	8.55
9	4.60	5.43	5.96	6.35	6.66	6.92	7.13	7.32	7.50	7.65	7.78	7.91	8.02	8.13
10	4.48	5.27	5.77	6.14	6.43	6.67	6.88	7.06	7.21	7.36	7.48	7.60	7.71	7.81
11	4.39	5.15	5.62	5.97	6.25	6.48	6.67	6.84	6.99	7.13	7.25	7.36	7.46	7.56
12	4.32	5.05	5.50	5.84	6.10	6.32	6.51	6.67	6.81	6.94	7.06	7.17	7.26	7.36
13	4.26	4.96	5.40	5.73	5.98	6.19	6.37	6.53	6.67	6.79	6.90	7.01	7.10	7.19
14	4.21	4.90	5.32	5.63	5.88	6.08	6.26	6.41	6.54	6.66	6.77	6.87	6.96	7.05
15	4.17	4.84	5.25	5.56	5.80	5.99	6.16	6.31	6.44	6.56	6.66	6.76	6.84	6.93
16	4.13	4.79	5.19	5.49	5.72	5.92	6.08	6.22	6.35	6.46	6.56	6.66	6.74	6.82
17	4.10	4.74	5.14	5.43	5.66	5.85	6.01	6.15	6.27	6.38	6.48	6.57	6.66	6.73
18	4.07	4.70	5.09	5.38	5.60	5.79	5.94	6.08	6.20	6.31	6.41	6.50	6.58	6.66
19	4.05	4.67	5.05	5.33	5.55	5.74	5.89	6.02	6.14	6.25	6.34	6.43	6.51	6.58
20	4.02	4.64	5.02	5.29	5.51	5.69	5.84	5.97	6.09	6.19	6.28	6.37	6.45	6.52
24	3.96	4.55	4.91	5.17	5.37	5.54	5.69	5.81	5.92	6.02	6.11	6.19	6.26	6.33
30	3.89	4.46	4.80	5.05	5.24	5.40	5.54	5.65	5.76	5.85	5.93	6.01	6.08	6.14
40	3.82	4.37	4.70	4.93	5.11	5.26	5.39	5.50	5.60	5.69	5.76	5.84	5.90	5.96
60	3.76	4.28	4.60	4.82	4.99	5.13	5.25	5.36	5.45	5.53	5.60	5.67	5.73	5.78
120	3.70	4.20	4.50	4.71	4.87	5.01	5.12	5.21	5.30	5.38	5.44	5.51	5.56	5.61
∞	3.64	4.12	4.40	4.60	4.76	4.88	4.99	5.08	5.16	5.23	5.29	5.35	5.40	5.45

TABLE H Critical values for the Mann-Whitney U test*

One-tailed test	Two-tailed test
$\alpha = .01$ (lightface)	$\alpha = .02$ (lightface)
$\alpha = .005$ (**boldface**)	$\alpha = .01$ (**boldface**)

N_2 \ N_1	1	2	3	4	5	6	7	8	9	10	11	12	13	14	15	16	17	18	19	20
1	—	—	—	—	—	—	—	—	—	—	—	—	—	—	—	—	—	—	—	—
2	—	—	—	—	—	—	—	—	—	—	—	—	0	0	0	0	0	0	1	1
	—	—	—	—	—	—	—	—	—	—	—	—	—	—	—	—	—	—	**0**	**0**
3	—	—	—	—	—	—	0	0	1	1	1	2	2	2	3	3	4	4	4	5
	—	—	—	—	—	—	—	—	**0**	**0**	**0**	**1**	**1**	**1**	**2**	**2**	**2**	**2**	**3**	**3**
4	—	—	—	—	0	1	1	2	3	3	4	5	5	6	7	7	8	9	9	10
	—	—	—	—	—	**0**	**0**	**1**	**1**	**2**	**2**	**3**	**3**	**4**	**5**	**5**	**6**	**6**	**7**	**8**
5	—	—	—	0	1	2	3	4	5	6	7	8	9	10	11	12	13	14	15	16
	—	—	—	—	**0**	**1**	**1**	**2**	**3**	**4**	**5**	**6**	**7**	**7**	**8**	**9**	**10**	**11**	**12**	**13**
6	—	—	—	1	2	3	4	6	7	8	9	11	12	13	15	16	18	19	20	22
	—	—	—	**0**	**1**	**2**	**3**	**4**	**5**	**6**	**7**	**9**	**10**	**11**	**12**	**13**	**15**	**16**	**17**	**18**
7	—	—	0	1	3	4	6	7	9	11	12	14	16	17	19	21	23	24	26	28
	—	—	—	**0**	**1**	**3**	**4**	**6**	**7**	**9**	**10**	**12**	**13**	**15**	**16**	**18**	**19**	**21**	**22**	**24**
8	—	—	0	2	4	6	7	9	11	13	15	17	20	22	24	26	28	30	32	34
	—	—	—	**1**	**2**	**4**	**6**	**7**	**9**	**11**	**13**	**15**	**17**	**18**	**20**	**22**	**24**	**26**	**29**	**30**
9	—	—	1	3	5	7	9	11	14	16	18	21	23	26	28	31	33	36	38	40
	—	—	**0**	**1**	**3**	**5**	**7**	**9**	**11**	**13**	**16**	**18**	**20**	**22**	**24**	**27**	**29**	**31**	**33**	**36**
10	—	—	1	3	6	8	11	13	16	19	22	24	27	30	33	36	38	41	44	47
	—	—	**0**	**2**	**4**	**6**	**9**	**11**	**13**	**16**	**18**	**21**	**24**	**26**	**29**	**31**	**34**	**37**	**39**	**42**
11	—	—	1	4	7	9	12	15	18	22	25	28	31	34	37	41	44	47	50	53
	—	—	**0**	**2**	**5**	**7**	**10**	**13**	**16**	**18**	**21**	**24**	**27**	**30**	**33**	**36**	**39**	**42**	**45**	**48**
12	—	—	2	5	8	11	14	17	21	24	28	31	35	38	42	46	49	53	56	60
	—	—	**1**	**3**	**6**	**9**	**12**	**15**	**18**	**21**	**24**	**27**	**31**	**34**	**37**	**41**	**44**	**47**	**51**	**54**
13	—	0	2	5	9	12	16	20	23	27	31	35	39	43	47	51	55	59	63	67
	—	—	**1**	**3**	**7**	**10**	**13**	**17**	**20**	**24**	**27**	**31**	**34**	**38**	**42**	**45**	**49**	**53**	**56**	**60**
14	—	0	2	6	10	13	17	22	26	30	34	38	43	47	51	56	60	65	69	73
	—	—	**1**	**4**	**7**	**11**	**15**	**18**	**22**	**26**	**30**	**34**	**38**	**42**	**46**	**50**	**54**	**58**	**63**	**67**
15	—	0	3	7	11	15	19	24	28	33	37	42	47	51	56	61	66	70	75	80
	—	—	**2**	**5**	**8**	**12**	**16**	**20**	**24**	**29**	**33**	**37**	**42**	**46**	**51**	**55**	**60**	**64**	**69**	**73**
16	—	0	3	7	12	16	21	26	31	36	41	46	51	56	61	66	71	76	82	87
	—	—	**2**	**5**	**9**	**13**	**18**	**22**	**27**	**31**	**36**	**41**	**45**	**50**	**55**	**60**	**65**	**70**	**74**	**79**
17	—	0	4	8	13	18	23	28	33	38	44	49	55	60	66	71	77	82	88	93
	—	—	**2**	**6**	**10**	**15**	**19**	**24**	**29**	**34**	**39**	**44**	**49**	**54**	**60**	**65**	**70**	**75**	**81**	**86**
18	—	0	4	9	14	19	24	30	36	41	47	53	59	65	70	76	82	88	94	100
	—	—	**2**	**6**	**11**	**16**	**21**	**26**	**31**	**37**	**42**	**47**	**53**	**58**	**64**	**70**	**75**	**81**	**87**	**92**
19	—	1	4	9	15	20	26	32	38	44	50	56	63	69	75	82	88	94	101	107
	—	**0**	**3**	**7**	**12**	**17**	**22**	**28**	**33**	**39**	**45**	**51**	**56**	**63**	**69**	**74**	**81**	**87**	**93**	**99**
20	—	1	5	10	16	22	28	34	40	47	53	60	67	73	80	87	93	100	107	114
	—	**0**	**3**	**8**	**13**	**18**	**24**	**30**	**36**	**42**	**48**	**54**	**60**	**67**	**73**	**79**	**86**	**92**	**99**	**105**

(*continued*)

* To be significant the U obtained from data must be equal to or **less than** the value shown in the table. Dashes in the body of the table indicate that no decision is possible at the stated level of significance.

SOURCE: From *Elementary Statistics*, Second Edition, by R. E. Kirk, Brooks/Cole Publishing, 1984.

T A B L E H (*continued*)

One-tailed test
α = .05 (lightface)
α = .025 (**boldface**)

Two-tailed test
α = .10 (lightface)
α = .05 (**boldface**)

N_2 \ N_1	1	2	3	4	5	6	7	8	9	10	11	12	13	14	15	16	17	18	19	20
1	—	—	—	—	—	—	—	—	—	—	—	—	—	—	—	—	—	—	0	0
2	—	—	—	—	0	0	0	1	1	1	1	2	2	2	3	3	3	4	4	4
	—	—	—	—	—	—	—	**0**	**0**	**0**	**0**	**1**	**1**	**1**	**1**	**1**	**2**	**2**	**2**	**2**
3	—	—	0	0	1	2	2	3	3	4	5	5	6	7	7	8	9	9	10	11
	—	—	—	—	**0**	**1**	**1**	**2**	**2**	**3**	**3**	**4**	**4**	**5**	**5**	**6**	**6**	**7**	**7**	**8**
4	—	—	0	1	2	3	4	5	6	7	8	9	10	11	12	14	15	16	17	18
	—	—	—	**0**	**1**	**2**	**3**	**4**	**4**	**5**	**6**	**7**	**8**	**9**	**10**	**11**	**11**	**12**	**13**	**13**
5	—	0	1	2	4	5	6	8	9	11	12	13	15	16	18	19	20	22	23	25
	—	—	**0**	**1**	**2**	**3**	**5**	**6**	**7**	**8**	**9**	**11**	**12**	**13**	**14**	**15**	**17**	**18**	**19**	**20**
6	—	0	2	3	5	7	8	10	12	14	16	17	19	21	23	25	26	28	30	32
	—	—	**1**	**2**	**3**	**5**	**6**	**8**	**10**	**11**	**13**	**14**	**16**	**17**	**19**	**21**	**22**	**24**	**25**	**27**
7	—	0	2	4	6	8	11	13	15	17	19	21	24	26	28	30	33	35	37	39
	—	—	**1**	**3**	**5**	**6**	**8**	**10**	**12**	**14**	**16**	**18**	**20**	**22**	**24**	**26**	**28**	**30**	**32**	**34**
8	—	1	3	5	8	10	13	15	18	20	23	26	28	31	33	36	39	41	44	47
	—	**0**	**2**	**4**	**6**	**8**	**10**	**13**	**15**	**17**	**19**	**22**	**24**	**26**	**29**	**31**	**34**	**36**	**38**	**41**
9	—	1	3	6	9	12	15	18	21	24	27	30	33	36	39	42	45	48	51	54
	—	**0**	**2**	**4**	**7**	**10**	**12**	**15**	**17**	**20**	**23**	**26**	**28**	**31**	**34**	**37**	**39**	**42**	**45**	**48**
10	—	1	4	7	11	14	17	20	24	27	31	34	37	41	44	48	51	55	58	62
	—	**0**	**3**	**5**	**8**	**11**	**14**	**17**	**20**	**23**	**26**	**29**	**33**	**36**	**39**	**42**	**45**	**48**	**52**	**55**
11	—	1	5	8	12	16	19	23	27	31	34	38	42	46	50	54	57	61	65	69
	—	**0**	**3**	**6**	**9**	**13**	**16**	**19**	**23**	**26**	**30**	**33**	**37**	**40**	**44**	**47**	**51**	**55**	**58**	**62**
12	—	2	5	9	13	17	21	26	30	34	38	42	47	51	55	60	64	68	72	77
	—	**1**	**4**	**7**	**11**	**14**	**18**	**22**	**26**	**29**	**33**	**37**	**41**	**45**	**49**	**53**	**57**	**61**	**65**	**69**
13	—	2	6	10	15	19	24	28	33	37	42	47	51	56	61	65	70	75	80	84
	—	**1**	**4**	**8**	**12**	**16**	**20**	**24**	**28**	**33**	**37**	**41**	**45**	**50**	**54**	**59**	**63**	**67**	**72**	**76**
14	—	2	7	11	16	21	26	31	36	41	46	51	56	61	66	71	77	82	87	92
	—	**1**	**5**	**9**	**13**	**17**	**22**	**26**	**31**	**36**	**40**	**45**	**50**	**55**	**59**	**64**	**67**	**74**	**78**	**83**
15	—	3	7	12	18	23	28	33	39	44	50	55	61	66	72	77	83	88	94	100
	—	**1**	**5**	**10**	**14**	**19**	**24**	**29**	**34**	**39**	**44**	**49**	**54**	**59**	**64**	**70**	**75**	**80**	**85**	**90**
16	—	3	8	14	19	25	30	36	42	48	54	60	65	71	77	83	89	95	101	107
	—	**1**	**6**	**11**	**15**	**21**	**26**	**31**	**37**	**42**	**47**	**53**	**59**	**64**	**70**	**75**	**81**	**86**	**92**	**98**
17	—	3	9	15	20	26	33	39	45	51	57	64	70	77	83	89	96	102	109	115
	—	**2**	**6**	**11**	**17**	**22**	**28**	**34**	**39**	**45**	**51**	**57**	**63**	**67**	**75**	**81**	**87**	**93**	**99**	**105**
18	—	4	9	16	22	28	35	41	48	55	61	68	75	82	88	95	102	109	116	123
	—	**2**	**7**	**12**	**18**	**24**	**30**	**36**	**42**	**48**	**55**	**61**	**67**	**74**	**80**	**86**	**93**	**99**	**106**	**112**
19	0	4	10	17	23	30	37	44	51	58	65	72	80	87	94	101	109	116	123	130
	—	**2**	**7**	**13**	**19**	**25**	**32**	**38**	**45**	**52**	**58**	**65**	**72**	**78**	**85**	**92**	**99**	**106**	**113**	**119**
20	0	4	11	18	25	32	39	47	54	62	69	77	84	92	100	107	115	123	130	138
	—	**2**	**8**	**13**	**20**	**27**	**34**	**41**	**48**	**55**	**62**	**69**	**76**	**83**	**90**	**98**	**105**	**112**	**119**	**127**

T A B L E J   **Critical values for the Wilcoxon matched-pairs signed-ranks T test***

No. of pairs N	α levels for a one-tailed test				N	α levels for a one-tailed test			
	.05	.025	.01	.005		.05	.025	.01	.005
	α levels for a two-tailed test					α levels for a two-tailed test			
	.10	.05	.02	.01		.10	.05	.02	.01
5	0	—	—	—	28	130	116	101	91
6	2	0	—	—	29	140	126	110	100
7	3	2	0	—	30	151	137	120	109
8	5	3	1	0	31	163	147	130	118
9	8	5	3	1	32	175	159	140	128
10	10	8	5	3	33	187	170	151	138
11	13	10	7	5	34	200	182	162	148
12	17	13	9	7	35	213	195	173	159
13	21	17	12	9	36	227	208	185	171
14	25	21	15	12	37	241	221	198	182
15	30	25	19	15	38	256	235	211	194
16	35	29	23	19	39	271	249	224	207
17	41	34	27	23	40	286	264	238	220
18	47	40	32	27	41	302	279	252	233
19	53	46	37	32	42	319	294	266	247
20	60	52	43	37	43	336	310	281	261
21	67	58	49	42	44	353	327	296	276
22	75	65	55	48	45	371	343	312	291
23	83	73	62	54	46	389	361	328	307
24	91	81	69	61	47	407	378	345	322
25	100	89	76	68	48	426	396	362	339
26	110	98	84	75	49	446	415	379	355
27	119	107	92	83	50	466	434	397	373

* To be significant the T obtained from the data must be equal to or **less than** the value shown in the table.
SOURCE: From *Elementary Statistics*, Second Edition, by R. E. Kirk, Brooks/Cole Publishing, 1984.

TABLE K Critical differences for the Wilcoxon-Wilcox multiple-comparisons test*
(Column N represents the number in one group.)

				$\alpha = .05$ (two-tailed)				
N	$K = 3$	$K = 4$	$K = 5$	$K = 6$	$K = 7$	$K = 8$	$K = 9$	$K = 10$
1	3.3	4.7	6.1	7.5	9.0	10.5	12.0	13.5
2	8.8	12.6	16.5	20.5	24.7	28.9	33.1	37.4
3	15.7	22.7	29.9	37.3	44.8	52.5	60.3	68.2
4	23.9	34.6	45.6	57.0	68.6	80.4	92.4	104.6
5	33.1	48.1	63.5	79.3	95.5	112.0	128.8	145.8
6	43.3	62.9	83.2	104.0	125.3	147.0	169.1	191.4
7	54.4	79.1	104.6	130.8	157.6	184.9	212.8	240.9
8	66.3	96.4	127.6	159.6	192.4	225.7	259.7	294.1
9	78.9	114.8	152.0	190.2	229.3	269.1	309.6	350.6
10	92.3	134.3	177.8	222.6	268.4	315.0	362.4	410.5
11	106.3	154.8	205.0	256.6	309.4	363.2	417.9	473.3
12	120.9	176.2	233.4	292.2	352.4	413.6	476.0	539.1
13	136.2	198.5	263.0	329.3	397.1	466.2	536.5	607.7
14	152.1	221.7	293.8	367.8	443.6	520.8	599.4	679.0
15	168.6	245.7	325.7	407.8	491.9	577.4	664.6	752.8
16	185.6	270.6	358.6	449.1	541.7	635.9	732.0	829.2
17	203.1	269.2	392.6	491.7	593.1	696.3	801.5	907.9
18	221.2	322.6	427.6	535.5	646.1	758.5	873.1	989.0
19	239.8	349.7	463.6	580.6	700.5	822.4	946.7	1072.4
20	258.8	377.6	500.5	626.9	756.4	888.1	1022.3	1158.1
21	278.4	406.1	538.4	674.4	813.7	955.4	1099.8	1245.9
22	298.4	435.3	577.2	723.0	872.3	1024.3	1179.1	1335.7
23	318.9	465.2	616.9	772.7	932.4	1094.8	1260.3	1427.7
24	339.8	495.8	657.4	823.5	993.7	1166.8	1343.2	1521.7
25	361.1	527.0	698.8	875.4	1056.3	1240.4	1427.9	1617.6

				$\alpha = .01$ (two-tailed)				
N	$K = 3$	$K = 4$	$K = 5$	$K = 6$	$K = 7$	$K = 8$	$K = 9$	$K = 10$
1	4.1	5.7	7.3	8.9	10.5	12.2	13.9	15.6
2	10.9	15.3	19.7	24.3	28.9	33.6	38.3	43.1
3	19.5	27.5	35.7	44.0	52.5	61.1	69.8	78.6
4	29.7	41.9	54.5	67.3	80.3	93.6	107.0	120.6
5	41.2	58.2	75.8	93.6	111.9	130.4	149.1	168.1
6	53.9	76.3	99.3	122.8	146.7	171.0	195.7	220.6
7	67.6	95.8	124.8	154.4	184.6	215.2	246.3	277.7
8	82.4	116.8	152.2	188.4	225.2	262.6	300.6	339.0
9	98.1	139.2	181.4	224.5	268.5	313.1	358.4	404.2
10	114.7	162.8	212.2	262.7	314.2	366.5	419.5	473.1
11	132.1	187.6	244.6	302.9	362.2	422.6	483.7	545.6
12	150.4	213.5	278.5	344.9	412.5	481.2	551.0	621.4
13	169.4	240.6	313.8	388.7	464.9	542.4	621.0	700.5
14	189.1	268.7	350.5	434.2	519.4	606.0	693.8	782.6
15	209.6	297.6	388.5	481.3	575.8	671.9	769.3	867.7
16	230.7	327.9	427.9	530.1	634.2	740.0	847.3	955.7
17	252.5	359.0	468.4	580.3	694.4	810.2	927.8	1046.5
18	275.0	391.0	510.2	632.1	756.4	882.6	1010.6	1140.0
19	298.1	423.8	553.1	685.4	820.1	957.0	1095.8	1236.2
20	321.8	457.6	597.2	740.0	885.5	1033.3	1183.3	1334.9
21	346.1	492.2	642.4	796.0	952.6	1111.6	1273.0	1436.0
22	371.0	527.6	688.7	853.4	1021.3	1191.8	1364.8	1539.7
23	396.4	563.8	736.0	912.1	1091.6	1273.8	1458.8	1645.7
24	422.4	600.9	784.4	972.1	1163.4	1357.6	1554.8	1754.0
25	449.0	638.7	833.8	1033.3	1236.7	1443.2	1652.8	1864.6

* To be significant the difference obtained from the data must be equal to or **larger than** the tabled value.
SOURCE: From *Some Rapid Approximate Statistical Procedures*, by F. Wilcoxon and R. Wilcox. Copyright ©
1964 Lederle Laboratories, a division of American Cyanamid Co. Reprinted with permission.

TABLE L Critical values for Spearman r_s^*

No. of pairs	α levels			
	One-tailed test		Two-tailed test	
N	.01	.05	.01	.05
4	—	1.000	—	—
5	1.000	.900	—	1.000
6	.943	.829	1.000	.886
7	.893	.714	.929	.786
8	.833	.643	.881	.738
9	.783	.600	.833	.700
10	.745	.564	.794	.648
11	.709	.536	.755	.618
12	.678	.503	.727	.587
13	.648	.484	.703	.560
14	.626	.464	.679	.538
15	.604	.446	.654	.521
16	.582	.429	.635	.503

For samples larger than 16, use Table A.

* To be significant, the r_s obtained from the data must be equal to or **larger than** the value shown in the table.

SOURCE: From "Testing the Significance of Kendall's T and Spearman's r_s," by M. Nijsse, 1988, *Psychological Bulletin*, *103*, 235–237. Copyright © 1988 American Psychological Association. Reprinted by permission.

Glossary of Words

abscissa The horizontal, or X, axis of a graph.

absolute value The value of a number without consideration of its algebraic sign.

alternative hypothesis A hypothesis about population parameters that remains tenable if the null hypothesis is rejected; symbolized H_1.

analysis of variance (ANOVA) An inferential statistics technique for comparing means, comparing variances, and assessing interactions.

***a priori* tests** A category of multiple comparison tests that must be planned before examining the data.

asymptotic A line that continually approaches but never reaches a specified limit.

bar graph A frequency distribution for nominal or qualitative data.

biased sample A sample selected in such a way that not all samples from the population have an equal chance of being chosen.

bimodal distribution A distribution with two modes.

binomial distribution A distribution of the frequency of events that have two possible outcomes.

bivariate distribution A joint distribution of two variables, the individual scores of which are paired in some logical way.

cell The portion of an ANOVA data array table containing the scores of subjects that are treated alike.

Central Limit Theorem The theorem in mathematical statistics that the sampling distribution of the mean approaches a normal curve as N gets larger. This normal curve has a mean equal to μ, and a standard deviation equal to $\sigma/\sqrt{N}$.

central tendency The mean, median, or mode; a statistic that best represents the scores in a distribution.

chi square distribution A theoretical sampling distribution of chi square values. Chi square distributions vary with degrees of freedom.

class interval A range of scores grouped together in a grouped frequency distribution.

coefficient of determination A squared correlation coefficient; an estimate of common variance.

common variance Variance held in common by two variables. Common variance is assumed to be related to the same factors.

confidence interval   An interval of scores which is expected, with specified confidence, to contain a parameter.

control group   A baseline group to which other groups are compared.

correlated-measures ANOVA   A technique that tests the hypothesis that two or more population means in a correlated-samples design are equal.

correlated-samples design   An experimental design in which measures from different groups can be logically paired.

correlation coefficient   A descriptive statistic calculated on bivariate data; expresses the degree of relationship between two variables.

critical region   Synonym for rejection region.

critical value   The value from a sampling distribution against which a computed statistic is compared to determine whether the null hypothesis may be rejected.

degrees of freedom   A concept in mathematical statistics that determines which distribution from a family of distributions is appropriate for a particular set of sample data.

dependent variable   The observed variable that is expected to change as a result of changes in the independent variable in an experiment.

descriptive statistic   A number that expresses some particular characteristic of a set of data. Graphs and tables are sometimes included in this category. (Congratulations to you if you have just looked up this entry after reading footnote 1 in Chapter 1. Very few students make the effort to check out their authors' claims, as you just did. You have one of the makings of a scholar.)

deviation score   A raw score minus the mean of the distribution it is in.

dichotomous variable   A variable that takes two, and only two, values.

distribution-free statistics   Inferential statistical methods that do not assume any particular population distribution.

effect size   The amount of difference that a variable makes.

empirical distribution   A set of scores that come from actual observations.

epistemology   The study or theory of the nature of knowledge.

error estimate (or error term)   Variance due to factors not controlled in the experiment; within-group or within-cell variance.

expected frequency   Value in a chi square analysis that is derived from the null hypothesis; theoretical frequency.

expected value   The mean value of a random variable over an infinite number of samplings. The expected value of a statistic is the mean of the sampling distribution of the statistic.

experimental group   A group that receives a treatment in an experiment and whose dependent-variable scores are compared with those of a control group.

extraneous variable   A variable other than the independent variable that may affect the dependent variable.

***F* distribution**   A theoretical sampling distribution of F values. There is a different F distribution for each combination of degrees of freedom.

***F* test**   A method of determining the significance of the difference among two or more means or between two variances.

factor   Independent variable.

factorial design   An experimental design with two or more independent variables; permits an analysis of interaction effects between independent variables.

frequency   The number of times a score occurs in a distribution.

frequency polygon   A graph with quantitative scores on the X axis and frequencies on the Y axis. Each point on the graph represents a score and the frequency of occurrence of that score.

goodness of fit   Degree to which observed data coincide with theoretical expectations.

grand mean   The mean of all the scores in an experiment.

grouped frequency distribution   An arrangement of scores from highest to lowest in which scores are grouped together into equal-sized ranges called class intervals. The number of scores that occur in each class interval is beside the appropriate class interval; may include midpoints of class intervals.

histogram   A graph with contiguous bars; quantitative scores are on the X axis and frequencies on the Y axis.

hypothesis testing   The process of hypothesizing a parameter, comparing (or testing) the parameter with a statistic from a sample, and deciding whether or not the parameter is reasonable.

independent   The occurrence or variation of one event does not affect the occurrence or variation of a second event.

independent-samples design   An experimental design using samples whose dependent-variable scores cannot logically be paired.

independent variable   A variable controlled by the experimenter; changes in this variable may produce changes in the dependent variable.

inferential statistics   A method of reaching conclusions about unmeasurable populations by using sample evidence and probability.

inflection point   A point on a curve that separates an arc that is concave upward from the part that is concave downward, and vice versa.

interaction   The situation in which the effect of one variable on the dependent variable depends on which level of a second variable is operative.

interpolation   A method for determining a value known to lie between two other values.

interval scale   A measurement scale in which equal differences between numbers represent equal differences in the thing measured. The zero point is arbitrarily defined.

J-curve   A severely skewed distribution with the mode at one extreme.

least squares solution   Method of fitting a regression line such that the sums of the squared deviations from the straight regression line will be a minimum.

level   One value of the independent variable.

Likert scale   A method of measuring attitudes; subjects indicate degree of approval or disapproval on a 5- or 7-point scale.

line graph   A graph presenting the relationship between two variables.

lower limit   The bottom of the range of possible values that a measurement on a quantitative variable can take.

main effect   The deviation of one or more treatment means from the grand mean.

Mann-Whitney U test   A nonparametric test of the hypothesis that two independent samples came from the same population.

matched pairs   A correlated-samples design in which individuals are paired by the researcher before the experiment.

mean   The arithmetic average; the sum of the scores divided by the number of scores.

mean square   An ANOVA term for the variance; a sum of squares divided by its degrees of freedom.

median   The point that divides a distribution of scores into two equal halves, so that half the scores are above the median and half are below it.

meta-analysis   A technique for reaching conclusions when the data consist of many separate studies done by different investigators.

mode   The score that occurs most frequently in a distribution.

multiple-comparison tests   Tests of differences between treatment means or combinations of means.

multiple correlation   A correlation method that combines intercorrelations among more than two variables into a single statistic.

natural pairs   A correlated-samples design in which pairing occurs naturally, prior to the experiment.

nominal scale   A scale of measurement in which numbers are used simply as names and not as quantities.

nonparametric statistics   Statistical methods that do not require assumptions about the parameters of the sampled populations.

normal distribution (or normal curve)   A bell-shaped, theoretical distribution that predicts the frequency of occurrence of chance events. An empirical distribution of similar shape.

NS   Not statistically significant.

null hypothesis   A hypothesis about a population or the relationship between populations; symbolized H_0.

one-sample t test   A statistical test of the hypothesis that a sample with mean, $\bar{X}$, came from a population with a mean, μ.

one-tailed test of significance   A statistical test in which the rejection region lies in one tail of the distribution.

one-way ANOVA   A technique that tests the hypothesis that two or more population means in an independent-samples design are equal.

operational definition   A definition that specifies how to measure a variable.

ordinal scale   Scale of measurement in which numbers are ranks but equal differences between numbers do not represent equal differences between the things measured.

ordinate   The vertical, or Y, axis of a graph.

parameter   A numerical or nominal characteristic of a population.

partial correlation   Technique that allows the separation or partialing out of the effects of one variable from the correlation of two other variables.

point estimation   Estimating one particular number to be the parameter of a population.

population   All members of a specified group.

post hoc **tests**   A category of multiple-comparison tests that can be made after examining the data.

power   power $= 1 - \beta$. Power is the probability of rejecting a false null hypothesis.

proportion   A part of a whole.

qualitative variable   A variable that exists in different kinds.

quantification   The idea that translating a phenomenon into numbers will promote a better understanding of the phenomenon.

quantitative variable   A variable that exists in different amounts.

random sample   A subset of a population chosen in such a way that all samples of the specified size have an equal probability of being selected.

range   The upper limit of the highest score minus the lower limit of the lowest score.

ratio scale   A scale that has all the characteristics of an interval scale; in addition, zero means that none of the thing measured is present.

raw score   A score as it is obtained in an experiment.

rectangular distribution   A distribution in which all scores have the same frequency; also called a uniform distribution.

regression coefficients   The values a (point where the regression line intersects the Y axis) and b (slope of the regression line).

regression equation   An equation used to predict particular values of Y for specific values of X.

regression line   The "line of best fit" that runs through a scatterplot.

rejection region   The area of the sampling distribution that includes values of the test statistic that lead to a decision to reject the null hypothesis.

reliability   The dependability or test-retest consistency of a measure.

repeated measures   An experimental design in which more than one dependent-variable measure is taken on each subject.

residual variance   ANOVA term for variability due to unknown or uncontrolled variables.

sample   A subset of a population.

sampling distribution   A theoretical distribution of a statistic based on all possible random samples drawn from the same population; used to determine probabilities.

scatterplot   The graph of a bivariate frequency distribution.

significance level   The probability of the observed data if the null hypothesis is true.

simple frequency distribution   Scores arranged from highest to lowest, with the frequency of each score placed in a column beside the score.

skewed distribution   An asymmetrical distribution. The skew may be positive or negative.

Spearman r_s   A correlation statistic for two sets of ranked data.

standard deviation   For a distribution of scores, a measure of the amount of variability around the mean of the distribution.

standard error   The standard deviation of a sampling distribution.

standard error of a difference   The standard deviation of a sampling distribution of differences between means.

standard error of estimate   The standard deviation of the differences between predicted outcomes and actual outcomes.

standard score   A score expressed in standard deviation units.

statistic   A numerical or nominal characteristic of a sample.

statistically significant   A conclusion based on observed data that would not be likely to occur if the null hypothesis is true; a reliable conclusion.

sum of squares   The sum of the squared deviations from the mean; the numerator of the formula for the standard deviation.

t distribution   Theoretical distribution used to determine probabilities when σ is not known or samples are small.

t test   Significance test that uses the t distribution.

***T* value**   Statistic used in the Wilcoxon matched-pairs signed-ranks *T* test.

theoretical distribution   Arrangement of hypothesized scores based on mathematical formulas and logic.

treatment   One level of the independent variable.

truncated range   The range of the sample is much smaller than the range of its population.

Tukey Honestly Significant Difference (HSD)   Significance test for all possible pairs of treatments in a multi-treatment experiment.

two-tailed test of significance   Any statistical test in which the rejection region is placed into the two tails of the distribution.

Type I error   Rejection of the null hypothesis when it is true.

Type II error   Retention of the null hypothesis when it is false.

univariate distribution   A frequency distribution of one variable.

upper limit   The top of the range of values that a measurement on a quantitative variable can take.

***U* value**   Statistic used in the Mann-Whitney *U* test.

variability   Characterized by having more than one value.

variable   Something that exists in more than one amount or in more than one form.

variance   The square of the standard deviation; also mean square in ANOVA.

very good idea   Examine a statistics problem until you understand it well enough to estimate the answer.

Wilcoxon matched-pairs signed-ranks *T* test   A nonparametric test of the hypothesis that two correlated samples came from the same population.

Wilcoxon rank-sum test   A nonparametric test of the difference between two independent samples.

Wilcoxon-Wilcox multiple-comparisons test   A nonparametric test of independent samples in which all possible pairs of treatments are tested.

Yates' correction   A correction for a 2×2 chi square with one small expected frequency.

***z* score**   A score expressed in standard deviation units; it is used to compare the relative standing of scores in two different distributions.

***z* test**   An inferential statistics test that uses the normal curve as a sampling distribution.

Glossary of Formulas

Analysis of Variance

degrees of freedom in one-way ANOVA

$$df_{treat} = K - 1$$
$$df_{error} = N_{tot} - K$$
$$df_{tot} = N_{tot} - 1$$

degrees of freedom in factorial ANOVA

$$df_A = A - 1$$
$$df_B = B - 1$$
$$df_{AB} = (A - 1)(B - 1)$$
$$df_{error} = N_{tot} - (A)(B)$$
$$df_{tot} = N_{tot} - 1$$

degrees of freedom in one-factor correlated-measures ANOVA

$$df_{subjects} = N_s - 1$$
$$df_{treat} = N_t - 1$$
$$df_{error} = (N_s - 1)(N_t - 1)$$
$$df_{tot} = N_{tot} - 1$$

F value in one-way ANOVA and correlated-measures ANOVA

$$F = \frac{MS_{treat}}{MS_{error}}$$

F values in factorial ANOVA

$$F_A = \frac{MS_A}{MS_{error}}$$
$$F_B = \frac{MS_B}{MS_{error}}$$
$$F_{AB} = \frac{MS_{AB}}{MS_{error}}$$

mean square

$$MS = \frac{SS}{df}$$

total sum of squares

$$SS_{tot} = \Sigma X_{tot}^2 - \frac{(\Sigma X_{tot})^2}{N_{tot}}$$

between-treatments sum of squares

$$SS_{treat} = \Sigma \left[\frac{(\Sigma X_t)^2}{N_t} \right] - \frac{(\Sigma X_{tot})^2}{N_{tot}}$$

between-cells sum of squares

$$SS_{cells} = \Sigma \left[\frac{(\Sigma X_c)^2}{N_c} \right] - \frac{(\Sigma X_{tot})^2}{N_{tot}}$$

between-subjects sum of squares

$$SS_{subjects} = \Sigma \left[\frac{(\Sigma X_s)^2}{N_s} \right] - \frac{(\Sigma X_{tot})^2}{N_{tot}}$$

error sum of squares for one-way ANOVA

$$SS_{error} = \Sigma \left[\Sigma X_t^2 - \frac{(\Sigma X_t)^2}{N_t} \right]$$

error sum of squares for factorial ANOVA

$$SS_{error} = \Sigma \left[\Sigma X_c^2 - \frac{(\Sigma X_c)^2}{N_c} \right]$$

error sum of squares for correlated-measures ANOVA

$$SS_{error} = SS_{tot} - SS_{subjects} - SS_{treat}$$

sum of squares for the interaction effect in factorial ANOVA

$$SS_{AB} = N_c \Sigma [(\bar{X}_{AB} - \bar{X}_A - \bar{X}_B + \bar{X}_{tot})^2]$$

Check:

$$SS_{AB} = SS_{cells} - SS_A - SS_B$$

sum of squares for a main effect in factorial ANOVA

$$SS = \frac{(\Sigma X_1)^2}{N_1} + \frac{(\Sigma X_2)^2}{N_2} + \cdots$$

$$+ \frac{(\Sigma X_K)^2}{N_K} - \frac{(\Sigma X_{tot})^2}{N_{tot}}$$

where 1 and 2 denote levels of a factor and K denotes the last level of a factor

Chi Square

basic formula

$$\chi^2 = \Sigma \left[\frac{(O - E)^2}{E} \right]$$

degrees of freedom for a chi square table

$$df = (R - 1)(C - 1)$$

where R = number of rows
C = number of columns

shortcut formula for a 2 × 2 table

$$\chi^2 = \frac{N(AD - BC)^2}{(A + B)(C + D)(A + C)(B + D)}$$

where A, B, C, and D designate the four cells of the table, moving left to right across the top row and then across the bottom row

Confidence Intervals

about a sample mean when
the normal curve is appropriate

$$LL = \bar{X} - z\sigma_{\bar{X}}$$
$$UL = \bar{X} + z\sigma_{\bar{X}}$$

about a sample mean when the
t distribution is appropriate

$$LL = \bar{X} - ts_{\bar{X}}$$
$$UL = \bar{X} + ts_{\bar{X}}$$

Correlation

Pearson product-moment
definition formula

$$r = \frac{\Sigma(z_X z_Y)}{N}$$

computation formulas

$$r = \frac{\dfrac{\Sigma XY}{N} - (\bar{X})(\bar{Y})}{(S_X)(S_Y)}$$

$$r = \frac{N\Sigma XY - (\Sigma X)(\Sigma Y)}{\sqrt{[N\Sigma X^2 - (\Sigma X)^2][N\Sigma Y^2 - (\Sigma Y)^2]}}$$

testing significance from .00
(or use Table A)

$$t = (r)\sqrt{\frac{N - 2}{1 - r^2}}$$
$$df = N - 2$$

Spearman r_s computation formula

$$r_s = 1 - \frac{6\Sigma D^2}{N(N^2 - 1)}$$

Degrees of Freedom

See specific statistical tests.

Deviation Score

$$x = X - \bar{X} \quad \text{or} \quad x = X - \mu$$

Effect Size

one-sample design

$$d = \frac{\bar{X} - \mu}{\hat{s}}$$

two independent samples,
$N_1 = N_2$

$$d = \frac{\bar{X}_1 - \bar{X}_2}{\sqrt{N_1}(s_{\bar{x}_1 - \bar{x}_2})}$$

two independent samples,
$N_1 \neq N_2$

$$d = \frac{\bar{X}_1 - \bar{X}_2}{\sqrt{\dfrac{\hat{s}_1^2(df_1) + \hat{s}_2^2(df_2)}{df_1 + df_2}}}$$

two correlated samples

$$d = \frac{\bar{X} - \bar{Y}}{\sqrt{N}(s_{\bar{D}})}$$

one-way ANOVA

$$f = \frac{\hat{s}_{\text{treat}}}{\hat{s}_{\text{error}}} = \frac{\sqrt{\dfrac{K-1}{N_{\text{tot}}}(MS_{\text{treat}} - MS_{\text{error}})}}{\sqrt{MS_{\text{error}}}}$$

Mann-Whitney U Test

value for U

$$U_1 = (N_1)(N_2) + \frac{N_1(N_1 + 1)}{2} - \Sigma R_1$$

where ΣR_1 = sum of the ranks of the N_1 group

testing significance for larger samples $N \geqslant 21$

$$z = \frac{(U_1 + c) - \mu_U}{\sigma_U}$$

where $c = 0.5$

$$\mu_U = \frac{N_1 N_2}{2}$$

$$\sigma_U = \sqrt{\frac{(N_1)(N_2)(N_1 + N_2 + 1)}{12}}$$

Mean

from a frequency distribution

$$\mu \text{ or } \bar{X} = \frac{\Sigma f X}{N}$$

from raw data

$$\mu \text{ or } \bar{X} = \frac{\Sigma X}{N}$$

of a set of means

$$\bar{\bar{X}} = \frac{\Sigma(N_1 \bar{X}_1 + N_2 \bar{X}_2 + \cdots + N_K \bar{X}_K)}{\Sigma N}$$

where N_1, N_2, and so on are the number of scores associated with their respective means and K is the last of the means being analyzed.

Range

$$\text{range} = X_H - X_L$$

where X_H = upper limit of highest score
X_L = lower limit of lowest score

Regression

for predicting Y from X

$$Y' = r \frac{S_Y}{S_X}(X - \bar{X}) + \bar{Y}$$

for a straight line

$$Y' = a + bX$$

where a = value at the Y intercept
b = slope of the regression line

the Y intercept of a regression line $\qquad a = \bar{Y} - b\bar{X}$

the slope of a regression line $\qquad b = r\dfrac{S_Y}{S_X}$

$$b = \frac{N\Sigma XY - (\Sigma X)(\Sigma Y)}{N\Sigma X^2 - (\Sigma X)^2}$$

Standard Deviation of a Population or Sample (for description)

by the raw-score method from ungrouped data

$$\sigma \quad \text{or} \quad S = \sqrt{\frac{\Sigma X^2 - \dfrac{(\Sigma X)^2}{N}}{N}}$$

by the raw-score method from grouped data

$$\sigma \quad \text{or} \quad S = \sqrt{\frac{\Sigma fX^2 - \dfrac{(\Sigma fX)^2}{N}}{N}}$$

by the deviation-score method from ungrouped data

$$\sigma \quad \text{or} \quad S = \sqrt{\frac{\Sigma x^2}{N}}$$

by the deviation-score method from grouped data

$$\sigma \quad \text{or} \quad S = \sqrt{\frac{\Sigma fx^2}{N}}$$

Standard Deviation of a Sample (an estimator of σ)

by the raw-score method from ungrouped data

$$\hat{s} = \sqrt{\frac{\Sigma X^2 - \dfrac{(\Sigma X)^2}{N}}{N - 1}}$$

$$\hat{s} = \sqrt{\frac{N\Sigma X^2 - (\Sigma X)^2}{N(N - 1)}}$$

by the raw-score method from grouped data

$$\hat{s} = \sqrt{\frac{\Sigma fX^2 - \dfrac{(\Sigma fX)^2}{N}}{N - 1}}$$

$$\hat{s} = \sqrt{\frac{N\Sigma fX^2 - (\Sigma fX)^2}{N(N - 1)}}$$

by the deviation-score method from ungrouped data

$$\hat{s} = \sqrt{\frac{\Sigma x^2}{N - 1}}$$

by the deviation-score method from grouped data

$$\hat{s} = \sqrt{\frac{\Sigma fx^2}{N - 1}}$$

for correlated samples
$$\hat{s}_D = \sqrt{\frac{\Sigma D^2 - \frac{(\Sigma D)^2}{N}}{N - 1}}$$

Standard Error

of the mean

where the population standard deviation is known
$$\sigma_{\bar{x}} = \frac{\sigma}{\sqrt{N}}$$

estimated from a single sample
$$s_{\bar{X}} = \frac{\hat{s}}{\sqrt{N}}$$

of a difference between means for equal-N samples
$$s_{\bar{x}_1 - \bar{x}_2} = \sqrt{s_{\bar{x}_1}^2 + s_{\bar{x}_2}^2}$$

$$= \sqrt{\left(\frac{\hat{s}_1}{\sqrt{N_1}}\right)^2 + \left(\frac{\hat{s}_2}{\sqrt{N_2}}\right)^2}$$

$$= \sqrt{\frac{\Sigma X_1^2 - \frac{(\Sigma X_1)^2}{N_1} + \Sigma X_2^2 - \frac{(\Sigma X_2)^2}{N_2}}{N_1(N_2 - 1)}}$$

for samples with unequal N's
$$s_{\bar{x}_1 - \bar{x}_2} = \sqrt{\frac{\Sigma X_1^2 - \frac{(\Sigma X_1)^2}{N_1} + \Sigma X_2^2 - \frac{(\Sigma X_2)^2}{N_2}}{N_1 + N_2 - 2}\left(\frac{1}{N_1} + \frac{1}{N_2}\right)}$$

for correlated samples by the direct-difference method
$$s_{\bar{D}} = \frac{\hat{s}_D}{\sqrt{N}}$$

where $s_D = \sqrt{\frac{\Sigma D^2 - \frac{(\Sigma D)^2}{N}}{N - 1}}$

for correlated samples when r is known
$$s_{\bar{D}} = \sqrt{s_{\bar{X}}^2 + s_{\bar{Y}}^2 - 2r_{XY}(s_{\bar{X}})(s_{\bar{Y}})}$$

t Test

as a test for whether a sample mean came from a population
$$t = \frac{\bar{X} - \mu}{s_{\bar{X}}}$$

with a mean μ
$$df = N - 1$$

for correlated samples where r is known

$$t = \frac{\bar{X} - \bar{Y}}{\sqrt{s_{\bar{X}}^2 + s_{\bar{Y}}^2 - 2r_{XY}(s_{\bar{X}})(s_{\bar{Y}})}}$$

for correlated samples where the direct-difference method is used

$$t = \frac{\bar{X} - \bar{Y}}{s_{\bar{D}}}$$

df = number of pairs minus one

for independent samples

$$t = \frac{\bar{X}_1 - \bar{X}_2}{s_{\bar{X}_1 - \bar{X}_2}}$$

$df = N_1 + N_2 - 2$

for testing whether a correlation coefficient is significantly different from .00

$$t = (r)\sqrt{\frac{N - 2}{1 - r^2}}$$

df = number of pairs minus two

Tukey HSD

$$HSD = \frac{\bar{X}_1 - \bar{X}_2}{s_{\bar{X}}}$$

where $s_{\bar{X}}$ (for $N_1 = N_2 = N_t$) = $\sqrt{\dfrac{MS_{error}}{N_t}}$

$$s_{\bar{X}} \text{ (for } N_1 \neq N_2) = \sqrt{\frac{MS_{error}}{2}\left(\frac{1}{N_1} + \frac{1}{N_2}\right)}$$

Variance

Use the formulas for the standard deviation. For $\hat{s}^2$ and σ^2, square $\hat{s}$ and σ.

Wilcoxon Matched-Pairs Signed-Ranks T Test

value for T

T = smaller sum of the signed ranks

testing significance for larger samples $N \geqslant 50$

$$z = \frac{(T + c) - \mu_T}{\sigma_T}$$

where $c = 0.5$

$$\mu_T = \frac{N(N + 1)}{4}$$

$$\sigma_T = \sqrt{\frac{N(N + 1)(2N + 1)}{24}}$$

N = number of pairs

z Score Formulas

descriptive
$$z = \frac{X - \bar{X}}{S}; \quad \text{also } z = \frac{X - \mu}{\sigma}$$

hypothesis testing for sample mean
$$z = \frac{\bar{X} - \mu}{\sigma_{\bar{X}}}$$

Answers to Problems

CHAPTER 1

1. **a.** 7.5–8.5 **b.** Qualitative **c.** 9.995–10.005
   d. Qualitative **e.** 4.45–4.55 **f.** 2.945–2.955
2. Many paragraphs can qualify as good answers to this question. Your paragraph should include variations of the following definitions:
   Population: a defined set of scores that is of interest to an investigator
   Sample: some subset of a population
   Statistic: a numerical or nominal characteristic of a sample
   Parameter: a numerical or nominal characteristic of a population
3. **a.** Inferential; sample data are being used to predict a future event.
   b. Descriptive; the population was available for counting.
   c. Descriptive; all the votes were counted and no inference was made.
   d. Inferential; past data are being used to predict a future event.
   e. Inferential; a conclusion about the generative approach is based on sample data.
4. Nominal, ordinal, interval, and ratio
5. *Nominal*: Different numbers are assigned to different classes of things.
   Ordinal: Nominal properties, plus the numbers carry information about greater than and less than.
   Interval: Ordinal properties, plus the distance between units is equal.
   Ratio: Interval properties, plus a zero point that means "none of the thing being measured."
6. **a.** Ordinal **b.** Ratio **c.** Nominal **d.** Nominal **e.** Ratio
   f. Ordinal **g.** Ratio **h.** Ordinal
7. **a.** *Independent variable*: gender of the candidate
   Dependent variable: rating
   Number and names of the levels of the independent variable: two; female and male
   b. *Independent variable*: identification of the spilled fruit in the discussion
   Dependent variable: degree of confidence that the spilled fruit was apples
   Number and names of the levels of the independent variable: three; correct (oranges), incorrect (apples), and not identified

    c. *Independent variable*: height
    Dependent variable: self-esteem scores
    Number and names of the levels of the independent variable: three; short, about average, and tall

8.   **a.** *Independent variable, number of levels, and their names*: hypnosis; two levels; yes and no
    Dependent variable: number of suggestions complied with
    i. Nominal variables: hypnosis, suggestions; statistic: mean number of suggestions complied with
    ii. Barber's study shows that the behavior of people under hypnosis can be equaled or exceeded by people who are not hypnotized (See Barber, 1976.)

    b. *Independent variable, number of levels, and their names*: mention of the barn in the question about how fast the car was going; two levels; mentioned and not mentioned
    Dependent variable: answer to the question, "Did you see the barn?"
    i. Population and parameter: Several answers can be correct here. The most general population answer is "people" or "people's memory." Less generally, "the memory of people given misinformation about an event." One parameter is the percent of all people who say they see a barn even though there was no barn in the film.
    ii. If people are given misinformation after they have witnessed an event, they sometimes incorporate the misinformation into their memory. [See Loftus's book, *Eyewitness Testimony* (1979), which relates this phenomenon to courts of law.]

    c. *Independent variable, number of levels, and their names*: the time shown on the clock on the wall; three levels; slow, on time, and fast
    Dependent variable: weight (grams) of crackers consumed
    i. Weight (grams)
    ii. Eating by obese males is affected by what time they think it is. That is, consumption by obese males does not depend entirely on internal physiological cues of hunger. (See Schachter and Gross, 1968.)

9. Besides differing in the textbook used, the two classes also differed in (1) the professor, one of whom might be a better teacher; (2) the time the class met; and (3) the length of the class period. A 10 A.M. class probably has younger students who have fewer outside commitments. Students may concentrate less during a 3-hour class than they do in three 1-hour classes. Any of these extraneous variables might be responsible for the difference in scores. Also, for the comparison to be valid, you would need assurance that the comprehensive test was fair to both textbooks.

CHAPTER 2

1.

Women				Men		
Height (in.)	Tally marks	f		Height (in.)	Tally marks	f
72	/	1		77	/	1
71		0		76	/	1
70	/	1		75	/	1
69	//	2		74	/	1
68	/	1		73	////	4
67	////	4		72	7#L	5
66	7#L	5		71	7#L //	7
65	7#L 7#L	10		70	7#L /	6
64	7#L ////	9		69	7#L ///	8
63	7#L //	7		68	7#L //	7
62	7#L	5		67	//	2
61	///	3		66	//	2
60	/	1		65	///	3
59	/	1		64	/	1
				63		0
				62	/	1

2. The order of candidates is arbitrary. They are listed and graphed in the order they are given in the problem; any order is correct.

Candidate	f	Candidate	f
Attila	13	Lenin	8
Gandhi	19	Mao	11
		Bolivar	5

3.

Temperature intervals	Tally marks	f
99.9–100.1	/	1
99.6–99.8		0
99.3–99.5	//	2
99.0–99.2	//	2
98.7–98.9	7#L /	6
98.4–98.6	7#L	5
98.1–98.3	7#L //	7
97.8–98.0	7#L 7#L	10
97.5–97.7	//	2
97.2–97.4	///	3
96.9–97.1	//	2
96.6–96.8		0
96.3–96.5	/	1

4.

Number of sentences heard before	Tally marks	f
16	//	2
15		0
14	//	2
13	///	3
12	///	3
11	7H/ ////	9
10	7H/ /	6
9	7H/ //	7
8	////	4
7	//	2
6	/	1

Interpretation: Here are some points that would be appropriate in your interpretation statement: (1) Surprisingly, no one recognized that all the statements were new. (2) Everyone thought they had heard more than a fourth of the statements before. (3) Most people thought that they had heard about half the statements before, but, as the number of statements increased or decreased from half, fewer and fewer thought that.

5. The range is $63.5-4.5 = 59$. $59 \div 3 = 19.67$, which is within the convention of 10–20 intervals. However, the lowest interval must begin with 3 if it is to be a multiple of 3 and also include the lowest score, 5. Starting with an interval of 3–5 leads to 21 intervals. The best i for these data is 5.

Two solutions using $i = 5$ are appropriate. One begins the intervals with multiples of 5. The other places multiples of 5 at the midpoints. Both are shown here.

Stat scores (class interval)	Tally marks	f	Stat scores (class interval)	Tally marks	f
60–64	/	1	63–67	/	1
55–59	//	2	58–62	/	1
50–54	////	4	53–57	///	3
45–49	////	4	48–52	////	4
40–44	7H/ /	6	43–47	////	4
35–39	7H/ //	7	38–42	7H/ //	7
30–34	7H/ ////	9	33–37	7H/ //	7
25–29	7H/ /	6	28–32	7H/ //	7
20–24	7H/	5	23–27	7H/ //	7
15–19	///	3	18–22	////	4
10–14	//	2	13–17	///	3
5–9	/	1	8–12	/	1
			3–7	/	1

6. a. The number 55 is the midpoint of the interval 54–56.
   b. Ten is a frequency number representing ten students.
   c. Fourteen

7.

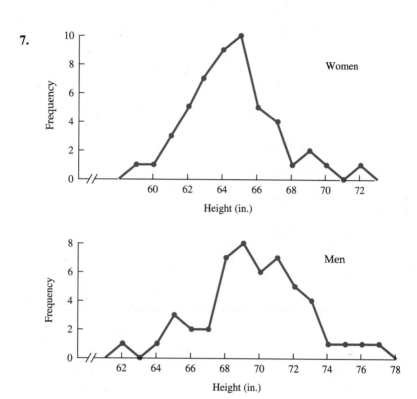

8. A bar graph is appropriate. Here are the thoughts that went into my design. I put the countries in the Y axis so the names would be easy to read. I gave some thought as to how to order the countries. I thought about trying to group them geographically, but finally decided on a least-to-most arrangement. Finally, a graph that was taller than it was wide was more pleasing to my eye.

Average industrial workweek in hours

9. They should be graphed as frequency polygons. Note that the midpoint of the class interval is used as the *X*-axis score.

a.

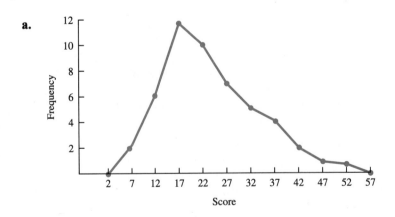

b.

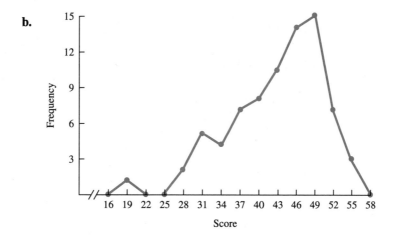

10. Problem 9a is positively skewed. Problem 9b is negatively skewed.
11. A bar graph is the proper graph for the qualitative data in problem 2.

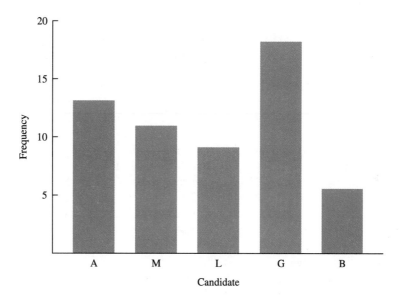

12. Check your sketches against Figures 2.7 and 2.10.
13. Right
14. In a frequency polygon the Y axis is frequency. In a line graph the Y axis is some variable other than frequency.
15. a. Positively skewed   **b.** Symmetrical   **c.** Positively skewed
d. Positively skewed   **e.** Negatively skewed   **f.** Symmetrical
16. a. 5   **b.** 12.5, halfway between 12 and 13
c. 9.5, halfway between 8 and 11

17. Only distribution c has two modes. They are 14 and 18.

18. *Mean* = 409/39 = 10.49

Median: There are 39 scores, so the halfway point will have 19.5 scores below and 19.5 above. Counting from the bottom of the distribution, there are 14 scores of 9 or less. Thus, you need 5.5 of the 6 scores in the interval of 9.5 to 10.5 (19.5 − 14 = 5.5). 5.5/6 = 0.92. 9.5 + 0.92 = 10.42 = median. *Mode* = 11

If a sentence has concepts similar to those in other material they have heard recently, people are poor at recognizing whether or not they have heard that particular sentence before. On the average, over half the sentences that subjects had never heard before were identified as having been heard before.

19. a. For these nominal data, only the mode is appropriate. The mode is Gandhi, which occurs on 34 percent of the signs.

b. Because only one precinct was covered and the student's interest was citywide, this is a sample mode, a statistic.

c. A simple interpretation would be: Because there are more Ghandhi yard signs than any other, I expect Gandhi to get the most votes.

20. In both cases the interest is in some larger group (the oral temperature of humans and five sections of Introduction to Sociology). Thus, both sets of numbers are samples.

21. a. $N = 19$ and half that is 9.5. Counting from the *top*, there are 7 when you include a score of 11. You need 2.5 of the 5 scores in the interval 9.5 to 10.5. 2.5/5 = 0.5, so *subtract* 0.5 from 10.5 to get the median, 10.0.

b. The easiest and quickest way to deal with 20 scores is to construct a simple frequency distribution. Counting from the top again, there are 6 when you include a score of 6. You need 4 of the 5 scores that lie in the interval 4.5 to 5.5. 4/5 = 0.80, so subtract 0.80 from 5.5 to find the median, 4.70.

22. $\Sigma(X - \bar{X}) = 0$ and $\Sigma(X - \bar{X})^2$ is a minimum.

23. *Mean*: $\bar{X} = \Sigma fX/N = 4026.2/41 = 98.2°$ F.

Median: $N/2 = 20.5$. The median will have 20.5 scores above it and 20.5 scores below it. From the bottom of the distribution, there are 18 scores in or below the interval, 97.8–98.0. The next interval, 98.1–98.3 contains the median. To get to 20.5, you need 2.5 scores (20.5 − 18 = 2.5). Thus, you need 2.5 scores of the 7 in the interval. 2.5/7 = 0.357. The interval is 0.3 score points wide. So, the point that is the median is 0.357 into an interval that is 0.3 points wide. 0.357 × 0.3 = 0.11. The lower limit of the interval 98.1–98.3 is 98.05. Finally, 0.11 + 98.05 = 98.16° F. Thus, the median = 98.2° F.

Mode: The midpoint of the interval with 10 frequencies, 97.8–98.0, is the mode. Thus, mode = 97.9° F.

24. *Mean*: $\bar{X} = \Sigma fX/N = 1725/50 = 34.50$, or $\bar{X} = \Sigma fX/N = 1720/50 = 34.40$. The small difference between the two means is due to the different grouping methods. The mean of the raw scores is 34.44, so neither grouped solution contains much error.

Median: $N/2 = 50/2 = 25$. The median will have 25 scores above it and 25 below it. Counting scores from the bottom of the first distribution, there are 17 through the interval 25–29 and 26 through the interval 30–34. Subtract the 17 scores below 30–34 from 25. Because 25 − 17 = 8, 8 more scores are needed. There are 9 in the interval containing the median, so you need 8/9, or 0.89 of the interval. The interval has 5 score points, so $0.889 \times 5 = 4.45$, the number of score points needed to reach the median. The lower limit of the interval 30–34 is 29.5. Adding 4.45 to 29.5, you get 33.95 as the median.

In the second distribution, 23 scores are below the interval 33–37. Two more are needed to reach the median. With 7 in the interval, 2/7, or 0.286, of the interval is needed, or 0.286 of 5. Multiply 0.286×5 to get 1.43. Add 1.43 points to the lower limit, 32.5, to find the median. Thus, $32.5 + 1.43 = 33.93$.

Mode: In the first distribution, the mode is the midpoint of the interval 30–34, because that interval has the greatest number of frequencies. The mode is 32. In the second distribution, four class intervals in a row have seven frequencies. The mode, in this case, is the point between intervals 28–32 and 33–37, or 32.5.

25. **a.** The mode is appropriate because the observations are of a nominal variable.
   b. The median or mode is appropriate because the observations are of an ordinal variable.
   c. The median is appropriate for data with an open-ended category.
   d. The median is the appropriate measure of central tendency to use. It is conventional to use the median for income data because the distribution is often severely skewed. (About half of the frequencies are in the $0–$15,000 range.)
   e. The mode is appropriate because these are nominal data.
   f. The mean is appropriate because the data are not severely skewed.

26. $12 \times 74 = 888$

        $31 \times 69 = 2139$

        $\dfrac{17 \times 75 = 1275}{\Sigma\ 60 \qquad 4302}$ *Overall mean:* $\dfrac{4302}{60} = 71.7$

27. The distribution in problem 9a is positively skewed (mean $= 23.80$, median $= 22.0$).
   The distribution in problem 9b is negatively skewed (mean $= 43.19$, median $= 44.61$).

28. This is not correct. To find his lifetime batting average he would need to add up his hits for the 3 years and divide by the total number of at-bats. From the description of the problem it appears that his average would be lower than .317.

CHAPTER 3

1. Range $17.5 - 0.5 = 17$

X	x
17	11
5	-1
1	-5
1	-5
$\Sigma X = 24$	$\Sigma x = 0$
μ or $\bar{X} = 6$	

2. Range $= 0.455 - 0.295 = 0.16$

X	x
0.45	0.10
0.30	-0.05
0.30	-0.05
$\Sigma X = 1.05$	$\Sigma x = 0$
μ or $\bar{X} = 0.35$	

3. σ is used to describe the variability of a population. $\hat{s}$ is used to estimate σ from a sample of the population. S is used to describe the variability of a sample when you have no desire to estimate σ.

4.

X	x	x^2
7	2	4
6	1	1
5	0	0
2	-3	9
Σ 20	0	14

μ or $\bar{X} = 5$

$$\sigma \text{ or } S = \sqrt{\frac{\Sigma x^2}{N}} = \sqrt{\frac{14}{4}} = \sqrt{3.5} = 1.87$$

5.

X	x	x^2
14	3.8	14.44
11	0.8	0.64
10	-0.2	0.04
8	-2.2	4.84
8	-2.2	4.84
Σ 51	0	24.80

μ or $\bar{X} = 10.2$

$$\sigma \text{ or } S = \sqrt{\frac{\Sigma x^2}{N}} = \sqrt{\frac{24.80}{5}} = \sqrt{4.96} = 2.23$$

6.

X	x	x^2
107	2	4
106	1	1
105	0	0
102	-3	9
Σ 420	0	14

μ or $\bar{X} = 105$

$$\sigma \text{ or } S = \sqrt{\frac{\Sigma x^2}{N}} = \sqrt{\frac{14}{4}} = \sqrt{3.5} = 1.87$$

7. Although the numbers in problem 6 are much larger than those in problem 4, the standard deviations are the same. Thus, the size of the numbers does not give you any information about the variability of the numbers.

8. Yes; the larger the numbers, the larger the mean.

9.

City	Mean	Standard deviation
San Francisco	56.75°	3.96°
Albuquerque	56.75°	16.24°

Although the mean temperature of the two cities is the same, Albuquerque has a wider variety of temperatures.

10. In eyeballing data for variability, use the range as a quick index.
   a. The second distribution is more variable.
   b. Equal variability
   c. The first distribution is more variable.
   d. Equal variability
   e. The second distribution is more variable; however, most of the variability is due to one extreme score, 15.

11. The second distribution (b) is more variable than the first.

a.

X	x	x^2	X^2
5	2.5	6.25	25
4	1.5	2.25	16
3	0.5	0.25	9
2	−0.5	0.25	4
1	−1.5	2.25	1
0	−2.5	6.25	0
Σ 15	0	17.50	55

$$S = \sqrt{\frac{\Sigma X^2 - \frac{(\Sigma X)^2}{N}}{N}} = \sqrt{\frac{55 - \frac{(15)^2}{6}}{6}} = \sqrt{2.917} = 1.71$$

$$S = \sqrt{\frac{\Sigma x^2}{N}} = \sqrt{\frac{17.50}{6}} = \sqrt{2.917} = 1.71$$

b.

X	x	x^2	X^2
5	2.5	6.25	25
5	2.5	6.25	25
5	2.5	6.25	25
0	−2.5	6.25	0
0	−2.5	6.25	0
0	−2.5	6.25	0
Σ 15	0	37.50	75

$$S = \sqrt{\frac{\Sigma X^2 - \frac{(\Sigma X)^2}{N}}{N}} = \sqrt{\frac{75 - \frac{(15)^2}{6}}{6}} = \sqrt{6.25} = 2.50$$

$$S = \sqrt{\frac{\Sigma x^2}{N}} = \sqrt{\frac{37.50}{6}} = \sqrt{6.25} = 2.5$$

12. For problem 11a, $6/1.71 = 3.51$. For problem 11b, $6/2.5 = 2.40$. Yes, these are between 2 and 5.

13. **a.** $\sigma = \sqrt{\dfrac{\Sigma X^2 - \dfrac{(\Sigma X)^2}{N}}{N}} = \sqrt{\dfrac{262 - \dfrac{(34)^2}{5}}{5}} = 2.48$

    b. $\sigma = \sqrt{\dfrac{294 - \dfrac{(38)^2}{5}}{5}} = 1.02$

14. For $\hat{s} = \$0.02$: Because you spend almost exactly \$1.78 each day in the Student Center, each day you don't go in will save you \$1.78.

    For $\hat{s} = \$2.00$: Because you spend widely varying amounts in the Student Center, restrict your buying of the large-ticket items, which will save you a bundle each time you spend. (The actual value of $\hat{s}$ for the data in Table 2.5 is \$2.03.)

15. The variance is the square of the standard deviation.

16. $\hat{s} = \sqrt{\dfrac{\Sigma X^2 - \dfrac{(\Sigma X)^2}{N}}{N - 1}} = \sqrt{\dfrac{5064 - \dfrac{(304)^2}{21}}{20}} = \sqrt{33.16} = 5.76$

    $\hat{s}^2 = 33.162$

17.

Females				Males			
Height (in.)	f	fX	fX^2	Height (in.)	f	fX	fX^2
72	1	72	5,184	77	1	77	5,929
70	1	70	4,900	76	1	76	5,776
69	2	138	9,522	75	1	75	5,625
68	1	68	4,624	74	1	74	5,476
67	4	268	17,956	73	4	292	21,316
66	5	330	21,780	72	5	360	25,920
65	10	650	42,250	71	7	497	35,287
64	9	576	36,864	70	6	420	29,400
63	7	441	27,783	69	8	552	38,088
62	5	310	19,220	68	7	476	32,368
61	3	183	11,163	67	2	134	8,978
60	1	60	3,600	66	2	132	8,712
59	1	59	3,481	65	3	195	12,675
	Sum	3,225	208,327	64	1	64	4,096
				62	1	62	3,844
					Sum	3,486	243,490

$\hat{s} = \sqrt{\dfrac{208{,}327 - \dfrac{3225^2}{50}}{49}} = 2.53$

$\hat{s} = \sqrt{\dfrac{243{,}490 - \dfrac{3486^2}{50}}{49}} = 3.02$

18. $\hat{s}$ is the more appropriate standard deviation; generalization to the manufacturing process would be expected.

$$\text{Process A: } \hat{s} = \sqrt{\frac{18 - \dfrac{0^2}{6}}{5}} = 1.90$$

$$\text{Process B: } \hat{s} = \sqrt{\frac{12 - \dfrac{0^2}{6}}{5}} = 1.55$$

Process B produces the more consistent doodads. Note that both processes have an average error of zero.

19. Before: $\hat{s} = \sqrt{1.429} = 1.20$, $\bar{X} = 5.0$. After: $\hat{s} = \sqrt{14.286} = 3.78$, $\bar{X} = 5.0$. It appears that, before studying poetry, students were homogeneous and neutral. After studying poetry for 9 weeks, some students were turned on and some were turned off; they were no longer neutral.

20.

Temperature intervals	Interval midpoint	f	fX	fX^2
99.9–100.1	100.0	1	100.0	10,000.00
99.6–99.8	99.7	0	0	0
99.3–99.5	99.4	2	198.8	19,760.72
99.0–99.2	99.1	2	198.2	19,641.62
98.7–98.9	98.8	6	592.8	58,568.64
98.4–98.6	98.5	5	492.5	48,511.25
98.1–98.3	98.2	7	687.4	67,502.68
97.8–98.0	97.9	10	979.0	95,844.10
97.5–97.7	97.6	2	195.2	19,051.52
97.2–97.4	97.3	3	291.9	28,401.87
96.9–97.1	97.0	2	194.0	18,818.00
96.6–96.8	96.7	0	0	0
96.3–96.5	96.4	1	96.4	9,292.96
		$\Sigma = 41$	4026.2	395,393.36

$$\hat{s} = \sqrt{\frac{\Sigma fX^2 - \dfrac{(\Sigma fX)^2}{N}}{N-1}} = \sqrt{\frac{395{,}393.36 - \dfrac{(4026.2)^2}{41}}{40}} = \sqrt{0.513} = 0.716$$

21. Zero

22. $\Sigma z = 0$. Because $\Sigma x = 0$, it follows that $\Sigma(x/S) = 0$.

23. Negative z scores are preferable when lower scores are preferable to higher scores. Examples include measures such as errors, pollution levels, and race times.

24.

Hattie	Missy
$z = \dfrac{37 - 39.5}{1.803} = -1.39$	$z = \dfrac{24 - 26.25}{1.479} = -1.52$

Of course, in timed events, the more negative the z score, the better the score. Missy's -1.52 is superior to Hattie's -1.39.

25.

Tobe's apple	Zeke's orange
$z = \dfrac{9 - 5}{1} = 4.00$	$z = \dfrac{10 - 6}{1.2} = 3.33$

Tobe's z score is larger, so the answer to Hamlet must be a resounding "To be." Notice that each fruit varies from its group mean by the same amount. It is the smaller variability of the apple weights that makes Tobe's fruit a winner.

26.

First test	Second test	Third test
$z = \dfrac{79 - 67}{4} = 3.00$	$z = \dfrac{125 - 105}{15} = 1.33$	$z = \dfrac{51 - 45}{3} = 2.00$

Milquetoast's performance was poorest on the second test.

CHAPTER 4

1. A bivariate distribution has two variables. The scores of the variables are paired in some logical way.
2. The statement means that variation in either variable is accompanied by predictable variation in the other. This statement says nothing about direction. The variables may vary in the same direction (positive correlation) or in the opposite direction (negative correlation).
3. In positive correlation, as X increases, Y increases. (It is equivalent to say that as X decreases, Y decreases.) In negative correlation, as X increases, Y decreases. (An equivalent statement is: As X decreases, Y increases in negative correlation.)
4. **a.** Yes; positive; taller people usually weigh more than shorter people.
   b. No; these scores cannot be correlated because there is no basis for pairing a particular oak tree with a particular pine tree.
   c. Yes, negative; as temperatures go up, less heat is needed and fuel bills go down.
   d. Yes, positive; people with higher IQs score higher on reading comprehension tests.
   e. Yes, positive; those who score well on quiz 1 are likely to score well on quiz 2. The fact that some students are in section 1 and some are in section 2 can be ignored.
   f. No; there is no basis for pairing the score of a student in section 1 with that of a student in section 2.

5.

	Fathers	Daughters
Mean	67.50	62.50
S	3.15	2.22
Σ (score)	405	375
Σ (score)^2	27,397	23,467
ΣXY		25,334
r		.513

r by the blanched procedure:

$$r = \frac{\dfrac{25,334}{6} - (67.5)(62.5)}{(3.15)(2.22)} = \frac{4,222.33 - 4,218.75}{6.98} = .51$$

By the raw-score procedure:

$$r = \frac{(6)(25,334) - (405)(375)}{\sqrt{[(6)(27,397) - (405)^2][(6)(23,467) - (375)^2]}} = .513$$

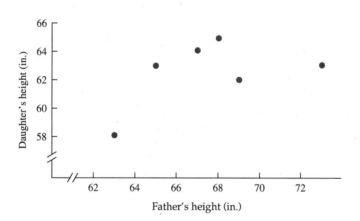

Father's height (in.)

6. $\bar{X} = 46.18$
$\bar{Y} = 30.00$

$$S_X = \sqrt{\frac{87,373 - \dfrac{(1,755)^2}{38}}{38}} = \sqrt{166.308} = 12.90$$

$$S_Y = \sqrt{\frac{37,592 - \dfrac{(1140)^2}{38}}{38}} = \sqrt{89.263} = 9.45$$

$$r = \frac{\dfrac{55,300}{38} - (46.18)(30)}{(12.90)(9.45)} = \frac{1,455.26 - 1,385.40}{121.91} = \frac{69.86}{121.91} = .57$$

7. $r = \dfrac{(50)(175{,}711) - (202)(41{,}048)}{\sqrt{[(50)(1{,}740) - (202)^2][(50)(35{,}451{,}830) - (41{,}048)^2]}} = .25$

$r = \dfrac{\dfrac{175{,}711}{50} - (4.04)(820.96)}{(4.299)(187.247)} = .25$

8. **a.** $-.60$ **b.** $-.95$ **c.** $.50$ **d.** $.25$

Would you be interested in knowing how statistics teachers did on this question? I asked 31 statistics teachers to estimate r for these four scatterplots. The means and standard deviations of their estimates were:

a. $\bar{X} = -.50$; $s = .19$ **b.** $\bar{X} = -.87$; $s = .06$
c. $\bar{X} = .42$; $s = .17$ **d.** $\bar{X} = .13$; $s = .11$

These data suggest to me that the question is a difficult one.

9. Coefficient of determination $= .32$. The two measures of self-esteem have about 32 percent of their variance in common. Although the two measures are to some extent measuring the same traits, they are, in large part, measuring different traits.

10. $r^2 = (.25)^2 = .0625$. This means that 6.25 percent of the variance in expenditure per pupil is related to population variability; 93.75 percent of the variance is related to factors other than population size. Thus, to "explain" the differences among the states, factors other than population differences will be needed.

11. Your shortest answer here is, No. Although you have two distributions, you do not have a bivariate distribution. There is no logical pairing of a height score for a woman with a height score for a man.

12. $(.10)^2 = .01$, or 1 percent; $(.40)^2 = .16$, or 16 percent. Note that a fourfold increase in the correlation coefficient produced a 16-fold increase in the common variance.

13.

	Cigarette consumption	Death rate
Mean	604.27	205.00
S	367.865	113.719
Σ (score)	6,647	2,255
Σ (score)^2	5,505,173	604,527
ΣXY	1,705,077	
N	11	
r	.74	

With $r = .74$, these data show that there is a fairly strong relationship between per capita cigarette consumption and male death rate 20 years later.

14. **a.** Vocational interests tend to remain similar from age 20 to age 40.
b. Identical twins raised together have very similar IQs.
c. There is a slight tendency for IQ to be lower as family size increases.
d. There is a slight tendency for taller men to have higher IQs than shorter men.

    e. The lower a person's income level is, the greater the probability that he or she will be diagnosed as psychotic.

    f. People who cannot tolerate ambiguity tend to be authoritarian.

15. a. There is some tendency for children with more older siblings to accept less credit or blame for their own successes and failures than children with fewer older siblings.

    b. Nothing; correlation coefficients do not permit you to make cause-and-effect statements.

    c. The coefficient of determination is .1369 $(-.37^2)$, so we can say that about 14 percent of the variance in acceptance of responsibility is predictable from knowledge of the number of older siblings; 86 percent is not. Thus, knowing the number of siblings is only somewhat helpful in predicting degree of acceptance of responsibility.

16. a. Because $.97^2 = .94$, these two tests have 94 percent of their variance in common. They must be measuring essentially the same trait. There is a very strong tendency for persons scoring high on one test to score high on the other.

    b. No; this is a cause-and-effect statement and is not justified on the basis of correlational data.

17. A Pearson r is not a useful statistic for data such as these because the relationship is a curved one. You might note that the scatterplot is similar to the curved relationship you saw in Figure 2.6.

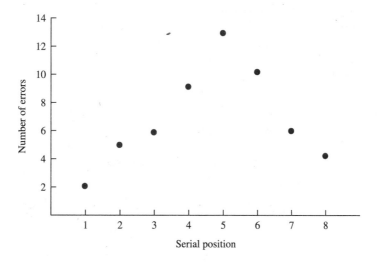

18. a. $b = r\dfrac{S_Y}{S_X} = (.513)\dfrac{2.217}{3.149} = (.513)(.704) = .361$

    $a = \bar{Y} - b\bar{X} = 62.5 - (.361)(67.5) = 38.12$

    b. To draw the regression line, I used two points: $(\bar{X}, \bar{Y})$ and (62, 60.5). I got the point (62, 60.5) by using the formula $Y' = a + bX$. Other X values would work just as well.

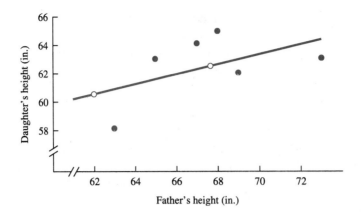

Father's height (in.)

19. a. $b = r\dfrac{S_Y}{S_X} = (.57)\dfrac{9.45}{12.90} = (.57)(.73) = .42$

$a = \bar{Y} - b\bar{X} = 30 - (.42)(46.18) = 30 - 19.40 = 10.60$

b. $Y' = a + bX = 10.60 + (.42)(42) = 10.60 + 17.64 = 28.24$

20.

	Advertising	Sales
Mean	4.286	108.571
S	1.030	20.304
Σ (score)	30	760
Σ (score)^2	136	85,400
ΣXY	3360	
r		.703

a. $Y' = a + bX = 49.176 + 13.858X$

b. One point that I used for the regression line on the scatterplot was that of the two means, $\bar{X} = 4.29$, $\bar{Y} = 108.57$. The other point was $X = 6.00$, $Y = 132.33$. Other X values would work just as well.

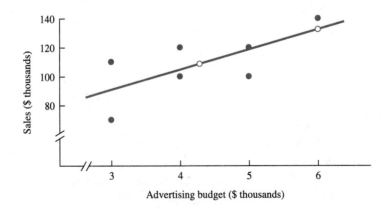

Advertising budget ($ thousands)

c. $Y' = a + bX$. For $X = 10$, $Y' = 187.756$, or, in terms of sales, \$187,756.

d. You were warned that the interpretation of r can be a tricky business. This problem leads you to one of those common errors I mentioned. With $r = .70$, you have a fair degree of confidence that the prediction will hold *in general*. However, if your thinking (and your answer) was that by increasing advertising, sales would increase, you made the error of inferring a causal relationship from a correlation coefficient. As a matter of fact, some analysts say that sales cause advertising, that advertising budgets are determined by last year's sales. In any case, don't infer more than the data permit. These data allow you to estimate with some confidence what sales were by knowing the amount spent for advertising. In this problem, the percent of sales spent on advertising (just less than 4 percent) is about the same as the percent spent nationally in the United States.

Also, \$10,000 for advertising is much more than any amount used to find the regression line. It is not safe to predict outcomes for points far from the data.

21. $Y' = (.80)\left(\dfrac{15}{16}\right)(65 - 100) + 100 - 73.75 = 74$

22. Here are the answers I asked you to find. It is important to designate the year variable as Y.

$$\Sigma X = 55.80 \qquad \Sigma Y = 477 \qquad\qquad N = 6$$
$$\Sigma X^2 = 519.02 \qquad \Sigma Y^2 = 37,939 \qquad \Sigma XY = 4,437.20$$
$$r = .930 \qquad\qquad a = -48.38 \qquad\qquad b = 13.75$$

For 1 million graduates ($10 \times 100,000$),

$$Y' = a + bX = -48.38 + 13.75(10) = 89.12$$

Thus, the predicted first year for 1 million graduates is 1990. This is a simple and straightforward prediction from a regression equation. Of course, at this time in history, the accuracy of this prediction can be checked: One million college graduates received baccalaureate degrees in 1989. The prediction was off by one year.

24. On the big, important questions like this one, you are on your own!

WHAT WOULD YOU RECOMMEND? Chapters 1–4

A. The mode is appropriate because this is a distribution of nominal data.

B. z scores can be used. The watermelon entries can be used to obtain a mean and a standard deviation. The z score of the largest watermelon can be calculated and then compared to the z score of the largest pumpkin, which is a z score based on the distribution of pumpkin weights.

C. A regression analysis would produce a prediction.

D. The median would be appropriate because the distribution is severely skewed (the class interval with zero has the highest frequency). Both the standard deviation and the range would be informative. The appropriate standard deviation is σ, because the 50 states represent the entire population of states.

E. A measure of variability *might* distinguish between the two applicants. A sample standard deviation would be described as "more consistent."

F. A line graph is appropriate for the relationship that is described. A Pearson product-moment correlation coefficient is not appropriate because the relationship is curvilinear (much forgetting at first, and much less after four days).

G. The median is the appropriate central tendency statistic when one or more of the categories is open-ended.

H. A correlation coefficient will describe this relationship. What direction and size would you guess for this coefficient? The answer (based on several studies) is −.10.

CHAPTER 5

1. There are seven cards between the 3 and jack, each with a probability of 4/52 = .077. So (7)(.077) = .539.

2. The probability of drawing a 7 is 4/52, and there are 52 opportunities to get a 7. Thus, (52)(4/52) = 4.

3. There are four cards that are higher than a jack or lower than a 3. Each has a probability of 4/52. Thus, (4)(4/52) = 16/52 = .308.

4. The probability of a 5 or 6 is 4/52 + 4/52 = 8/52 = .154. In 78 draws, (78)(.154) = 12 cards that are 5s or 6s.

5. a. .7500; adding the probability of one head (.3750) to that of two heads (.3750), you get a figure of .7500.

   b. .1250 + .1250 = .2500

6. 16(.1250) = 2 times

7. a. .0668   b. The test scores are normally distributed.

8. a. .0832   b. .2912   c. .4778

9. Empirical; the scores are based on observations.

10. $z = (X − \mu)/\sigma$. For IQ = 55, z score = −3.00; for IQ = 110, z score = .67; for IQ = 103, z score = .20; for IQ = 100, z score = .00.

11. A quick check of your answer can be made by comparing it with the proportion having IQs of 120 or higher, .0918. The proportion with IQs of 70 or lower will be less than the proportion with IQs of 120 or higher. Is your calculated proportion less? Following is a picture of a normal distribution in which the proportion of the

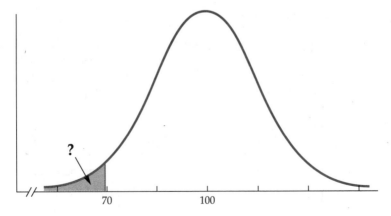

population with IQs of 70 or lower is shaded. For IQ $= 70$, $z = (70 - 100)/15 = -30/15 = -2.00$. The proportion beyond $z = 2.00$ is .0228 (column C in Table C). Thus, .0228 or 2.28 percent of the population would be expected to be in special education classes.

12. $(.0228)(4000) = 91.2$, or 91 students.

13. $z = (110 - 100)/15 = 10/15 = .67$. The proportion beyond $z = .67$ is .2514, which is the expected proportion of people with IQs of 110 or higher.

14. **a.** $.2514 \times 250 = 62.85$, or 63 students

   b. $250 - 62.85 = 187.15$, or 187 students

   c. $1/2 \times 250 = 125$. I hope you were able to get this one immediately by thinking about the symmetrical nature of the normal distribution.

15. The z score that includes a proportion of .02 is 2.06. If $z = 2.06$, $X = 100 + (2.06)(15) = 100 + 30.9 = 130.9$. In fact, Mensa requires an IQ of 130 on tests that have a standard deviation of 15.

16. **a.** The z score you need is 1.65. (1.64 will include more than the tallest 5 percent.)
   $1.65 = (X - 64.5)/2.5$; $X = 64.5 + 4.1 = 68.6$ inches.

   b. $z = (58 - 64.5)/2.5 = -6.5/2.5 = -2.60$. The proportion excluded is .0047.

17. **a.** $z = (60 - 69.7)/3.0 = -9.7/3.0 = -3.23$. The proportion excluded is .0006.

   b. $z = (62 - 69.7)/3.0 = -7.7/3.0 = -2.57$. The proportion taller than Napoleon is .9949 (.5000 + .4949).

18. **a.** $z = \dfrac{3.20 - 3.11}{.05} = 1.80$, proportion $= .0359$

   b. The z score corresponding to a proportion of .1000 is 1.28. Thus,
   $X = 3.11 + (1.28)(.05)$
   $\quad = 3.11 + .064$
   $\quad = 3.17$
   $X = 3.11 - .064$
   $\quad = 3.05$
   Thus, the middle 80 percent of the pennies weigh between 3.05 and 3.17 grams.

19.

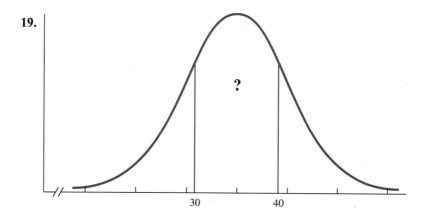

a. $z = \dfrac{30 - 35}{6} = -.83$, proportion $= .2967$; $z = \dfrac{40 - 35}{6} = .83$,

proportion $= .2967$ $(2)(.2967) = .5934 =$ the proportion of students with scores between 30 and 40.

b. The probability is also .5934.

20. No, because the distance between 30 and 40 straddles the mean, where scores that occur frequently are found. The distance between 20 and 30 is all in one tail of the curve, where scores occur less frequently.

21.

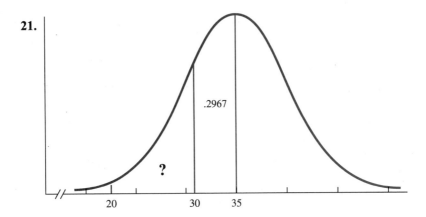

$z = \dfrac{20 - 35}{6} = -2.50$, proportion $= .4938$

$z = \dfrac{30 - 35}{6} = -.83$, proportion $= .2967$

$.4938 - .2967 = .1971$. The proportion of students with scores between 35 and 30 is subtracted from the proportion with scores between 35 and 20. There are also other correct ways to set up this problem.

22. 800 people $\times$ .1971 $= 157.7 = 158$ people

23.

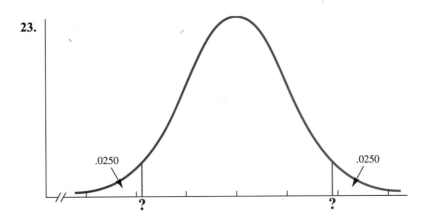

The z score corresponding to a proportion of .0250 is 1.96 (column C). Thus,

$$X = \mu + (z)(\sigma)$$
$$= 100 + (1.96)(15)$$
$$= 129.4 \text{ or } 129$$

and

$$X = 100 + (-1.96)(15)$$
$$= 70.6 \text{ or } 71$$

IQ scores of 129 or higher or 71 or lower are achieved by 5 percent of the population.

24. Because .05 of the curve lies outside the limits of the scores 71 and 129, the probability is .05. The probability is .025 that the randomly selected person has an IQ of 129 or higher and .025 that it is 71 or lower.

25. $z = \dfrac{.85 - .99}{.17} = \dfrac{-.14}{.17} = -.82$. The proportion associated with

$z = -.82$ is .2061 (column C), and $(.2061)(185,822) = 38,298$ truck drivers. The usual procedure would be to report this result as "about 38,000 truck drivers," because the normal curve corresponds only approximately to the empirical curve.

26. This problem has two parts: determining the number of trees over 8 inches DBH on 1 acre and multiplying by 100:

$$z = \frac{8 - 13.68}{4.83} = -1.18, \text{ proportion} = .5000 + .3810 = .8810$$

$.8810(199) = 175.319$ trees on 1 acre

$175.319(100) = 17,532$ trees on 100 acres (about 17,500 trees)

27. $z = (300 - 268)/14 = 32/14 = 2.29$. The proportion of gestation periods expected to last 10 months or longer is .0110. In modern obstetrical care, pregnancies are normally brought to term (by drugs or surgery) earlier than 300 days.

28. Converting this problem to inches, the z score is

$$z = \frac{80 - 69.7}{3.0} = 3.43$$

The proportion of the curve above 3.43 is .0003, so only 3 in 10,000 will have to duck to enter a residential room.

29. $\mu = \dfrac{3600}{100} = 36.0$ inches

$$\sigma = \sqrt{\frac{130,000 - \dfrac{(3600)^2}{100}}{100}} = 2.00 \text{ inches}$$

a. $z = \dfrac{32 - 36}{2} = -2.00$, proportion $= .5000 + .4772 = .9772$

b. $z = \dfrac{39 - 36}{2} = 1.50$, proportion $= .5000 + .4332 = .9332$

$.9332(300) = 279.96$ or 280 hobbits

c. $z = \dfrac{46 - 36}{2} = 5.00$, proportion $<.00003$ or, practically speaking, zero

CHAPTER 6

1. $\Sigma X = 180$ $\mu = 9.0$

   $\Sigma X^2 = 1700$ $\sigma = 2.0$

   The numbers in this population are derived from a data set in Dudek (1981).
2. Standard error, expected value
3. **a.** Select a set of numbers as the population. Draw many samples of the same size, find the range for each sample, and arrange those ranges into a frequency distribution.

   b. The standard error of the range

   c. The expected value of the range, which is the mean of the sampling distribution of the range, will be *less* than the population range. Only a very few sample ranges (maybe only one) can have a value equal to the population range. Thus, almost all the sample ranges will be less than the population range. Statistics (such as the range) whose expected value is not equal to the corresponding population parameter are called *biased statistics* and are not popular among mathematical statisticians.
4. Please check your answer with the definition and explanation in the chapter.
5. If this standard error (whose name is the standard error of estimate) were very small, it would mean that you would be confident that the actual Y would be close to the predicted Y. A very large standard error would indicate that you could not put much confidence in a predicted Y, that the actual Y would be subject to lots of other influences besides X.

6.

			σ	
N	1	2	4	8
1	1	2	4	8
4	0.50	1	2	4
16	0.25	0.50	1	2
64	0.125	0.25	0.50	1

7. As N increases by a factor of 4, $\sigma_{\bar{X}}$ is reduced by one-half. More simply but less accurately, as N increases, $\sigma_{\bar{X}}$ decreases.
8. 16 times
9. $z = \dfrac{8.5 - 9}{2/\sqrt{8}} = \dfrac{.5}{.707} = -.71$ Proportion $= .2389$

10. For sample means of 8.5 or less,

$$z = \dfrac{8.5 - 9}{2/\sqrt{16}} = \dfrac{-.5}{.5} = -1.00$$ Proportion $= .1587$

For sample means of 10 or greater,

$$z = \frac{10 - 9}{2/\sqrt{16}} = \frac{1}{.5} = 2.00 \qquad \text{Proportion} = .0228$$

11. For mean IQs of 105 or greater,

$$z = \frac{105 - 100}{15/\sqrt{25}} = \frac{5}{3} = 1.67 \qquad \text{Probability} = .0475$$

For mean IQs of 90 or less,

$$z = \frac{90 - 100}{15/\sqrt{25}} = \frac{-10}{3} = -3.33 \qquad \text{Probability} = .0004$$

Classrooms with 25 students will have a mean IQ fairly close to 100; departures of as much as 5 points will happen less than 5 percent of the time; departures of as much as 10 points will be very rare, occurring less than one time in a thousand. This analysis is based on the premise that no systematic factors are influencing a child's assignment to a classroom.

12. $\sigma_{\bar{X}} = \sigma/\sqrt{N} = 12,000/\sqrt{400} = 600$

$\text{LL} = \bar{X} - z(\sigma_{\bar{X}}) = 48,300 - 1.96(600) = \$47,124$

$\text{UL} = \bar{X} + z(\sigma_{\bar{X}}) = 48,300 + 1.96(600) = \$49,476$

With a sample mean of \$48,300 based on a random sample of 400 State U. students, we are 95 percent confident that the interval \$47,124 to \$49,476 captures the State U. mean. However, because the interval also captures the U.S. mean of \$49,000, we do not have any strong evidence that the State U. mean is less than the national mean.

13. Estimation and hypothesis testing

14. Wider

15. $\sigma_{\bar{X}} = \sigma/\sqrt{N} = 12,000/\sqrt{400} = 600$

$\text{LL} = \bar{X} - z(\sigma_{\bar{X}}) = 46,500 - 2.58(600) = \$44,952$

$\text{UL} = \bar{X} + z(\sigma_{\bar{X}}) = 46,500 + 2.58(600) = \$48,048$

The conclusion is that we are more than 99 percent confident that the mean family income for State U. students is less than the mean for all U.S. students.

16. For 90 percent confidence intervals, $z = \pm 1.64$ or $z = \pm 1.65$ ($z = \pm 1.65$ is traditionally used). For 98 percent confidence intervals, $z = \pm 2.33$.

17. $\bar{X} = \dfrac{\Sigma X}{N} = \dfrac{58}{7} = 8.286$

18. $\sigma_{\bar{X}} = \dfrac{\sigma}{\sqrt{N}} = \dfrac{2}{\sqrt{7}} = 0.756$

$\text{LL} = \bar{X} - z(\sigma_{\bar{X}}) = 8.29 - 1.96(.756) = 6.81$

$\text{UL} = \bar{X} + z(\sigma_{\bar{X}}) = 8.29 + 1.96(.756) = 9.77$

You have 95 percent confidence that the interval 6.81 to 9.77 contains the population mean. In this case, because you know that the sample was drawn from a population with a mean of 9, your confidence is well founded.

19. For this problem, it is not the numbers in your sample that are "right" or "wrong" but the method you used to get them. Using the table of random numbers, the proper method is to create an identifying number (01–20) for every population member. (Table 6.2 will work fine.) Haphazardly find a starting place in Table B and record in sequence 10 two-digit identifying numbers (ignoring numbers over 20 and duplications). Write the 10 population members that correspond to the 10 selected identifying numbers.

20. As in problem 19, the method, not the numbers, determines a correct answer. You should give each number an identifying number of 01 to 38, begin at some chance starting place, and select 12 numbers. The starting place for every random sample taken from Table B should be haphazardly chosen. If you start at the same place in the table every time, you will have the same identifying numbers every time.

21. Those with greater educational accomplishments are more likely to return the questionnaire. Such a biased sample will overestimate the accomplishments of the population.

22. The sample will be biased. Many of the general population will not even read the article and thus cannot be in the sample. Of those who read it, there will be those who have no opinion and those who will not trouble to mark the ballot and return it. In short, the results of such a questionnaire cannot be generalized to the larger population.

23. $\bar{X} = \dfrac{\Sigma X}{N} = \dfrac{15,600}{16} = 975$

$\hat{s} = \sqrt{\dfrac{15,247,500 - \dfrac{(15,600)^2}{16}}{15}} = 50.0$

$s_{\bar{X}} = \dfrac{\hat{s}}{\sqrt{N}} = \dfrac{50}{\sqrt{16}} = 12.5$

24. The normal curve is appropriate when you know σ (the standard deviation of the population) and when sample size is adequate. The t distribution is appropriate in cases where sample size is small or σ is unknown. In both cases, samples that are representative of the population are required.

25. $\bar{X} = 27.00$ $\hat{s} = 3.863$ $s_{\bar{X}} = 1.032$ $df = 13$

The t value for 13 df and 95 percent confidence is 2.160.

LL $= \bar{X} - t(s_{\bar{X}}) = 27 - 2.16(1.032) = 24.77$

UL $= \bar{X} + t(s_{\bar{X}}) = 27 + 2.16(1.032) = 29.23$

The workshop is effective. A 95 percent confidence interval about the mean of all who might take the workshop, based on a sample of 14 who completed the workshop, gives a range of 24.77 to 29.23. Because the national mean for the test is *less* than the lower limit of the confidence interval of those taking the workshop, conclude that clients are more assertive than the national average after completing the workshop. (This conclusion assumes that the clients were not better than average in assertiveness before the workshop.)

26. a. $\pm 3.182(2) = \pm 6.364$ (12.728 units wide)

    b. $\pm 2.131(1) = \pm 2.131$ (4.262 units wide). The fourfold increase in N produces a confidence interval about one-third as wide. The effect of changing sample size depends on the relative change in N, not the absolute change.

    c. $\pm 3.182(1) = \pm 3.182$ (6.364 units wide). Reducing $\hat{s}$ by half reduces the size of the confidence interval by half. This relationship is true regardless of the size of $\hat{s}$. Data gatherers reduce $\hat{s}$ by taking measurements carefully, using precise instruments, and increasing N.

27. The size of the standard deviation of each sample. If you happen to get a sample that produces a large standard deviation, a large standard error will result and the confidence interval will be wide.

28. a. 19

    b. Narrower; because N was increased by four, $s_{\bar{X}}$ would be reduced to half of its size. In addition, increasing N will decrease the t value from Table D. Therefore, the lines would be less than half as long.

29. a. 18

    b. Narrower; the t value is smaller.

30. The standard error of the mean for each of the four trial means is 1.0; the t value (24 df) for the 95 percent confidence interval is 2.064. The lower and upper limits for each mean are ± 2.064.

    Look at the graph following this problem. The graph in problem 30 has been modified by adding 95 percent confidence intervals to each trial mean. The overlap of the trial 2 confidence interval with that of trial 1 indicates that the dip at trial 2 may not be reliable. That is, you are 95 percent confident that the population mean for trial 2 is in the range of about 3.94 to 8.06. Because the upper half of this range is also included in the trial 1 confidence interval, you aren't confident that a second sample of data would show the dip at trial 2. [Note that the figure that shows only the means invites you to conclude that performance (always) goes down on trial 2. The figure with the confidence intervals leads to a more cautious interpretation.]

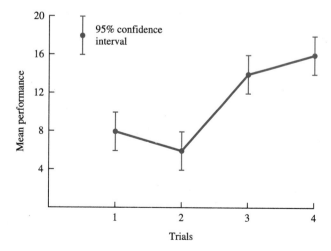

Looking at the *lack* of overlap between the confidence intervals of trial 3 and trial 2 leads you to conclude that the trial 3 performance is *reliably* better than that of trial 2; that is, you are over 95 percent confident that a second sample of data would also show that the trial 3 mean is higher than the trial 2 mean.

CHAPTER 7

1. The null hypothesis is a statement about a parameter.
2. The null hypothesis is a tentative hypothesis about the value of some parameter (often the mean) of the population the sample is taken from. The alternative hypothesis specifies other values for the parameter. The symbols are: null hypothesis, H_0; alternative hypothesis, H_1.
3. Your outline should include the following points:
   a. Gather sample data from the population you are interested in and calculate a statistic.
   b. Recognize two logical possibilities for the population:
      H_0: a statement specifying an exact value for the parameter of the population
      H_1: a statement specifying all other values for the parameter
   c. Using a sampling distribution that assumes H_0 is correct, find the probability of the statistic you calculated.
   d. If the probability is very small, reject H_0 and accept H_1. If the probability is large, retain both H_0 and H_1, acknowledging that the data do not allow you to reject H_0.
4. Disagree
5. False. Although the researcher probably thinks that H_0 is false, the sampling distribution is based on the assumption that H_0 is true.
6. An α level is chosen by the researcher as the point that separates "This is an unlikely event if H_0 is true; reject H_0" from "This could reasonably happen if H_0 is true; retain H_0." Significance level is the probability produced by the data—the probability of obtaining such data when H_0 is true. The largest acceptable α level without justification is .05.
7. Phrases that should be circled are "p is small" and "Accept H_1."
8. a. $t_{.01}(17\ df) = 2.898$
   b. $t_{.001}(40\ df) = 3.551$
   c. $t_{.05}(\infty\ df) = 1.96$
9. α is the probability of a Type I error; a Type I error occurs when a researcher mistakenly rejects H_0. To reduce Type I errors, researchers set α at a low level such as .05 or less. A statistical test on the data produces a p value, which is the probability of obtaining the sample statistic actually obtained, if H_0 is true. The significance level is determined by the p value produced by the statistical test.
10. A Type II error is retaining a false null hypothesis.
11. The probability of making a Type I error is zero. If, at first, you don't agree with this answer, keep thinking until you understand.
12. The probability of a Type I error decreases from .05 to .01 and the probability of a Type II error increases.

13. The null hypothesis is $H_0: \mu_0 = 100$.

$$\bar{X} = \frac{5250}{50} = 105.00 \qquad \hat{s} = \sqrt{\frac{560,854 - \frac{(5250)^2}{50}}{49}} = 14.00$$

$$s_{\bar{X}} = \frac{14.00}{\sqrt{50}} = 1.98 \qquad t = \frac{\bar{X} - \mu_0}{s_{\bar{X}}} = \frac{105 - 100}{1.98} = 2.53; \ 49 \ df$$

$t_{.02}(40 \ df) = 2.423$. The null hypothesis can be rejected; $p < .02$. The conclusion is that the Head Start participants had IQs that were significantly above average.

The actual studies that evaluated Head Start were more sophisticated than the one described here, but the conclusions were the same. The IQ differences, however, weren't lasting, but other benefits were.

14. $H_0: \mu_0 = 58.7 \qquad \Sigma X = 360.8 \qquad \Sigma X^2 = 21,703.86 \qquad N = 6$

$\bar{X} = 60.1333 \qquad s_{\bar{X}} = 0.5084$

$$t = \frac{\bar{X} - \mu_0}{s_{\bar{X}}} = \frac{60.133 - 58.7}{0.5084} = 2.82; \ 5 \ df$$

$t_{.05}(5 \ df) = 2.571$

The statistical decision is to reject the null hypothesis. Conclude that Snickers candy bars weigh more than the company claims, $p < .05$.

15. $t = \dfrac{\bar{X} - \mu_0}{s_{\bar{X}}} = \dfrac{56.10 - 50.00}{2.774} = 2.199; \ df = 12$

$t_{.05}(12 \ df) = 2.179$

Reject the null hypothesis and conclude that those successful in used car sales have significantly higher extroversion scores than those in the general population.

16. I suggest that you check each part of this problem as soon as you do it. If you have already worked them all, just look at each answer until you miss one. If an answer here surprises you, make a note of what surprised you. (Students, like researchers, have expectations and surprises. A surprise often means that you are about to learn something.)

	Critical value	t-test value	Reject or retain H_0	p values
a.	2.262	2.25	retain	$p > .05$
b.	2.539	2.57	reject	$p < .01$
c.	2.042	2.03	retain	$p > .05$
d.	6.869	6.72	retain	$p > .001$
e.	1.714	1.72	reject	$p < .05$
f.	2.423	2.41	retain	$p > .02$

17. $d = \dfrac{\bar{X} - \mu}{\sigma} = \dfrac{270.675 - 269.3}{0.523} = 2.63$

The effect size index, 2.63, is much larger than the conventional designation of "large." Even though the sample mean was only 1.375 grams more than the Frito-Lay company claimed, the effect was large because the bag-filling machinery was so precise (the standard deviation was small).

18. $d = \dfrac{\bar{X} - \mu}{\sigma} = \dfrac{56.1 - 50.0}{10} = 0.61$

The difference in extroversion between those successful in used car sales and the population mean, although statistically significant, has an effect size index that is medium in size.

19. $d = \dfrac{\bar{X} - \mu}{\sigma} = \dfrac{98.2 - 98.6}{0.7} = -0.57; \ |-0.57| = 0.57$

An effect size index of .57 shows that a medium-size effect has existed undetected for a long time.

20. $t = (r)\sqrt{\dfrac{N - 2}{1 - r^2}}$

  a. $t = 2.24$; $df = 8$. Because $t_{.05}$ (8 df) $= 2.31$, $r = .62$ is not significantly different from .00.
  b. $t = -2.12$; $df = 120$. Because $t_{.05}$ (120 df) $= 1.98$, $r = -.19$ is significantly different from .00.
  c. $t = 2.08$; $df = 13$; $t_{.05}$ (13 df) $= 2.16$. Retain the null hypothesis.
  d. $t = -2.85$; $df = 62$; $t_{.05}$ (60 df) $= 2.00$. Reject the null hypothesis.

21. Here's my response: "Chance is not a likely explanation for a correlation coefficient of .74 based on an N of 11 pairs. If there is really no relationship between cigarette smoking and lung cancer, chance would produce an $r = \pm.7348$ only 1 time in 100 (Table A, $df = 9$). An $r = .74$ is less likely than an $r = .7348$, so the probability that chance produced this result is less than 1 in 100. I don't believe chance did it."

22. $r = \dfrac{N\Sigma XY - (\Sigma X)(\Sigma Y)}{\sqrt{[N\Sigma X^2 - (\Sigma X)^2][N\Sigma Y^2 - (\Sigma Y)^2]}}$

$= \dfrac{42(46{,}885) - (903)(2079)}{\sqrt{[42(25{,}585) - (903)^2][42(107{,}707) - (2079)^2]}} = .40$

From Table A, $r_{.01}$ (40 df) $= .3932$. Because $.40 > .3932$, reject the null hypothesis and conclude that there is a positive correlation between the time a student works on a test and the grade, $p < .01$.

If, as you finished up this problem, you thought about waiting until the end of the period to turn in your next test (so that you would get a better grade), you drew a cause-and-effect conclusion from a correlation coefficient. Better go back and reread page 92.

CHAPTER 8

1. Your outline should include the following points:
   a. Two random samples are selected.
   b. The samples are treated the same except for one thing.
   c. The samples are measured, and the difference found is attributed to chance or to the difference in treatments.
2. Random assignment of participants to groups
3. **a.** Independent variable: personality type
   Dependent variable: satisfaction in life
   Null hypothesis: The mean satisfaction in life is the same for people with Type A personalities as those with Type B personalities.
   b. Independent variable: type of crime
   Dependent variable: years served in prison
   Null hypothesis: convicted robbers and convicted embezzlers spend, on the average, the same number of years in jail.
4. Your outline should include the following points:
   a. Recognize two logical possibilities:
      1. The treatment had no effect (null hypothesis, H_0).
      2. The treatment had an effect (alternative hypothesis, H_1).
   b. Assume H_0 to be correct and establish an α level.
   c. Calculate a test statistic from the sample data. Compare the test statistic to a sampling distribution (which is based on the assumption that H_0 is true).
   d. If the test statistic has a probability less than α, reject H_0, adopt H_1, and write a conclusion about the effects of the treatment. If the test statistic has a probability greater than α, retain H_0 and write a conclusion that says the data did not allow you to choose between H_0 and H_1.
5. Experiments are conducted so that researchers can write a conclusion about the effect of an independent variable on a dependent variable.
6. Degrees of freedom are always equal to the number of observations minus the (a) number of necessary relationships that exist among the observations or (b) number of parameters that are estimated from the observations.
7. **a.** Correlated samples. With twins divided between the two groups, there is a logical pairing (natural pairs); independent variable: environment, dependent variable: attitude toward education.
   b. Correlated samples. This is a before-and-after experiment. For each group, the amount of aggression before watching the other children is paired with the amount afterward (repeated measures); independent variable: whether children watched others with better toys, dependent variable: number of aggressive encounters.
   c. Correlated samples. The depression score after treatment is compared to the depression score before treatment (repeated measures); independent variable: bright artificial light, dependent variable: depression score.
   d. Correlated samples (yoked control design). Each participant is paired with another so that the amount of interrupted sleep is equal for the two. One,

however, is deprived of REM sleep, the other is not; independent variable: REM deprivation, dependent variable: mood questionnaire score.

e. Independent samples. The random assignment of individuals and the subsequent procedures give you no reason to pair two particular participants' mood questionnaire scores; independent variable: REM deprivation, dependent variable: mood questionnaire score.

f. Independent samples. The dean randomly assigned individuals to one of the two groups; independent variable: sophomore honors course, dependent variable: grade point average.

This paragraph is really not about statistics, and you may skip it if you wish. Were you somewhat more anxious about your decisions on parts c and d than on parts a and b? If so, and if this anxiety was based on the expectation that surely it was time for an answer to be "independent samples" and if you based your answer on this expectation rather than on the problem, you were exhibiting a *response bias*. A response bias occurs when a current response is made on the basis of previous responses rather than on the basis of the current stimulus. Such response biases often lead to a correct answer in textbooks, and you may learn to make some decisions (those you are not sure about) on the basis of irrelevant cues (such as what your response was on the last question). To the extent it rewards your response biases, a textbook is doing you a disservice. So, be forewarned; recognize response bias and resist it. (I'll try to do my part.)

8. A two-tailed test is appropriate; if one of the two kinds of animals is superior, you would want to know which one.

$H_0: \mu_1 = \mu_2$
$H_1: \mu_1 \neq \mu_2$

	Fellow humans	Pigeons
ΣX	147	75
ΣX^2	7515	1937
N	3	3
$\bar{X}$	49	25
$\hat{s}$	12.4900	5.5678

$$t = \frac{\bar{X}_1 - \bar{X}_2}{\sqrt{\left(\frac{\hat{s}_1}{\sqrt{N_1}}\right)^2 + \left(\frac{\hat{s}_2}{\sqrt{N_2}}\right)^2}} = \frac{49 - 25}{\sqrt{\left(\frac{12.49}{\sqrt{3}}\right)^2 + \left(\frac{5.5678}{\sqrt{3}}\right)^2}} = \frac{24}{7.895} = 3.04;$$

$df = 4$

Reject the null hypothesis and conclude that pigeons can spot a person in the ocean more quickly than humans can, $p < .05$.

9. Independent variable: percent of cortex removed; dependent variable: number of errors

$$\hat{s}_{0\,percent} = \sqrt{\frac{1706 - \dfrac{(208)^2}{40}}{40 - 1}} = 4.00$$

$$\hat{s}_{20\,percent} = \sqrt{\frac{2212 - \dfrac{(252)^2}{40}}{40 - 1}} = 4.00$$

$$s_{\bar{X}_1 - \bar{X}_2} = \sqrt{\left(\frac{4.00}{\sqrt{40}}\right)^2 + \left(\frac{4.00}{\sqrt{40}}\right)^2} = 0.894$$

$$t = \frac{\bar{X}_0 - \bar{X}_{20}}{s_{\bar{X}_1 - \bar{X}_2}} = \frac{5.2 - 6.3}{0.894} = -1.23; \; df = 78$$

$t_{.05}(60 \; df) = 2.00$. Retain the null hypothesis and conclude that a 20 percent loss of cortex does not reduce significantly a rat's memory for a simple maze. *Note*: Some brain operations *improve* performance on some tasks, so a two-tailed test is appropriate.

10.

	New package	Old package
ΣX	45.9	52.8
ΣX^2	238.73	354.08
$\bar{X}$	5.1	6.6
N	9	8
$\hat{s}$	0.76	0.89

$$t = \frac{6.6 - 5.1}{\sqrt{\left(\dfrac{10.24}{15}\right)(0.24)}} = \frac{1.5}{0.40} = 3.75; \; df = 15$$

$t_{.005}(15 \; df) = 2.947$ (one-tailed test). Thus, $p < .005$; therefore, reject the null hypothesis and conclude that the test procedure can be worked more quickly with the new package. A one-tailed test is appropriate here because the only interest is whether the new package is better than the one on hand. (See the section "One-Tailed and Two-Tailed Tests" in Chapter 7.)

11. First of all, you might point out that if the populations are those particular freshman classes, no statistics are necessary; you have the population data and there is no sampling error. State U. is one-tenth of a point higher than The U. If the question is not about those freshman classes but about the two schools, and the two freshman classes can be treated as representative samples, a two-tailed test is called for

because superiority of either school would be of interest.

$$s_{\bar{X}_1-\bar{X}_2} = \sqrt{\left(\frac{\hat{s}_1}{\sqrt{N_1}}\right)^2 + \left(\frac{\hat{s}_2}{\sqrt{N_2}}\right)^2} = 0.0474$$

$$t = \frac{\bar{X}_1 - \bar{X}_2}{s_{\bar{X}_1-\bar{X}_2}} = \frac{21.4 - 21.5}{0.0474} = -2.11; \; df = 120+$$

$t_{.05}(120 \; df) = 1.980$. Students at State U. have statistically significantly higher ACT admission scores than do students at The U. You may have noted how small the difference actually is—only one-tenth of a point, an issue that is discussed in a later section in the text.

12. a. $s_D = \sqrt{\dfrac{\Sigma D^2 - \dfrac{(\Sigma D)^2}{N}}{N-1}}$

s_D is the standard deviation of the distribution of differences between correlated scores.

b. $D = X - Y$. D is the difference between two correlated scores.

c. $s_{\bar{D}} = s_D/\sqrt{N}$. $s_{\bar{D}}$ is the standard error of the difference between means for a correlated set of scores.

d. t is the name of a theoretical probability distribution. In this chapter you have used it to determine the probability that two samples came from populations with the same mean.

e. $\bar{Y} = \Sigma Y/N$. $\bar{Y}$ is the mean of a set of scores that is correlated with another set.

13.

	X	Y	D	D^2
	16	18	-2	4
	10	11	-1	1
	17	19	-2	4
	4	6	-2	4
	9	10	-1	1
	12	14	-2	4
Σ	68	78	-10	18
Mean	11.3	13.0		

$$s_D = \sqrt{\frac{\Sigma D^2 - \frac{(\Sigma D)^2}{N}}{N-1}} = \sqrt{\frac{18 - \frac{(-10)^2}{6}}{5}} = \sqrt{0.2667} = 0.5164$$

$$s_{\bar{D}} = \frac{s_D}{\sqrt{N}} = \frac{0.5164}{\sqrt{6}} = 0.2108$$

$$t = \frac{11.3333 - 13.0000}{0.2108} = \frac{-1.6667}{0.2108} = -7.91; \; df = 5$$

$t_{.001}(5 \ df) = 6.869$. Therefore, $p < .001$. Notice that the small difference between means (1.67) is highly significant even though the data consist of only six pairs of scores. This illustrates the power that a large correlation can have in reducing the standard error. The conclusion reached by the researchers was that frustration (produced by seeing others treated better) leads to aggression.

14. On this problem you cannot use the direct-difference method, because you do not have the raw data to work with. This leaves you with the definition formula, $t = (\bar{X} - \bar{Y})/\sqrt{s_{\bar{X}}^2 + s_{\bar{Y}}^2 - 2r_{XY}(s_{\bar{X}})(s_{\bar{Y}})}$, which requires a correlation coefficient. Fortunately, you have the data necessary to calculate r. (The initial clue for many students is the term ΣXY.)

Females	Males
$\bar{X} = \$12,782.25$	$\bar{Y} = \$15,733.25$
$\hat{s}_X = 2,358.99$	$\hat{s}_Y = 3,951.27$
$s_{\bar{X}} = 589.75$	$s_{\bar{Y}} = 987.82$
$r = .693$	

$$t = \frac{12,782.25 - 15,733.25}{\sqrt{(589.75)^2 + (987.82)^2 - (2)(.693)(589.75)(987.82)}} = -4.11$$

$t_{.001}(15 \ df) = 4.073$. The null hypothesis can be rejected at the .001 level. Here is the conclusion about these data written for the court.

> Chance is not a likely explanation for the \$2951 annual difference in favor of men. Chance would be expected to produce such a difference less than one time in a thousand. In addition, the difference cannot be attributed to education because both groups were equally educated. Likewise, the two groups were equal in their work experience at the rehabilitation center. One explanation that has not been eliminated is discrimination, based on sex.

(These data are 1979 salary data submitted in Hartman and Hobgood *v.* Hot Springs Rehabilitation Center, HS-76-23-C.)

15. This is a correlated-samples study.

$$\Sigma D = -.19 \qquad \Sigma D^2 = .0101 \qquad N = 11 \qquad \bar{X} - \bar{Y} = \frac{\Sigma D}{N} = -.0173$$

$$s_D = \sqrt{\frac{\Sigma D^2 - \dfrac{(\Sigma D)^2}{N}}{N - 1}} = \sqrt{0.00068} = 0.0261$$

$$s_{\bar{D}} = \frac{0.026}{\sqrt{11}} = 0.00787$$

$$t = \frac{-0.0173}{0.00787} = -2.19; \ df = 10$$

$t_{.05}(10\ df) = 2.23$. The obtained t does not indicate statistical significance. (This is a case in which a small sample led to a Type II error. On the basis of larger samples, auditory RT has been found to be faster than visual RT—approximately 0.14 second for auditory RT and 0.18 second for visual RT, using practiced subjects.)

16. I hope that you did not treat these data as correlated samples. If you did, you exhibited a response bias that led you astray. If you treated these data as the independent samples they are, you are on your (mental) toes. So much for verbal reinforcement; here is the data analysis.

Recency	Primacy
$\Sigma X_1 = 86$	$\Sigma X_2 = 104$
$\Sigma X_1^2 = 934$	$\Sigma X_2^2 = 1312$
$N_1 = 9$	$N_2 = 9$

$$s_{\bar{X}_1 - \bar{X}_2} = \sqrt{\frac{\Sigma X_1^2 - \dfrac{(\Sigma X_1)^2}{N} + \Sigma X_2^2 - \dfrac{(\Sigma X_2)^2}{N}}{N(N-1)}} = 1.758$$

$$t = \frac{\bar{X}_1 - \bar{X}_2}{s_{\bar{X}_1 - \bar{X}_2}} = \frac{9.556 - 11.556}{1.758} = -1.14;\ df = 16$$

$t_{.05}(16\ df) = 2.12$. Therefore, retain the null hypothesis. The conclusion is: This experiment does not provide evidence that primacy or recency is more powerful than the other.

17. $t = \dfrac{\bar{X}_1 - \bar{X}_2}{s_{\bar{X}_1 - \bar{X}_2}} = \dfrac{60 - 36}{\sqrt{\left(\dfrac{12}{\sqrt{45}}\right)^2 + \left(\dfrac{6}{\sqrt{45}}\right)^2}} = \dfrac{24}{2} = 12.00;\ df = 88$

$t_{.001}(60\ df) = 3.460$. The meaning of very large t-test values is quite clear; they indicate a very, very small probability. Thus, conclude that children whose illness begins before age 3 exhibit symptoms longer than do those whose illness begins after age 6, $p < .001$.

18. You can be completely confident that the probabilities are accurate when the dependent-variable scores are normally distributed and have equal variances, and when the samples are randomly selected from the populations.

19. $d = \dfrac{\bar{X}_1 - \bar{X}_2}{\hat{s}} = \dfrac{49 - 25}{\sqrt{3}\,(7.895)} = 1.76$

An effect size index of 1.76 indicates that pigeons are quite superior to humans at spotting people lost at sea.

20. $d = \dfrac{\bar{X}_1 - \bar{X}_2}{\hat{s}} = \dfrac{5.2 - 6.3}{\sqrt{40}\,(0.894)} = -0.19;\ |-0.19| = 0.19$

An effect size index of 0.19 is small. Thus, d is small, and a t test using large samples did not find a significant difference. This combination leads to the conclusion that the effect (if any) of removal of 20 percent of the cortex is small. If there is an effect, it will be difficult to detect.

21. $d = \dfrac{\bar{X} - \bar{Y}}{\hat{s}} = \dfrac{28.36 - 34.14}{5.74} = -1.01; \ |-1.01| = 1.01$

An effect size index of 1.01 is large; thus, the effect of the biracial camp on racial attitudes was both reliable (statistically significant) and sizable. Note that the effect size index gives you information that the means do not. That is, $d = 1.01$ tells you much more than does the fact that after the camp, the girls scored an average of 5.78 points higher.

22. Power is the probability of rejecting the null hypothesis when it is false.

23. My version of the list is (1) effect size, (2) sample size, (3) α level, and (4) the preciseness of measuring the dependent variable.

24. Here is one version: "Significant differences are due to the independent variable and not to chance; important differences are ones that change our understanding about something. Important differences are significant, but significant differences may or may not be important."

25. The p refers to the probability of obtaining such a difference in samples (or a larger difference) if the two populations they come from (holistic scores and lecture-discussion scores) are identical. That is, if the populations are the same, then chance would produce such a difference (or one larger) in sample means less than 1 time in 100.

26. $\Sigma D = -8; \ \Sigma D^2 = 18$

$$s_D = \sqrt{\dfrac{\Sigma D^2 - \dfrac{(\Sigma D)^2}{N}}{N-1}} = \sqrt{\dfrac{18 - \dfrac{(-8)^2}{8}}{7}} = 1.1952$$

$$s_{\bar{D}} = \dfrac{s_D}{\sqrt{N}} = \dfrac{1.1952}{\sqrt{8}} = 0.4226$$

$$t = \dfrac{\bar{X} - \bar{Y}}{s_{\bar{D}}} = \dfrac{2.5 - 3.5}{0.4226} = -2.37; \ df = 7$$

$t_{.05}(7 \ df) = 2.365$. Participants remembered significantly more CVCs when they spent the time between learning and recall asleep (as compared to being awake).

$$d = \dfrac{\bar{X} - \bar{Y}}{\hat{s}} = \dfrac{2.5 - 3.5}{1.1952} = -0.84; \ |-0.84| = 0.84$$

Thus, the effect of sleep (as compared to being awake) is to significantly improve retention by a large amount.

27. The null hypothesis is that experience has no effect on performance. A two-tailed test would allow a conclusion about the advantage *or* the disadvantage of experience. The experimental design is independent samples.

$$t = \frac{7.40 - 5.05}{\sqrt{\left(\frac{2.13}{\sqrt{20}}\right)^2 + \left(\frac{2.31}{\sqrt{20}}\right)^2}} = \frac{2.35}{0.7026} = 3.34; \; df = 38$$

$t_{.01}(30 \; df) = 2.750$. Because the experienced group took longer, you should conclude that the previous experience with the switch *retarded* the subject's ability to recognize the solution.

$$d = \frac{\bar{X}_1 - \bar{X}_2}{\hat{s}} = \frac{7.40 - 5.05}{\sqrt{20}\,(0.7026)} = 0.75$$

An effect size index of 0.75 borders on being large.

28.

| | Listened to tape | | | | |
| | In car | | In laboratory | | | |
	Rank	Errors	Rank	Errors	D	D^2
	1	7	2	7	0	0
	4	7	3	15	−8	64
	5	12	6	22	−10	100
	8	12	7	24	−12	144
	9	21	10	32	−11	121
Sum		59		100	−41	429
Mean		11.8		20.0	−8.2	

$$s_D = \sqrt{\frac{\Sigma D^2 - \frac{(\Sigma D)^2}{N}}{N-1}} = \sqrt{\frac{429 - \frac{(-41)^2}{5}}{4}} = 4.817$$

$$s_{\bar{D}} = \frac{s_D}{\sqrt{N}} = \frac{4.817}{\sqrt{5}} = 2.154$$

$$t = \frac{\bar{X} - \bar{Y}}{s_{\bar{D}}} = \frac{11.8 - 20.0}{2.154} = \frac{-8.2}{2.154} = -3.81; \; df = 4$$

$t_{.02}(4 \; df) = 3.747$. Reject the null hypothesis at the .02 level. Because the car group made fewer errors, conclude that concrete, immediate experience facilitates learning vocabulary. A two-tailed test is appropriate because *disproving* a claim is always of interest to the scientifically minded. To use your data to support *or* refute a claim requires you to choose a two-tailed test (before the data are gathered).

I hope you recognized that these data could be arranged as correlated samples (matched pairs). Each pair represents a level of achievement in French II.

$$d = \frac{\bar{X} - \bar{Y}}{\hat{s}} = \frac{11.8 - 20.0}{4.817} = -1.70; \; |-1.70| = 1.70$$

An effect size index of 1.70 is very large. (It is true that concrete, immediate experience has a great effect on learning foreign vocabulary words.)

WHAT WOULD YOU RECOMMEND? Chapters 5–8

A. A correlated-samples t test will allow you to determine whether the number of smiles (dependent variable) is related to whether or not there is an audience (independent variable). The samples are correlated because each infant contributes data to both the audience and the no audience condition.

B. A z-score analysis using the normal curve will give the percent of the population that is less gregarious than the social worker.

C. A one-sample t test will give the probability of obtaining a sample with a mean of 55 from a population with a mean of 50. If this probability is very low, you might conclude that the workshop resulted in increased gregariousness scores.

D. The lower and upper limits of a confidence interval will capture a population mean with a certain level of confidence (usually 95 or 99 percent).

E. The state's population mean is not captured in problem D because the sample is not a random one (or representative either) from the state. A better interpretation is that the interval captures the mean of "those students planning on going to college."

F. To determine the probability that an $r = .20$ could be the result of chance, test the null hypothesis, H_0: $\rho = 0$. If H_0 is rejected, conclude that there is a reliable relationship between the two variables. (The relationship between illness and stress is reliable though the r is small, about .20 to .30—Cohen and Williamson, 1991.)

G. The probability that a student, chosen at random, comes from one of the six categories, is found by dividing the number of students in that category by the total number of students. (See the most recent issue of *Statistical Abstract of the United States* for current data.)

H. A correlated-samples t test will allow you to determine whether activity (the dependent variable) depends on whether or not a child ate sugar for breakfast (the independent variable). The samples are correlated because each child served in both conditions.

CHAPTER 9

1. *Example 2:* The independent variable is the schedule of reinforcement; there are four levels. The dependent variable is persistence. The null hypothesis is that the mean persistence is the same for the four populations; that is, the schedule of reinforcement has no effect on persistence.

Example 3: The independent variable is degree of modernization; there are three levels. The dependent variable is suicide rate. The null hypothesis is that suicide rates are not dependent on degree of modernization, that the mean suicide rates for countries with low, medium, and high degrees of modernization are the same. In formal terms, the null hypothesis is H_0: $\mu_{low} = \mu_{med} = \mu_{high}$.

Example 4: The independent variable is the injection; there are three levels. The dependent variable is memory. The null hypothesis is that the three populations of memory scores are the same (the effect of the three injections will be the same).

2. Larger
3. One estimate is obtained from variability among the sample means, and another estimate is obtained by pooling the sample variances.
4. F is a ratio of the two estimates of the population variance.
5. **a.** Retain the null hypothesis and conclude that the data do not provide evidence that social classes differ in their attitude toward religion.
   b. Reject the null hypothesis and conclude that there is a relationship between degree of modernization and suicide rate.
6. There are several ways to craft a fine answer to this question. Some ways use graphs, some use algebra, and some use only words. a good answer includes such elements as: (a) F is a ratio of two variances—a between-treatments variance over an error variance; (b) when the null hypothesis is true, both variances are good estimators of the population variance, which produces a ratio of about 1.00; (c) when the null hypothesis is false, the between-treatments variance overestimates the population variance, resulting in a ratio that is larger than 1.00; (d) the F value calculated from the data is evaluated by comparing it to a critical value from the F distribution, a distribution that assumes the null hypothesis is true.

7.

	X_1	X_2	X_3	Σ
ΣX	12	12	24	48
ΣX^2	50	56	224	330
$\bar{X}$	4	4	8	

$$SS_{tot} = 330 - \frac{48^2}{9} = 330 - 256 = 74.00$$

$$SS_{treat} = \frac{12^2}{3} + \frac{12^2}{3} + \frac{24^2}{3} - \frac{48^2}{9} = 48 + 48 + 192 - 256$$
$$= 288 - 256 = 32.00$$

$$SS_{error} = \left(50 - \frac{12^2}{3}\right) + \left(56 - \frac{12^2}{3}\right) + \left(224 - \frac{24^2}{3}\right)$$
$$= 2 + 8 + 32 = 42.00$$

Check: $74.00 = 32.00 + 42.00$

8. Independent variable: degree of modernization; dependent variable: suicide rate.

	Degree of modernization			
	Low	Medium	High	Σ
ΣX	24	48	84	156
ΣX^2	154	614	1522	2290
$\bar{X}$	6.0	12.0	16.8	

$$SS_{tot} = 2290 - \frac{(156)^2}{13} = 418.00$$

$$SS_{mod} = \frac{(24)^2}{4} + \frac{(48)^2}{4} + \frac{(84)^2}{5} - \frac{(156)^2}{13} = 259.20$$

$$SS_{error} = \left(154 - \frac{(24)^2}{4}\right) + \left(614 - \frac{(48)^2}{4}\right) + \left(1522 - \frac{(84)^2}{5}\right) = 158.80$$

Check: $418.00 = 259.20 + 158.80$

Incidentally, the rate per 100,000 is about 18 in the United States and about 17 in Canada.

9. Independent variable: kind of therapy; dependent variable: snake-approach responses.

	Model	Film	Desensitization	Control	Σ
ΣX	104	72	68	40	284
ΣX^2	2740	1322	1182	430	5674
$\bar{X}$	26	18	17	10	
N	4	4	4	4	16

$$SS_{tot} = 5674 - \frac{284^2}{16} = 5674 - 5041 = 633$$

$$SS_{treat} = \frac{104^2}{4} + \frac{72^2}{4} + \frac{68^2}{4} + \frac{40^2}{4} - \frac{284^2}{16}$$

$$= 2704 + 1296 + 1156 + 400 - 5041$$

$$= 5556 - 5041 = 515$$

$$SS_{error} = \left(2740 - \frac{104^2}{4}\right) + \left(1322 - \frac{72^2}{4}\right)$$

$$+ \left(1182 - \frac{68^2}{4}\right) + \left(430 - \frac{40^2}{4}\right)$$

$$= 36 + 26 + 26 + 30 = 118$$

Check: $633 = 515 + 118$

10. Six groups. The critical values of F are based on 5, 70 *df* and are 2.35 and 3.29 at the .05 and .01 levels, respectively. If $\alpha = .01$, retain the null hypothesis; if $\alpha = .05$, reject the null hypothesis.

11. Data from problem 7:

a. Source	SS	df	MS	F	p
Treatments	32.00	2	16.00	2.29	>.05
Error	42.00	6	7.00		
Total	74.00	8			

$F_{.05}(2, 6 \, df) = 5.14$. Because 2.29 does not reach the critical value, retain the null hypothesis and conclude that you do not have evidence against the hypothesis that these three groups came from the same population.

b. $SS_{tot} = 280 - \dfrac{36^2}{6} = 64.00$

$SS_{treat} = \dfrac{12^2}{3} + \dfrac{24^2}{3} - \dfrac{36^2}{6} = 24.00$

$SS_{error} = \left(56 - \dfrac{12^2}{3}\right) + \left(224 - \dfrac{24^2}{3}\right) = 40.00$

Source	SS	df	MS	F	p
Treatments	24.00	1	24.00	2.40	>.05
Error	40.00	4	10.00		
Total	64.00	5			

$F_{.05}(1, 4 \, df) = 7.71$. Retain the null hypothesis.

$t = \dfrac{\bar{X}_1 - \bar{X}_2}{s_{\bar{X}_1 - \bar{X}_2}} = \dfrac{8 - 4}{2.582} = 1.55; df = 4$

Note that $(1.55)^2 = 2.40$; that is, $t^2 = F$.

12. Durkheim's modernization and suicide data from problem 8. For *df*, use

$df_{tot} = N_{tot} - 1 = 13 - 1 = 12$
$df_{mod} = K - 1 = 3 - 1 = 2$
$df_{error} = N_{tot} - K = 13 - 3 = 10$

Source	SS	df	MS	F	p
Modernization	259.20	2	129.60	8.16	<.01
Error	158.80	10	15.88		
Total	418.00	12			

Because $F_{.01}(2, 10 \, df) = 7.56$, reject the null hypothesis and conclude that there is a relationship between suicide rate and a country's degree of modernization.

13.

Source	SS	df	MS	F	p
Treatments	515	3	171.667	17.46	<.01
Error	118	12	9.833		
Total	633	15			

$F_{.01}(3, 12 \, df) = 5.95$. Because $17.46 > 5.95$, reject the null hypothesis at the .01 level and conclude that the treatment method had an effect on the number of snake-approach responses. By the end of this chapter, you will be able to make comparisons among the pairs of means to determine which method is best (or worst).

14. *A priori* tests are those planned in advance. They are tests based on the logic of the design of an experiment. *Post hoc* tests are chosen after examining the data.

15. Recall that the mean suicide rates for the different degrees of modernization were: low—6.0; medium—12.0; high—16.8. Comparing countries with low and medium degrees of modernization:

$$HSD = \frac{12.0 - 6.0}{\sqrt{\dfrac{15.88}{4}}} = \frac{6.0}{1.99} = 3.02 \qquad NS \qquad HSD_{.05} = 3.88$$

Comparing countries with medium and high degrees of modernization:

$$HSD = \frac{16.8 - 12.0}{\sqrt{\dfrac{15.88}{2}\left(\dfrac{1}{4} + \dfrac{1}{5}\right)}} = \frac{4.8}{1.890} = 2.54 \qquad NS \qquad HSD_{.05} = 3.88$$

Comparing countries with low and high degrees of modernization:

$$HSD = \frac{16.8 - 6.0}{1.890} = \frac{10.8}{1.890} = 5.71 \qquad p < .01 \qquad HSD_{.01} = 5.27$$

Suicide rates are significantly higher in countries with a high degree of modernization than they are in countries with a low degree of modernization. Although the other differences were not significant, the trend is in the direction predicted by Durkheim's hypothesis.

16. I calculated HSDs for each of the six pairwise differences. For some of these HSDs the equal N formula was appropriate, and for some the unequal N formula was necessary. The two values of $s_{\bar{X}}$ are

$$s_{\bar{X}} = \sqrt{\frac{MS_{error}}{N_t}} = \sqrt{\frac{3.93}{5}} = 0.887$$

$$s_{\bar{X}} = \sqrt{\frac{MS_{error}}{2}\left(\frac{1}{N_1} + \frac{1}{N_2}\right)} = \sqrt{\frac{3.93}{2}\left(\frac{1}{5} + \frac{1}{7}\right)} = 0.821$$

The table shows all six HSD values. The significance level of each comparison is indicated with one or two asterisks. $HSD_{.05} = 4.00$, $HSD_{.01} = 5.09$.

	crf	FR2	FR4
FR2	4.63*		
FR4	7.44**	3.41	
FR8	9.47**	5.60**	2.03

* $p < .05$
** $p < .01$

The trend of the means is consistent; the higher the ratio of response to reinforcement during learning, the greater the persistence during extinction. The crf schedule produced the least persistence, significantly less than any other schedule. The FR2 schedule produced less persistence than the FR4 schedule (NS) and significantly less than the FR8 schedule ($p < .01$).

17. Comparing the best treatment (model) with the second best (film),

$$HSD = \frac{\bar{X}_1 - \bar{X}_2}{\sqrt{\dfrac{MS_{error}}{N_t}}} = \frac{26 - 18}{1.568} = 5.10; \quad p < .05; \quad HSD_{.05} = 4.20$$

Comparing the poorest treatment (desensitization) with the control group,

$$HSD = \frac{\bar{X}_3 - \bar{X}_4}{s_{\bar{X}}} = \frac{17 - 10}{1.568} = 4.46; \quad p < .05; \quad HSD_{.05} = 4.20$$

For treatment of phobias, using a live model and encouraging participation is significantly better than using a film or desensitization. All three of the treatments produced significant improvement compared to an untreated control group.

18. Normally distributed dependent variable, homogeneity of variance, and random sampling

19. $f = \dfrac{\sqrt{\dfrac{K-1}{N_{tot}}(MS_{treat} - MS_{error})}}{\sqrt{MS_{error}}} = \dfrac{\sqrt{\dfrac{4-1}{16}(171.667 - 9.833)}}{\sqrt{9.833}} = 1.76$

An effect size index of 1.76 is much larger than the $f = .40$ that Cohen identifies as a large effect; thus, the type of therapy given to those suffering from a phobia has a large effect on improvement.

20.

	Normals	Schizophrenics	Depressives	Σ
ΣX	178	138	72	388
ΣX^2	4162	2796	890	7848
$\bar{X}$	22.25	19.71	12.00	
N	8	7	6	21

Source	SS	df	MS	F	p
Disorder categories	376.31	2	188.16	11.18	<.01
Error	302.93	18	16.83		
Total	679.24	20			

$F_{.01}(2, 18) = 6.01$

Depressive and schizophrenic:

$$HSD = \frac{12.00 - 19.71}{\sqrt{\frac{16.83}{2}\left(\frac{1}{7}+\frac{1}{6}\right)}} = \frac{-7.71}{1.614} = -4.78; \; p < .01; \quad HSD_{.01} = 4.70$$

Depressive and normal:

$$HSD = \frac{12.00 - 22.25}{\sqrt{\frac{16.83}{2}\left(\frac{1}{6}+\frac{1}{8}\right)}} = \frac{-10.25}{1.567} = -6.54; \; p < .01; \quad HSD_{.01} = 4.70$$

Normal and schizophrenic:

$$HSD = \frac{22.25 - 19.71}{\sqrt{\frac{16.83}{2}\left(\frac{1}{8}+\frac{1}{7}\right)}} = \frac{2.54}{1.501} = 1.69; \; NS; \quad HSD_{.05} = 3.61$$

The R scores on the Rorschach test do not distinguish schizophrenics from normals. Both schizophrenics and normals score significantly higher than depressives.

21.

	ACS	4MAT	Control
$\bar{X}$	8.9	18.1	3.1

Source	SS	df	MS	F	p
Instruction methods	2288.533	2	1144.267	20.81	<.01
Error	3134.400	57	54.989		
Total	5422.933	59			

$F_{.01}(2, 55 \; df) = 5.01$
HSD values calculated for the three pairwise comparisons are:

ACS and 4MAT:	HSD = 5.55**
ACS and Control:	HSD = 3.50*
4MAT and Control:	HSD = 9.05**

$* \; p < .05 \quad ** p < .01 \quad HSD_{.01} = 4.37$
$HSD_{.05} = 3.44$

$$f = \frac{\sqrt{\frac{K-1}{N_{tot}}(MS_{treat} - MS_{error})}}{\sqrt{MS_{error}}} = \frac{\sqrt{\frac{3-1}{60}(1144.267 - 54.989)}}{\sqrt{54.989}} = 0.81$$

Three months after BSE training, women retained significantly more information if they learned with the 4MAT method than if they were taught with the ACS method. Either method is better than no training. The effect size index of .81 shows that BSE training has a very large effect on information retained three months later.

22.

Source	SS	df	MS	F	p
Injections	235.64	2	117.82	7.98	<.01
Error	339.40	23	14.76		
Total	575.04	25			

$F_{.01}(2, 23\ df) = 5.66$
HSD values calculated for the three pairwise comparisons are:

Strychnine and saline:	HSD = 4.00*
Strychnine and no injection:	HSD = 5.36**
Saline and no injection:	HSD = 1.29

$*p < .05$ $**p < .01$ $HSD_{.05} = 3.58$ $HSD_{.01} = 4.64$

Strychnine injections *improved* memory; significantly fewer trials were necessary to relearn the avoidance task for rats given a strychnine injection after learning.

CHAPTER 10

1. **a.** 3 × 2   **b.** 4 × 3   **c.** 2 × 2
   d. 4 × 3 × 3. This last is a factorial design in which there are *three* independent variables. This text does not cover the analysis of three-way ANOVAs. If you are interested in such complex designs, see footnote 1 in Chapter 10.
2. A main effect is the effect that the different levels of one factor have on the dependent variable.
3. An interaction occurs when the effect of one factor on the dependent variable depends on which level of the other factor is being administered.
4. Main effect
5. Main effect
6. **a.** The number of scores in each group (cell) must be the same.
   b. The samples in each cell must be independent.
   c. The levels of both factors must be chosen by the experimenter and not left to chance.
   d. The samples are drawn from normally distributed populations.
   e. The variances of the populations are equal.
   f. The samples are drawn randomly from the populations, or the participants are assigned randomly to the group.

7. i. a.

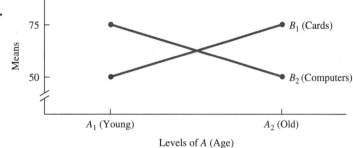

Levels of A (Age)

b. There appears to be an interaction.

c. Attitude toward playing the games depends on age. The young would rather play computer games, and the old would rather play cards.

d. There appear to be no main effects. Margin means for games are cards = 62.5 and computers = 62.5. Margin means for age are young = 62.5 and old = 62.5.

e. Later in this chapter, you will learn that when there is a significant interaction, the interpretation of main effects isn't simple when there is an interaction. Although all four means are exactly the same, there is a difference in attitudes toward the two games, but this difference depends on the age of the participant.

ii. a.

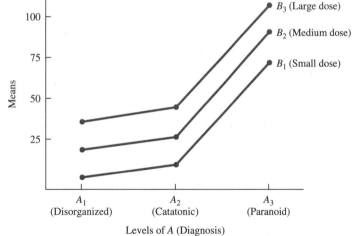

Levels of A (Diagnosis)

b. There appears to be no interaction.

c. The effect of the drug dosage does not depend on the diagnosis of the patient.

d. There appear to be main effects for both factors.

e. For all schizophrenics, the larger the dose, the higher the competency score (Factor B, main effect). Paranoids have higher competency scores than the others at all dose levels (Factor A, main effect).

iii. a.

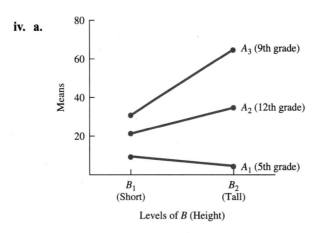

b. There appears to be no interaction.

c. The improvement observed with the three kinds of therapy does not depend on whether the patient is high or low in verbal ability.

d. There appear to be main effects for both factors.

e. High-verbal patients showed greater improvement than low-verbal patients. Patients who received cognitive therapy showed the most improvement. Psychoanalytic patients showed the least improvement; behavior modification patients were just above psychoanalytic patients.

iv. a.

b. There appears to be an interaction.

c. The effect of a person's height on self-consciousness depends on the person's grade in school. Height seems to make little difference in the 5th grade. In the 9th grade, tall students are much more self-conscious than short students, but this difference is much smaller in the 12th grade.

d. The main effects appear to be significant. The margin means are 20 and 35 for height; 7.5, 47.5, and 27.5 for grade level.

e. Here's an interpretation that is appropriate for the information you've been given so far. (More on interpretation of main effects when the interaction is significant will come later.) Height: tall students ($\bar{X} = 35$) are more self-conscious than short students ($\bar{X} = 20$). Grade level: 9th graders ($\bar{X} = 47.5$) are more self-conscious than 12th graders ($\bar{X} = 27.5$) or 5th graders ($\bar{X} = 7.5$).

v. a.

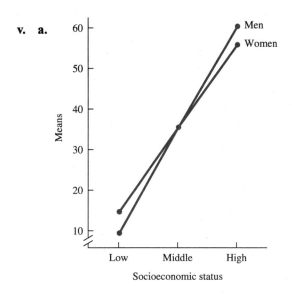

b. There appears to be no interaction. Although the lines in the graph are not parallel—they even cross—the departure from parallel is slight and probably should be attributed to sampling variation.

c. Gender differences in attitudes toward lowering taxes on investments do not depend on socioeconomic status.

d. There appears to be a main effect for socioeconomic status but not for gender.

e. Men and women do not differ in their attitudes toward lowering taxes on investments (Factor B). The higher the socioeconomic status, the more positive are attitudes toward lowering taxes on investments.

8. a. 2×2

b. Independent variables: gender and handedness; dependent variable: age at death

c. $SS_{tot} = 61,455 - \dfrac{855^2}{12} = 61,455 - 60,918.75 = 536.25$

$$SS_{cells} = \frac{219^2}{3} + \frac{186^2}{3} + \frac{234^2}{3} + \frac{216^2}{3} - \frac{855^2}{12}$$

$$= 61,323 - 60,918.75 = 404.25$$

$$SS_{gender} = \frac{453^2}{6} + \frac{402^2}{6} - 60,918.75 = 61,135.50 - 60,918.75$$

$$= 216.75$$

$$SS_{\text{handedness}} = \frac{405^2}{6} + \frac{450^2}{6} - 60{,}918.75 = 61{,}087.50 - 60{,}918.75$$
$$= 168.75$$

$$SS_{AB} = 3[(73 - 75.5 - 67.5 + 71.25)^2 + (62 - 67.5 - 67 + 71.25)^2$$
$$+ (78 - 75 - 75.5 + 71.25)^2 + (72 - 75 - 67 + 71.25)^2]$$
$$= 3(6.25) = 18.75$$

Check: $SS_{AB} = 404.25 - 216.75 - 168.75 = 18.75$

$$SS_{\text{error}} = \left[16{,}013 - \frac{219^2}{3}\right] + \left[11{,}574 - \frac{186^2}{3}\right]$$
$$+ \left[18{,}284 - \frac{234^2}{3}\right] + \left[15{,}584 - \frac{216^2}{3}\right]$$
$$= 26.00 + 42.00 + 32.00 + 32.00 = 132.00$$

Check: $536.25 = 404.25 + 132.00$

9. **a.** 2×3

b. Teacher response and gender

c. Errors on a comprehensive examination in arithmetic

d. $SS_{\text{tot}} = 13{,}281 - \dfrac{845^2}{60} = 1380.58$

$$SS_{\text{cells}} = \frac{(169)^2}{10} + \frac{(136)^2}{10} + \frac{(118)^2}{10} + \frac{(146)^2}{10} + \frac{(167)^2}{10} + \frac{(109)^2}{10} - \frac{(845)^2}{60}$$
$$= 306.28$$

$$SS_{\text{response}} = \frac{(315)^2}{20} + \frac{(303)^2}{20} + \frac{(227)^2}{20} - \frac{(845)^2}{60} = 227.73$$

$$SS_{\text{gender}} = \frac{(423)^2}{30} + \frac{(422)^2}{30} - \frac{(845)^2}{60} = 0.02$$

$$SS_{AB} = 10[(16.90 - 15.75 - 14.10 + 14.0833)^2$$
$$+ (13.60 - 15.15 - 14.10 + 14.0833)^2$$
$$+ (11.80 - 11.35 - 14.10 + 14.0833)^2$$
$$+ (14.60 - 15.75 - 14.0667 + 14.0833)^2$$
$$+ (16.70 - 15.15 - 14.0667 + 14.0833)^2$$
$$+ (10.90 - 11.35 - 14.0667 + 14.0833)^2]$$
$$= 78.53$$

Check: $SS_{AB} = 306.28 - 227.73 - 0.02 = 78.53$

$$SS_{error} = \left(3129 - \frac{(169)^2}{10}\right) + \left(2042 - \frac{(136)^2}{10}\right) + \left(1620 - \frac{(118)^2}{10}\right)$$

$$+ \left(2244 - \frac{(146)^2}{10}\right) + \left(2955 - \frac{(167)^2}{10}\right) + \left(1291 - \frac{(109)^2}{10}\right)$$

$$= 1074.30$$

Check: $1380.58 = 306.28 + 1074.30$

10. $SS_{tot} = 1405 - \dfrac{145^2}{20} = 353.75$

$$SS_{cells} = \frac{25^2}{5} + \frac{57^2}{5} + \frac{43^2}{5} + \frac{20^2}{5} - \frac{145^2}{20} = 173.35$$

$$SS_{learn} = \frac{68^2}{10} + \frac{77^2}{10} - \frac{145^2}{20} = 4.05$$

$$SS_{recall} = \frac{63^2}{10} + \frac{82^2}{10} - \frac{145^2}{20} = 18.05$$

$$SS_{AB} = 5[(8.60 - 6.80 - 6.30 + 7.25)^2 + (4.00 - 7.70 - 6.30 + 7.25)^2$$

$$+ (5.00 - 6.80 - 8.20 + 7.25)^2 + (11.40 - 7.70 - 8.20 + 7.25)^2]$$

$$= 151.25$$

Check: $SS_{AB} = 173.35 - 4.05 - 18.05 = 151.25$

$$SS_{error} = \left(400 - \frac{43^2}{5}\right) + \left(100 - \frac{20^2}{5}\right) + \left(155 - \frac{25^2}{5}\right) + \left(750 - \frac{57^2}{5}\right)$$

$$= 180.40$$

Check: $353.75 = 173.35 + 180.40$

11. There is not an equal number of scores in each cell.

12.

Source	SS	df	MS	F	p
A (gender)	216.75	1	216.75	13.14	<.01
B (handedness)	168.75	1	168.75	10.23	<.05
A × B	18.75	1	18.75	1.14	>.05
Error	132.00	8	16.50		
Total	536.25	11			

$F_{.05}(1, 8 \, df) = 5.32;\quad F_{.01}(1, 8 \, df) = 11.26$

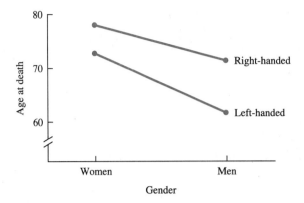

Interpretation: The interaction is not significant, so the main effects can be interpreted directly. Both gender and handedness produced significant differences. Right-handed people live longer than left-handed people ($p < .05$) and women live longer than men ($p < .01$).

13.

Source	SS	df	MS	F	p
A (response)	227.73	2	113.87	5.72	<.01
B (gender)	0.02	1	0.02	0.00	>.05
AB	78.53	2	39.27	1.97	>.05
Error	1,074.30	54	19.89		
Total	1,380.58	59			

$F_{.05}$ (2, 50 *df*) = 3.18; $F_{.01}$ (2, 50 *df*) = 5.06

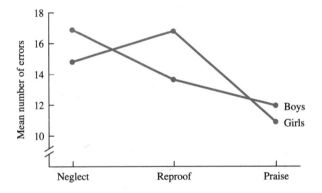

Interpretation: The interaction was not statistically significant. Only the teacher-response variable produced results that were statistically significant ($p < .01$). The null hypothesis $\mu_{\text{neglect}} = \mu_{\text{reproof}} = \mu_{\text{praise}}$ is rejected. The three kinds of teacher response produced different numbers of errors. The hypothesis that girls and boys were affected in the same way could not be rejected.

14.

Source	SS	df	MS	F	p
A (learning)	4.05	1	4.05	0.36	>.05
B (recall)	18.05	1	18.05	1.60	>.05
AB	151.25	1	151.25	13.41	<.01
Error	180.40	16	11.28		
Total	353.75	19			

$F_{.05}(1, 16\ df) = 4.49;\quad F_{.01}(1, 16\ df) = 8.53$

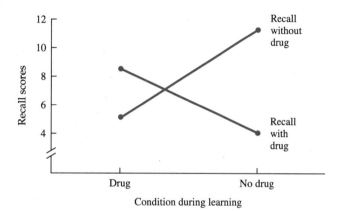

15. i. a. Independent variables: (1) treatment the participant received (insulted or treated neutrally) and (2) administrator of test (same or different). Dependent variable: number of captions produced.
   b. $F_{\text{treatments}} = 9.08;\ p < .01;\ F_{\text{experimenters}} = 6.82;\ p < .05;\ F_{AB} = 4.88;\ p < .05.$
   c. The significant interaction means that the effect of the experimenter *depended* on whether the participant was insulted or treated neutrally. If the participant had been insulted, then taking the humor test from the person who had done the insulting caused larger number of captions to be written—larger, that is, than if a new experimenter had administered the test. If the participant had not been insulted, there was little difference between the two kinds of experimenters.

Both main effects are statistically significant. For both main effects, the difference in margin means depends heavily on one cell, Cell A_1B_1. Interpreting the main effects for this problem isn't advisable.

ii. a. Independent variables: (1) name and (2) occupation of father. Dependent variable: score assigned to the theme.
   b. $F_{\text{name}} = 0.44;\quad p > .05;\quad F_{\text{occupation}} = 0.05;\quad p > .05;\quad F_{AB} = 8.20;\ p < .01.$
   c. The significant interaction means that the effect of the name depends on the occupation of the father. The effect of the name David is good if his father

is a research chemist (Cell A_1B_1) and bad if his father is unemployed (Cell A_1B_2). The effect of the name Elmer is good if his father is unemployed (Cell A_2B_2) and bad if his father is a research chemist (Cell A_2B_1).

The main effects are not significant. The average "David" theme is about equal to the average "Elmer" theme. Similarly, the themes do not differ significantly according to whether the father was unemployed or a research chemist. Notice that this interpretation of the main effects is misleading; that is, *names do have an effect*, but the effect depends on the occupation of the father. The appropriate response to these main effects is to ignore them.

iii. a. Independent variables: (1) gender of speaker and (2) job experience of speaker. Dependent variable: number of students looking at the speaker.

b. $F_{gender} = 4.59$; $p < .05$; $F_{experience} = 9.80$; $p < .01$; $F_{AB} = 0.24$; $p > .05$.

c. Because the interaction is not significant, the interpretations of the main effects are straightforward. The gender of the speaker was significant; students paid more attention to men than to women. The experience of the speaker was significant; students paid more attention to those who had been on the job for longer than two years than to those who had been on the job less than six months.

16. *Interpretation:* Only the interaction is significant. The conclusion is that when conditions of learning and recall are the same, memory is good. When conditions of learning and recall are different, memory is poorer.

17. No, because the interaction in that study was significant.

18.

Source	SS	df	MS	F	p
A (rewards)	199.31	3	66.44	3.90	<.05
B (classes)	2.25	1	2.25	<1	>.05
AB	0.13	3	0.04	<1	>.05
Error	953.25	56	17.02		
Total	1154.94	63			

$F_{.05}(3, 55\ df) = 2.78$

There is no significant interaction between rewards and class standing, and freshmen and seniors do not differ significantly. There is a significant difference among the rewards; that is, rewards have an effect on attitudes toward the police. Pairwise comparisons are called for.

The size of a mean difference that will be significant at the .05 level with $N = 16$ is 3.91, which was determined by the formula:

$$(HSD_{.05})(s_{\bar{X}}) = (3.79)\sqrt{\frac{17.02}{16}} = 3.91$$

To facilitate an overall view, I arranged the four means and the six mean differences into a table, marking with an asterisk the one that is larger than the critical value of 3.91.

	$10	$5	$1	50¢
Means	8.938	10.250	12.125	13.563
$10		1.31	3.19	4.63*
5			1.88	3.31
1				1.44

The less students were paid to defend the police, the more positive their attitude toward the police. Those paid 50 cents had significantly more positive attitudes toward the police than those paid ten dollars, $p < .05$.

19. $SS_{tot} = 817 - \dfrac{(129)^2}{24} = 123.625$

$SS_{cells} = \dfrac{(22)^2}{3} + \dfrac{(9)^2}{3} + \dfrac{(23)^2}{3} + \dfrac{(15)^2}{3} + \dfrac{(11)^2}{3} + \dfrac{(26)^2}{3} + \dfrac{(9)^2}{3}$
$\qquad + \dfrac{(14)^2}{3} - \dfrac{(129)^2}{24} = 104.29$

$SS_{gender} = \dfrac{(69)^2}{12} + \dfrac{(60)^2}{12} - \dfrac{(129)^2}{24} = 3.375$

$SS_{pictures} = \dfrac{(33)^2}{6} + \dfrac{(35)^2}{6} + \dfrac{(32)^2}{6} + \dfrac{(29)^2}{6} - \dfrac{(129)^2}{24} = 3.125$

$SS_{G \times P} = 3[(7.333 - 5.5 - 5.75 + 5.375)^2 + (3.667 - 5.50 - 5.0 + 5.375)^2$
$\qquad + (3.0 - 5.833 - 5.75 + 5.375)^2$
$\qquad + (8.667 - 5.833 - 5.0 + 5.375)^2$
$\qquad + (7.667 - 5.333 - 5.75 + 5.375)^2$
$\qquad + (3.0 - 5.333 - 5.0 + 5.375)^2$
$\qquad + (5.0 - 4.833 - 5.75 + 5.375)^2$
$\qquad + (4.667 - 4.833 - 5.0 + 5.375)^2]$
$\qquad = 3(32.60) = 97.79$

Check: $SS_{cells} = 3.375 + 3.125 + 97.79 = 104.29$

$SS_{error} = \left(166 - \dfrac{(22)^2}{3}\right) + \left(41 - \dfrac{(11)^2}{3}\right) + \left(29 - \dfrac{(9)^2}{3}\right) + \left(226 - \dfrac{(26)^2}{3}\right)$
$\qquad + \left(179 - \dfrac{(23)^2}{3}\right) + \left(29 - \dfrac{(9)^2}{3}\right) + \left(77 - \dfrac{(15)^2}{3}\right)$
$\qquad + \left(70 - \dfrac{(14)^2}{3}\right)$

$\qquad = 19.33$

Check: $123.62 = 104.29 + 19.33$

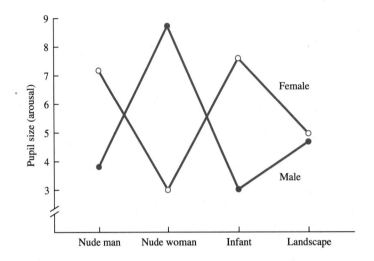

Source	MS	df	F	p
Gender	3.375	1	2.79	NS
Pictures	1.042	3	0.86	NS
Gender × Pictures	32.597	3	26.98	<.01
Error	1.208	16		
Total		23		

$F_{.05}(1, 16) = 4.49; \quad F_{.01}(3, 16) = 5.29$

Only the interaction is significant. Based on the graph, males were aroused by nude women and not by nude men or infants. Females, however, were aroused by nude men and infants and not by nude women. Responses toward the landscape were about the same for both genders.

CHAPTER 11

1.

SS_{tot}

$SS_{subjects}$ | SS_{treat} | SS_{error}

2.

Subject	Rest interval (weeks)			Σ
	2	4	8	
1	40	50	58	148
2	58	56	65	179
3	44	70	69	183
4	57	61	74	192
Σ	199	237	266	702
$\bar{X}$	49.75	59.25	66.50	

Source	SS	df	MS	F
Subjects	365.667	3		
Intervals	564.500	2	282.250	7.21
Error	234.833	6	39.139	
Total	1165.000	11		

Tukey HSD tests $HSD_{.05} = 4.34$ $s_{\bar{X}} = \sqrt{\dfrac{39.139}{4}} = 3.128$

$HSD_{8v4} = 2.32$ NS
$HSD_{8v2} = 5.35$ $p < .05$
$HSD_{4v2} = 3.04$ NS

These data show that as the rest interval increases from 2 weeks to 4 weeks to 8 weeks, there is an increase in percent of recall. That is, the longer the rest interval, the better the percent of recall. Only the difference between 2 weeks and 8 weeks was statistically significant, $p < .05$.

3. $HSD_{1st \ v. \ cont} = \dfrac{75 - 75}{\sqrt{\dfrac{1.333}{4}}} = \dfrac{-1}{.577} = -1.73$ $p > .05$

$HSD_{.05} = 4.34$. Although always staying with your first choice resulted in an average loss of one point, the difference was not statistically significant.

4. The comparison of the three therapy techniques does not lend itself to a correlated-measures design using the same participants. Clearly, it doesn't make sense to provide therapy for someone for a period sufficient to help and then use the next therapy on that same person.

5. 1. When the same subjects are serving at all treatment levels, you should be sure there are no carryover effects (unless you are looking for carryover effects).

   2. To use the analysis presented in this chapter, the levels of the independent variable must be chosen by the researcher and the subjects must be randomly assigned to rows.

6. The assumptions for a repeated measures ANOVA are (1) normal distribution of scores, (2) random sampling, and (3) the covariance matrix is spherical.

7.

Source	SS	df	MS	F'	p
Subjects	34543.33	4			
Periods	5890.00	2	2945.00	44.73	<.01
Error	526.67	8	65.83		
Total	40960.00	14			

Tukey HSD tests $\quad HSD_{.01} = 5.64 \quad s_{\bar{X}} = \sqrt{\dfrac{65.833}{5}} = 3.629$

$HSD_{before\ v\ during} = 11.30; \quad p < .01$
$HSD_{before\ v\ after} = 0.55; \quad NS$
$HSD_{after\ v\ during} = 11.85; \quad p < .01$

During meditation, oxygen consumption drops to about 80 percent of the level observed before and after meditation. The drop is statistically significant, $p < .01$. The oxygen consumption rates before and after meditation are not significantly different.

8. If the conclusion that meditation lowers oxygen consumption is wrong, it is a Type I error. If the conclusion that oxygen consumption before and after meditation is not significantly different is wrong, it is a Type II error.

9.

Matching characteristic	No therapy	Behavior therapy	Cognitive therapy	Drug therapy	ΣX_s
Females, at age 20	18	11	8	10	47
Males, at age 30	21	9	15	10	55
Females, at age 45	24	16	10	16	66
ΣX_t	63	36	33	36	168
$\bar{X}_t$	21	12	11	12	

Source	SS	df	MS	F
Subjects	45.500	2		
Therapies	198.000	3	66.000	8.16*
Error	48.500	6	8.083	
Total	292.000	11		

$F_{.05}(3, 6) = 4.76 \qquad\qquad *p < .05$

Tukey HSD tests $\quad HSD_{.05} = 4.90 \quad s_{\bar{X}} = \sqrt{\dfrac{8.083}{3}} = 1.641$

$HSD_{no\ v\ behavior} = 5.48; \quad p < .05$
$HSD_{no\ v\ drug} = 5.48; \quad p < .05$
$HSD_{no\ v\ cognitive} = 6.09; \quad p < .05$
$HSD_{drug\ v\ cognitive} = .61; \quad NS$

Each of the three therapies is significantly better than the no therapy control group. There are no significant differences among the three therapy groups.

CHAPTER 12

1. $\chi^2 = \Sigma\left[\dfrac{(O-E)^2}{E}\right] = \dfrac{(316-300)^2}{300} + \dfrac{(84-100)^2}{100}$

$= 0.8533 + 2.5600$

$= 3.4133; \quad df = 1$

$\chi^2_{.05}$ (1 df) = 3.84. Retain the null hypothesis and conclude that these data are consistent with a 3 : 1 hypothesis.

2. The expected frequencies are 20 correct and 40 incorrect: (1/3)(60) = 20 and (2/3)(60) = 40.

$\chi^2 = \dfrac{(32-20)^2}{20} + \dfrac{(28-40)^2}{40} = 10.80; \quad df = 1; \quad p < .01$

Thus, these three emotions can be distinguished from each other. Subjects did not respond in a chance fashion.

There is an interesting sequel to this experiment. Subsequent researchers did not find the simple, clear-cut results that Watson reported. One experimenter (Sherman, 1927) found that if only the infant's reactions were observed, there was a great deal of disagreement. However, if the observers also knew the stimuli (dropping, stroking, and so forth), they agreed with each other. This seems to be a case in which Watson's design did not permit him to separate the effect of the infant's reaction (the independent variable) from the effect of knowing what caused the reaction (an extraneous variable).

3.

	Received program	Control	Total
Police record	114	101	215
No police record	211	224	435
Total	325	325	650

The expected frequencies are:

$\dfrac{(215)(325)}{650} = 107.5; \quad \dfrac{(215)(325)}{650} = 107.5; \quad \dfrac{(435)(325)}{650} = 217.5;$

$\dfrac{(435)(325)}{650} = 217.5$

O	E	$O-E$	$(O-E)^2$	$\dfrac{(O-E)^2}{E}$
114	107.5	6.5	42.25	0.393
101	107.5	−6.5	42.25	0.393
211	217.5	−6.5	42.25	0.194
224	217.5	6.5	42.25	0.194
			$\chi^2 =$	1.174
			$df = 1$	

By the shortcut method:

$$\chi^2 = \frac{650[(114)(224) - (101)(211)]^2}{(215)(435)(325)(325)} = 1.17$$

$\chi^2_{.05}$ (1 df) = 3.84. Therefore, retain the null hypothesis and conclude that the two groups did not differ significantly in the number who had police records. The Cambridge-Somerville Youth Study, discussed in many sociology texts, was first reported by Powers and Witmer (1951).

Did you notice that the proportion of those with police records was *higher* for the boys who participated in the program?

4. $\chi^2 = \frac{100[(9)(34) - (26)(31)]^2}{(35)(65)(40)(60)} = 4.579$

Calculations of the expected frequencies are as follows.

$$\frac{(35)(40)}{100} = 14 \qquad \frac{(35)(60)}{100} = 21$$

$$\frac{(65)(40)}{100} = 26 \qquad \frac{(65)(60)}{100} = 39$$

O	E	$O - E$	$(O - E)^2$	$\dfrac{(O - E)^2}{E}$
9	14	−5	25	1.786
26	21	5	25	1.191
31	26	5	25	0.962
34	39	−5	25	0.641
				$\chi^2 = 4.58$
				$df = 1$

$\chi^2_{.05}$ (1 df) = 3.84. Reject the hypothesis that group size and joining are independent. Conclude that passersby are more likely to join a group of five than a group of two.

5. a.

		Recapture site	
		Issaquah	East Fork
Capture site	Issaquah	46 (34.027)	0 (11.973)
	East Fork	8 (19.973)	19 (7.027)

$$\chi^2 = \frac{73[(46)(19) - (0)(8)]^2}{(46)(27)(54)(19)} = 43.76$$

$\chi^2_{.001}$ (1 df) = 10.83. Reject the null hypothesis (which is that the second choice is independent of the first) and conclude that choices are very consistent; salmon tend to choose the same stream each time.

b.

Recapture site

		Issaquah	East Fork
Capture site	Issaquah	39 (40.071)	12 (10.929)
	East Fork	16 (14.929)	3 (4.071)

$$\chi^2 = \frac{70[(39)(3) - (12)(16)]^2}{(51)(19)(55)(15)} = 0.49$$

$\chi^2_{.05}$ (1 *df*) = 3.84. Hasler's hypothesis is supported; fish with plugged nasal openings do not make a consistent choice of streams, but those that get olfactory cues consistently choose the same stream. For further confirmation and a short summary of Hasler's work, see Hasler, Scholz, and Horrall (1978).

6. The work that needs to be done on the data is to divide each set of applicants into two independent categories, hired and not hired. The following table results. Expected values are in parentheses.

	White	African–American	Total
Hired	390 (339.327)	18 (68.673)	408
Not hired	3810 (3860.673)	832 (781.327)	4642
Total	4200	5050	5050

$$\chi^2 = \frac{5050[(390)(832) - (3810)(18)]^2}{(4200)(850)(408)(4642)} = \frac{3.3070 \times 10^{14}}{6.7514 \times 10^{12}} = 48.91$$

Given that $\chi^2_{.001}$ (1 *df*) = 10.83, the null hypothesis can be rejected. Because 9.29 percent of the white applicants were hired compared to 2.12 percent of the African-American applicants, the statistical consultant and the company in question concluded that discrimination had occurred.

7. To get the expected frequencies, multiply the percentages given by Professor Stickler by the 340 students. Then enter these expected frequencies in the usual table.

O	E	O − E	(O − E)²	$\frac{(O - E)^2}{E}$
20	23.8	−3.8	14.44	0.607
74	81.6	−7.6	57.76	0.708
120	129.2	−9.2	84.64	0.655
88	81.6	6.4	40.96	0.502
38	23.8	14.2	201.64	0.472

$$\chi^2 = 10.94$$
$$df = 5 - 1 = 4$$

(In this problem, the only restriction on the theoretical frequencies is that $\Sigma E = \Sigma O$.) $\chi^2_{.05}$ (4 df) = 9.49. Therefore, reject the contention of the professor that his grades conform to "the curve." By examining the data, you can also reject the contention of the colleague that the professor is too soft. The primary reason the data do not fit the curve is that there were too many "flunks."

8. Goodness of fit
9. Each of the six sides of a die is equally likely. With the total number of throws as 1200, the expected value for any one of them is $\frac{1}{6}$ times 1200, or 200. Thus, retain the null hypothesis. The results of the evening do not differ significantly from the "unbiased dice" model.

O	E	$O - E$	$(O - E)^2$	$\dfrac{(O - E)^2}{E}$
195	200	-5	25	0.125
200	200	0	0	0.000
220	200	20	400	2.000
215	200	15	225	1.125
190	200	-10	100	0.500
180	200	-200	400	2.000
Σ 1200	1200			$\chi^2 = 5.75$

$\chi^2_{.05}$ (5 df) = 11.07

10. Goodness of fit

11. O	E	$O - E$	$(O - E)^2$	$\dfrac{(O - E)^2}{E}$
21	12.735	8.265	68.310	5.365
5	13.265	-8.265	68.310	5.150
27	28.408	-1.408	1.982	0.070
31	29.592	1.408	1.982	0.067
0	6.857	-6.857	47.018	6.857
14	7.143	6.857	47.018	6.583

$\chi^2_{.001}$ (2 df) = 13.82 $\qquad \chi^2 = 24.09$

Reject the null hypothesis and conclude that Madison used the word *by* significantly more frequently than did Hamilton. The stage is now set to examine the 12 disputed papers for their use of *by*, an examination that Mosteller and Wallace carried out. They describe the rates they found as "Madisonian." In addition, other words that distinguished between Madison and Hamilton (*upon*, *also*, and others) were found in Madisonian rates in the 12 papers.

12. Independence
13. To convert the information you have into frequencies that can be analyzed with a χ^2, multiply the percentages (as proportions) by the number of observations. Thus, for *observed frequencies*:

.193 (140) = 27 schizophrenic offspring; .807 (140) = nonschizophrenic offspring

For the *expected frequencies*:

.25 (140) = 35 schizophrenic offspring; .75 (140) = 105 nonschizophrenic offspring

O	E	$O - E$	$(O - E)^2$	$\dfrac{(O - E)^2}{E}$
27	35	−8	64	1.829
113	105	−8	64	0.610
				$\chi^2 = 2.439$
				$df = 2 - 1 = 1$

$\chi^2_{.05}$ (1 df) = 3.84. Thus, retain the null hypothesis and conclude that these data are consistent with the 1 : 3 ratio predicted by the theory.

This problem illustrates the point that data that are consistent with the theory do not prove that the theory is true. Current explanations of schizophrenia include a genetic component, but not the simple one envisioned by the early hypothesis.

14. The difference between the tests for goodness of fit and for independence is in how the expected frequencies are obtained. In the goodness-of-fit test, expected frequencies are predicted by a theory, whereas in the independence test, they are obtained from the data.

15. The first task in this problem is to determine the expected frequencies. To do this, multiply each color's population percentage by the total number of M&Ms, 23. The summary table with the steps for calculating χ^2 is

Color	O	E	$O - E$	$(O - E)^2$	$\dfrac{(O - E)^2}{E}$
Brown	4	6.9	−2.9	8.41	1.219
Red	6	4.6	1.4	1.96	0.426
Yellow	8	4.6	3.4	11.56	2.513
Green	2	2.3	−0.3	0.09	0.039
Orange	3	2.3	0.7	0.48	0.213
Blue	0	2.3	−2.3	5.29	2.300
					$\chi^2 = 6.71$

Degrees of freedom is number of categories minus one; $df = 6 - 1 = 5$. The critical value is $\chi^2_{.05}$ (5 df) = 11.07. Retain the null hypothesis. The observed frequencies are not significantly different from the expected frequencies, which are based on the company's claims.

Perhaps you have some complaints about this problem. (If not, reread the problem and its answer before reading further.)

Perhaps your complaint was that an N of 23 isn't large enough for a goodness-of-fit problem with six categories. A more sophisticated version of the complaint is that five of the six categories have expected values that are less than 5. The most sophisticated version is that such a small sample carries a very high probability of making a Type II error.

Fortunately, this is the kind of problem you can gather your own data on. Of course, you *must* use a larger sample (and having a research partner would be nice, too). Should you and your partner find yourselves having to test peanut M&Ms, the percentages are the same except that brown is 20 percent and blue is 20 percent. Finally, I want to acknowledge my debt to Randolph A. Smith, who explained to me the value of M&Ms as a statistical tool.

16. You should be gentle with your friend but explain that the observations in his study are not independent and cannot be analyzed with χ^2. Explain that, because one person is making five observations, the observations are not independent; the choice of one female candidate may cause the subject to pick a male candidate next (or vice versa).

17.

Candidates

		Hill	Dale	Σ
Signaled turn	Yes	57 (59.248)	31 (28.752)	88
	No	11 (8.752)	2 (4.248)	13
	Σ	68	33	101

$$\chi^2 = \frac{101[(57)(2) - (31)(11)]^2}{(88)(13)(68)(33)} = 2.03$$

$\chi^2_{.05}$ (1 df) = 3.84. Cars with bumper stickers for Hill and for Dale were observed as they turned left at a busy intersection. Sixteen percent of the 68 Hill cars failed to signal their left turn. Six percent of the 33 Dale cars failed to signal. This difference was not statistically significant; that is, a difference this big would be expected more than 5 times in 100 if there is really no difference in percentages among the supporters of the two candidates.

18. I hope you would say something like, "Gee, I've been able to analyze data like these since I learned the t test. I will need to know the standard deviations for those means, though." If you attempted to analyze these data using χ^2, you erred (which, according to Alexander Pope, is human). If you erred, forgive yourself and reread pages 288 and 305.

19.

Houses

		Brick	Frame	Σ
Candidates	Hill	17 (37.116)	59 (38.884)	76
	Dale	88 (67.884)	51 (71.116)	139
	Σ	105	110	215

$$\chi^2 = \frac{N(AD - BC)^2}{(A + B)(C + D)(A + C)(B + D)}$$

$$= \frac{215(867 - 5192)^2}{(76)(139)(105)(110)} = 32.96; \quad df = 1$$

$\chi^2_{.001}$ (1 df) = 10.83; $p < .001$. Houses with yard signs for Hill and for Dale were classified as brick (affluent) or one-story frame (less affluent). Eighty-four percent of the 105 affluent houses had Dale signs; 16 percent had Hill signs. The relationship was in the other direction at the 110 less affluent houses: 46 percent had Dale signs and 54 percent had Hill signs. These differences, which are statistically significant, support the conclusion that Dale's supporters are more affluent than Hill's supporters.

20. $df = (R - 1)(C - 1)$, except for a table with only one row.

a. 3   **b.** 12   **c.** 3   **d.** 10

CHAPTER 13

1. Sum of the ranks of those given estrogen = 76
Sum of ranks of the control animals = 44

Smaller $U = (7)(8) + \dfrac{(7)(8)}{2} - 76 = 8$

Larger $U = 48$

To find the critical value, look in Table H for a two-tailed test with $\alpha = .05$ (the second page). At the intersection of $N_1 = 7$, $N_2 = 8$, you find a U value of 10. The obtained U of 8 is smaller than the critical value, so you should reject the null hypothesis. By examining the data and noting that the estrogen-injected animals were the lowest ranking ones, you can conclude that estrogen causes rats to be less dominant.

If you gave the rat that won 0 bouts a rank of 1, the sum of ranks for the estrogen-injected group is 36; the sum for the sesame oil–injected group is 84. You get the same U values, however, from these sums.

2. If you gave 0 parasites a rank of 2, and 110 parasites a rank of 40, then the sum of the ranks for Present Day is 409.5 and the sum of the ranks for Ten Years Earlier is 410.5. If you ranked the number of parasites in reverse order, the sum of the ranks for Present Day is 574.5 and the sum of the ranks for Ten Years Earlier is 245.5. In either case, the smaller U is 109.5 and the larger U is 274.5.

$$U = (24)(16) + \frac{(24)(25)}{2} - 574.5 = 109.5$$

$$\mu_U = \frac{(24)(16)}{2} = 192$$

$$\sigma_U = \sqrt{\frac{(24)(16)(41)}{12}} = 36.22$$

$$z = \frac{(109.5 + 0.5) - 192}{36.22} = -2.26$$

The critical value of z is ± 1.96, so you may reject the null hypothesis that the distributions are the same. By examining the average ranks (23.9 for present-day birds and 15.3 for 10-years-ago birds), you should conclude that present-day birds have significantly fewer brain parasites. (If you gave the 0 birds a rank of 2, the average ranks are 17.1 for present-day birds and 25.7 for 10-years-ago birds. The same conclusion is reached, however.)

3. With $\alpha = .01$, there are *no* possible results that would allow H_0 to be rejected. We must find more cars.

4. $\Sigma R_Y = 13$, $\Sigma R_Z = 32$.

$$\text{Smaller } U = (4)(5) + \frac{(5)(6)}{2} - 32 = 3$$

Even with $\alpha = .05$ and a two-tailed test, a U of 1 or less is required to reject H_0. With an obtained $U = 3$, you must conclude that the quietness test did not produce evidence that Y cars are quieter than Z cars.

5. $$\mu_T = \frac{N(N + 1)}{4} = \frac{112(113)}{4} = 3164$$

$$\sigma_T = \sqrt{\frac{N(N + 1)(2N + 1)}{24}} = \sqrt{\frac{112(113)(225)}{24}} = 344.46$$

$$z = \frac{(4077 + 0.5) - 3164}{344.46} = 2.65$$

Because 2.65 is greater than 1.96, reject H_0 and conclude that incomes were significantly different after the program. Because you do not have the actual data, you cannot tell whether the incomes were higher or lower than before.

6.

Worker	Without rests	With rests	D	Signed ranks
1	2240	2421	181	2
2	2069	2260	191	4
3	2132	2333	201	5
4	2095	2314	219	6
5	2162	2297	135	1
6	2203	2389	186	3

Check: $21 + 0 = 21$

$$\frac{(6)(7)}{2} = 21$$

Σ positive $= 21$
Σ negative $= 0$
$T = 0$
$N = 6$

The critical value of T at the .05 level for a two-tailed test is 0. Because 0 (obtained) is equal to or less than 0 (table), reject the null hypothesis and conclude that the output with rests is greater than the output without rests. Here is the story behind this study.

From 1927 to 1932 the Western Electric Company conducted a study on a group of six workers at their Hawthorne Works near Chicago. The final conclusion reached by this study has come to be called the Hawthorne effect. In the study, six workers who assembled telephone relays were separated from the rest of the workers. A variety of changes in their daily routine followed, one at a time, with the following results. Five-minute rest periods morning and afternoon increased output; 10-minute rest periods increased output; company-provided snacks during rest periods increased output; and quitting 30 minutes early increased output. Going back to the original no-rest schedule increased output again, as did the reintroduction of the rest periods. Finally, the management concluded that it was the special attention paid to the workers, rather than the rest periods, that increased output. Thus, the Hawthorne effect is an improvement in performance that is due just to being in an experiment (getting special attention) rather than to any specific manipulation in the experiment. For a summary of this study, see Mayo (1946).

7. The appropriate test for this study is one for two independent samples. A Mann-Whitney U test will work. If high scores are given high rank (that is, 39 ranks 1):

Σ (Canadians) $= 208$

Σ (U.S.) $= 257$

(If low scores are given high ranks, the sums are reversed.)

$$U = (15)(15) + \frac{(15)(16)}{2} - 257 = 88$$

By consulting the second page of Table H (boldface type), you will find that a U value of 64 is required in order to reject the null hypothesis ($\alpha = .05$, two-tailed test). The lower of the two calculated U values for this problem is 88. (The U value using the Canadian sum is 137.) Because 88 is larger than the tabled value of 64, the null hypothesis must be retained. Thus, our political science student must conclude that there is no strong evidence that Canadians and people from the United States have different attitudes toward the regulation of business.

Once again you must be cautious in writing the conclusion when the null hypothesis is retained. You have not demonstrated that the two groups are the same; you have shown only that the groups were not significantly different, which is a rather unsatisfactory way to leave things.

8.

Student	Before	After	D	Signed rank
1	18	4	14	13
2	14	14	0	1.5
3	20	10	10	9
4	6	9	−3	−4
5	15	10	5	6
6	17	5	12	11
7	29	16	13	12
8	5	4	1	3
9	8	8	0	−1.5
10	10	4	6	7
11	26	15	11	10
12	17	9	8	8
13	14	10	4	5
14	12	12	0	Eliminated

Check: $85.5 + 5.5 = 91$

$$\frac{(13)(14)}{2} = 91$$

Σ positive $= 85.5$

Σ negative $= -5.5$

$T = 5.5$

$N = 13$

The tabled value for T for a two-tailed test with $\alpha = .01$ is 9, so you may reject H_0 and conclude that the after-distribution is from a different population than the before-distribution. Except for one person, the number of misconceptions stayed the same or decreased, so you may conclude that the course *reduced* the number of misconceptions.

I would like to remind you here of the distinction made in Chapter 8 between statistically significant and important. There is a statistically significant decrease in the number of misconceptions, but a professor might be quite dismayed at the number of misconceptions that remain.

9. Σ positive $= 81.5$ Check: $81.5 + 54.5 = 136$

Σ negative $= -54.5$ $$\frac{(16)(17)}{2} = 136$$

$T = 54.5$

$N = 16$

A $T \le 29$ is required for rejection at the .05 level. Thus, there is no significant difference in the weight 10 months later. Put in the most positive language, there is no statistically significant evidence that the participants gained back the weight lost during the workshop.

10. Arranging the data into a summary table:

Groups

		1	2	3	4	5
	2	85				
	3	31	54			
Groups	4	18	67	13		
	5	103	188*	134	121	
	6	31	116	62	49	72

* $p < .05$

For $K = 6$, $N = 8$, differences of 159.6 and 188.4 are required to reject the null hypothesis at the .05 and .01 levels, respectively. Thus, the only significant difference is between the means of groups 2 and 5 at the .05 level. (With data such as these, the novice investigator may be tempted to report "almost significant at the .01 level." Resist that temptation.)

11.

Authoritarian		Democratic		Laissez-faire	
X	Rank	X	Rank	X	Rank
77	5	90	10.5	50	1
86	9	92	12	62	2
90	10.5	100	16	69	3
97	13	105	21	76	4
100	16	107	22	79	6
102	19	108	23	82	7
120	27.5	110	24	84	8
121	29	118	26	99	14
128	32	125	30	100	16
130	33	131	34	101	18
135	37	132	35.5	103	20
137	38	132	35.5	114	25
141	39.5	146	41	120	27.5
147	42	156	44	126	31
153	43	161	45	141	39.5
Σ	393.5		419.5		222.0

	Authoritarian (393.5)	Democratic (419.5)	Laissez-faire (222)
Democratic (419.5)	26		
Laissez-faire (222)	171.5*	197.5*	

$* p < .05$

For $K = 3$, $N = 15$, a difference in the sum of the ranks of 168.6 is required to reject H_0 at the .05 level. Therefore, both the authoritarian leadership and the democratic leadership resulted in higher personal satisfaction scores than did laissez-faire leadership, but the authoritarian and democratic types of leadership did not differ significantly from each other.

12. The Mann-Whitney U test can be performed on ranked data from two independent samples with equal or unequal N's. The Wilcoxon matched-pairs signed-ranks T test can be performed on ranked data from two correlated samples. The Wilcoxon–Wilcox multiple-comparisons test makes all possible pairwise comparisons among K independent equal-N groups of ranked data.

13. **a.** Wilcoxon matched-pairs signed-ranks test; this is a before-and-after study.

   b. Wilcoxon matched-pairs signed-ranks test; pairs are formed by family.

   c. Spearman's r_s; the degree of relationship is desired.

   d. Wilcoxon-Wilcox multiple-comparisons test; the experiment has three independent groups.

   e. Wilcoxon matched-pairs signed-ranks test; again, this is a before-and-after design.

14.

	Symbol of statistic	Appropriate for what design?	Calculated statistics must be (larger, smaller) than the tabled statistic to reject H_0
Mann-Whitney test	U	Two independent samples	Smaller
Wilcoxon matched-pairs signed-ranks test	T	Two correlated samples	Smaller
Wilcoxon-Wilcox multiple-comparisons test	None	More than two independent samples, equal N's	Larger

15.

Year	Rank in marriages (X)	(X^2)	Rank in grain (Y)	(Y^2)	$(X)(Y)$
1930	1	1.00	3	9.00	3.0
1931	2	4.00	1	1.00	2.0
1927	3	9.00	4	16.00	12.0
1925	4	16.00	2	4.00	8.0
1926	5	25.00	5	25.00	25.0
1923	6	36.00	6	36.00	36.0
1922	7	49.00	7	49.00	49.0
1919	8.5	72.25	11	121.00	93.5
1924	8.5	72.25	9	81.00	76.5
1928	10	100.00	9	81.00	90.0
1929	11	121.00	9	81.00	99.0
1918	12	144.00	12.5	156.25	150.0
1921	13	169.00	12.5	156.25	162.5
1920	14	196.00	14	196.00	196.0
Σ	105	1014.5	105	1012.5	1002.5

This is a case of tied ranks, so I will compute a Pearson r on the ranks.

$$r = \frac{N\Sigma XY - (\Sigma X)(\Sigma Y)}{\sqrt{[N\Sigma X^2 - (\Sigma Y)^2][N\Sigma Y^2 - (\Sigma Y)^2]}}$$

$$= \frac{14(1002.5) - (105)(105)}{\sqrt{[(14)(1014.5) - (105)^2][(14)(1012.5) - (105)^2]}} = .95$$

Consulting Table A for the significance of correlation coefficients, we find that for 12 df, a coefficient of .78 is significant at the .001 level. The sentence of explanation might be: There is a highly significant relationship between the number of marriages and the value of the grain crop; as one goes up the other does too.

16. $r_s = 1 - \dfrac{6\Sigma D^2}{N(N^2 - 1)} = 1 - \dfrac{6(308)}{16(16^2 - 1)} = .55$

From Table L, a coefficient of .503 is required for significance (two-tailed test). Thus, an $r_s = .55$ is significantly different from .00.

17. If the correlation is low, the philosophers probably have different sets of criteria of what is important. On the other hand, a high correlation probably means they have about the same set of criteria.

Candidates	Locke	Kant	D	D^2
A	7	8	−1	1
B	10	10	0	0
C	3	5	−2	4
D	9	9	0	0
E	1	1	0	0
F	8	7	1	1
G	5	3	2	4
H	2	4	−2	4
I	6	6	0	0
J	4	2	2	4
				$\Sigma D^2 = 18$

$$r_s = 1 - \frac{6\Sigma D^2}{N(N^2 - 1)} = 1 - \frac{6(18)}{10(99)} = 1 - .109 = .891$$

The two professors seem to be in pretty close agreement on the selection criteria.

18. Unfortunately, with only four pairs of scores there is no possible way to reject the hypothesis that the population correlation is .00 (see Table L). Advise your friend that more data must be obtained before any inference can be made about the population.

WHAT WOULD YOU RECOMMEND? CHAPTERS 9–13

A. A Mann-Whitney U test should be used to compare flooding and systematic desensitization. The two groups are independent samples—there is no reason to pair the score of a trainee who experienced flooding with that of a trainee who experienced systematic desensitization. A nonparametric test is called for because the data were skewed.   *goodness of fit test or indendence*

B. The proper analysis for these data is a chi square ~~test of independence~~. The score that a participant received was counted as a frequency in a category (comply or not comply).

C. These data should be analyzed with a factorial ANOVA. The two independent variables are the looks and intentionality of the defendant. The dependent variable was the likelihood estimate (a quantitative measure). (The factorial ANOVA of these data produced a significant interaction. Baby-faced defendants were less likely to be convicted of an intentional offense than a mature-faced defendant, but more likely to be convicted of a negligent offense.)

D. A Spearman r_s gives the degree of relationship between these two variables. A Spearman r_s is more appropriate than a Pearson r because one of the variables is a set of ranks.

E. The theory, which predicts the number of participants on subsequent occasions of the exercise, can be evaluated using a chi square goodness-of-fit test. (The expected values would be 50, 25, and 12.5.)

F. This experiment has one independent variable that has three levels. Because the variance of the "no cues" condition was much larger than that of the other two conditions, a Wilcoxon-Wilcox multiple-comparisons test is appropriate.

G. A Wilcoxon matched-pairs signed-ranks T test is appropriate for these skewed data (ounces eaten). This is a correlated-samples design because a before-meal score is paired with an after-meal score. (Both scores were obtained from the same dog.)

H. To compare each pair of recipes, a Tukey HSD would be appropriate. (No evidence in the problem suggests that the data are not normally distributed or have unequal variances.)

I. A one-factor correlated-measures ANOVA will help evaluate the program. Tukey HSDs can determine the immediate effect of the program (posttest vs. pretest), the long-term effect (follow-up vs. pretest), and effect of time on retention (follow-up vs. posttest).

CHAPTER 14

Set A

1. A mean of a set of means is needed, and to find it you need to know the number of dollars invested in each division.

2. Wilcoxon matched-pairs signed-ranks T test. You have evidence that the population of dependent-variable scores (reaction time) is not normally distributed.

3. A one-factor correlated-measures ANOVA for four treatments would be appropriate.

4. A median is appropriate because the scores are skewed.

5. Assume that these measures are normally distributed and use a normal curve to find the proportion. (.0228 is the answer.)

6. Either an independent-measures t test or a Mann-Whitney U test will do.

7. A 95 or 99 percent confidence interval about the mean reading achievement of those 50 sixth graders will answer the question.

8. χ^2 goodness of fit

9. Mode

10. Neither will do. The relationship described is nonlinear—one that first increases and then decreases. Neither r nor r_s is appropriate. A statistic, *eta*, is appropriate for this.

11. Correlated-samples t test or Wilcoxon matched-pairs signed-ranks T test

12. A regression equation will provide the predictions and a correlation coefficient will indicate how accurate the predictions will be.

13. A 2×2 factorial ANOVA

14. The student's wondering may be translated into a question of whether a correlation of .20 is a statistically significant one. Use a t test to find whether $r = .20$ for that class is significantly different from $r = .00$. Or, look in Table A in Appendix B.

15. This is a χ^2 problem but it cannot be worked using the techniques in this text because these before-and-after data are correlated, not independent. Intermediate texts describe appropriate techniques for correlated χ^2 problems.

16. Finding the probability that the sample mean came from a population in which $\mu = 1000$ (a one-sample t test) will answer the question.

17. Pearson product-moment correlation coefficient

18. Wilcoxon-Wilcox multiple-comparisons test (if the reaction-time scores are skewed) or one-way analysis of variance.

19. A χ^2 test of independence will determine whether a decision to report shoplifting is influenced by gender (one χ^2 test) and by the dress of the shoplifter (a second χ^2 test).

20. A line graph with serial position on the X axis and number of errors on the Y axis will illustrate this relationship.

21. A Mann-Whitney U test would be preferred, because the dependent variable is skewed. The question is whether there is a relationship between diet and cancer. (Note that if the two groups differ, two interpretations are possible. For example, if the incidence is higher among the red meat cultures, it might be because of the red meat or because of the lack of cereals.)

22. An r_s will give the degree of relationship for these two ranked variables.

23. For each species of fish, a set of z scores may be calculated. The fish with the largest z score should be declared the overall winner.

24. Calculating an effect size for the experiments mentioned in the first part of this question will tell you how big a difference psychotherapy makes.

Set B

25. The critical values are 377.6 at the .05 level and 457.6 at the .01 level. You can conclude that Herbicide B is significantly better than A, C, or D at the .01 level and that D is better than C at the .05 level.

26. An effect size of .10 is quite small. Using Cohen's (1969) conventions, effect sizes of .20, .50, and .80 are classified as small, medium, and large. In the case of a company offering instruction, a large number of clients provides a large N, which can ensure a "statistically significant improvement," even if the effect size is quite small.

27. The screening process is clearly worthwhile in making each batch profitable. Of course, the cost of screening will have to be taken into account.

28. $t_{.001}$ (40 df) = 3.55; therefore, reject the null hypothesis and conclude that extended practice *improved* performance. Because this is just the opposite of the theory's prediction, we may conclude that the theory is not adequate to explain the serial position effect.

29. Because $\chi^2_{.05}$ (2 df) = 5.99, the null hypothesis for this goodness-of-fit test is retained. The data do fit the theory; the theory is adequate.

30. The overall F is significant ($p < .01$), so conclude that the differences should not be attributed to chance. The low dose is significantly better than the placebo ($HSD_{.01}$ = 4.70) and significantly better than the high dose (which has a smaller mean than the placebo). The placebo and high dose are not significantly different.

31. The critical value of U for a two-tailed test with α = .01 is 70. You may conclude that the difference between the two methods is statistically significant at the .01 level. You cannot, from the information supplied, tell which of the two methods is superior.

32. The critical value at the .001 level of an r_s with N = 18 is .708 (Table A). Attitudes of college students and people in business toward the 18 groups are similar. There is little likelihood that the similarity of attitudes found in the sample is due to chance.

33. $t_{.05}(60\ df) = 2.00$; $F_{.01}(30, 40\ df) = 2.20$; the poetry unit appears to have no significant effect on mean attitudes toward poetry. There is a very significant effect on the variability of the attitudes of those who studied poetry. It appears that the poetry unit turned some students on and some students off, thus causing a great deal of variability in the attitude scores.

34. $F_{.05}(1, 44\ df) = 4.06$; $F_{.01}(1, 44\ df) = 7.24$; only the interaction is significant. A graph, as always, will help in the interpretation. See the graph here. The question of whether to present one or both sides to get the most attitude change depends on the level of education of the audience. If the members of a group have less than a high school education, present one side. If they have some college education, present both sides.

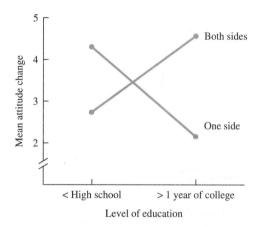

35. The tabled value of $F_{.05}(2, 60\ df) = 3.15$, so the null hypothesis is retained. There is no significant difference in the verbal ability scores of third graders who are left- or right-handed or mixed.

36. Because $\chi^2_{.02}(2\ df) = 7.82$, the null hypothesis for this test of independence is rejected at the .02 level. Alcoholism and toilet-training are not independent; they are related. Experiments such as this do not allow you to make cause-and-effect statements, however.

APPENDIX A

Part 1 (Pretest)

1. About 16	**2.** About 16	**3.** About 16	**4.** 6.1
5. 0.4	**6.** 10.3	**7.** 11.424	**8.** 27.141
9. 19.85	**10.** 2.065	**11.** 49.972	**12.** 0.0567
13. 3.02	**14.** 17.11	**15.** 1.375	**16.** 0.83
17. 0.1875	**18.** 0.056	**19.** 0.60	**20.** 0.10
21. 1.50	**22.** 3.11	**23.** 6	**24.** −22
25. −7	**26.** −6	**27.** 25	**28.** −24
29. 3.33	**30.** −5.25	**31.** 33%	**32.** 22.68
33. 0.19	**34.** 60 birds	**35.** 5	**36.** 4

37. 6, 10	**38.** 4, 22	**39.** 16	**40.** 6.25
41. 0.1225	**42.** 3.05	**43.** 0.96	**44.** 0.30
45. 4.20	**46.** 4.00	**47.** 4.50	**48.** 0.45
49. 15.25	**50.** 8.00	**51.** 16.25	**52.** 1.02
53. 12.00	**54.** 10, 30	**55.** 12, 56	**56.** 13.00
57. 11.00	**58.** 2.50	**59.** −2.33	

Part 2 (Review)

1. a. A sum is the answer to an addition problem.

   b. A quotient is the answer to a division problem.

   c. A product is the answer to a multiplication problem.

   d. A difference is the answer to a subtraction problem.

2. a. About 10   **b.** About 7   **c.** About 100   **d.** About 0.2

   e. About 0.16   **f.** About 60   **g.** Just less than 1   **h.** About 0.1

3. a. 14   **b.** 126   **c.** 9   **d.** 0   **e.** 128

   f. 13   **g.** 12   **h.** 13   **i.** 9

4. a. 6.33   **b.** 13.00   **c.** 0.05   **d.** 0.97   **e.** 2.61

   f. 0.34   **g.** 0.00   **h.** 0.02   **i.** 0.99

5.
```
 0.001
10.000
 3.652
 2.500
16.153
```

6.
```
 14.20
 −7.31
  6.89
```

7.
```
  1.26
× 0.04
0.0504
```

8.
```
143.300
 16.920
  2.307
  8.100
170.627
```

9.
```
        76.5
0.04)3.06
      28
      26
      24
      20
      20
```

10.
```
          2.04
11.75)24.00
      23 50
       5000
       4700
        300
```

11.
```
  152.12
− 127.40
   24.72
```

12.
```
   0.07
 × 0.5
 0.035
```

13. $\dfrac{9}{10} + \dfrac{1}{2} + \dfrac{2}{5} = 0.90 + 0.50 + 0.40 = 1.80$

14. $\dfrac{9}{20} \div \dfrac{19}{20} = 0.45 \div 0.95 = 0.47$

15. $\left(\dfrac{1}{3}\right)\left(\dfrac{5}{6}\right) = 0.333 \times 0.833 = 0.28$ **16.** $\dfrac{4}{5} - \dfrac{1}{6} = 0.80 - 0.167 = 0.63$

17. $\dfrac{1}{3} \div \dfrac{5}{6} = 0.333 \div 0.833 = 0.40$ **18.** $\dfrac{3}{4} \times \dfrac{5}{6} = 0.75 \times 0.833 = 0.62$

19. $18 \div \dfrac{1}{3} = 18 \div 0.333 = 54$

20. a. $(24) + (-28) = -4$ **b.** -23

   c. $(-8) + (11) = 3$ **d.** $(-15) + (8) = -7$

21. -40

22. 24

23. 48

24. -33

25. $(-18) - (-9) = -9$

26. $14 \div (-6) = -2.33$

27. $12 - (-3) = 15$

28. $(-6) - (-7) = 1$

29. $(-9) \div (-3) = 3$

30. $(-10) \div 5 = -2$

31. $4 \div (-12) = -0.33$

32. $(-7) - 5 = -12$

33. $6 \div 13 = .46$

34. $.46 \times 100 = 46$ percent

35. $18 \div 25 = .72$

36. $85 \div 115 = 0.74 \times 100 = 74$ percent

37. 25.92   **38.** .31   **39.** 128

40. 187   **41.** $|-31| = 31$

42. $|21 - 25| = |-4| = 4$

43. $12 \pm (2)(5) = 12 \pm 10 = 2, 22$

44. $\pm(5)(6) + 10 = \pm 30 + 10 = -20, 40$

45. $\pm(2)(2) - 6 = \pm 4 - 6 = -10, -2$

46. $(2.5)^2 = 2.5 \times 2.5 = 6.25$ **47.** $9^2 = 9 \times 9 = 81$

48. $0.3 \times 0.3 = 0.09$ **49.** $(1/4)^2 = (0.25)(0.25) = 0.0625$

50. a. 25.00   **b.** 2.50   **c.** 0.79   **d.** 0.25   **e.** 4.10
     f. 0.0548   **g.** 426.00   **h.** 0.50   **i.** 4.74

51. $\dfrac{(4-2)^2 + (0-2)^2}{6} = \dfrac{2^2 + (-2)^2}{6} = \dfrac{4+4}{6} = \dfrac{8}{6} = 1.33$

52. $\dfrac{(12-8)^2 + (8-8)^2 + (5-8)^2 + (7-8)^2}{4-1}$

$$= \dfrac{4^2 + 0^2 + (-3)^2 + (-1)^2}{3}$$

$$= \dfrac{16 + 0 + 9 + 1}{3} = \dfrac{26}{3} = 8.67$$

53. $\left(\dfrac{5+6}{3+2-2}\right)\left(\dfrac{1}{3} + \dfrac{1}{2}\right) = \left(\dfrac{11}{3}\right)(0.333 + 0.500) = (3.667)(0.833) = 3.05$

54. $\left(\dfrac{13+18}{6+8-2}\right)\left(\dfrac{1}{6} + \dfrac{1}{8}\right) = \left(\dfrac{31}{12}\right)(0.167 + 0.125) = (2.583)(0.292) = 0.75$

55. $\dfrac{8[(6-2)^2 - 5]}{(3)(2)(4)} = \dfrac{8[4^2 - 5]}{24} = \dfrac{8[16-5]}{24} = \dfrac{8[11]}{24} = \dfrac{88}{24} = 3.67$

56. $\dfrac{[(8-2)(5-1)]^2}{5(10-7)} = \dfrac{[(6)(4)]^2}{5(3)} = \dfrac{[24]^2}{15} = \dfrac{576}{15} = 38.40$

57. $\dfrac{6}{1/2} + \dfrac{8}{1/3} = (6 \div 0.5) + (8 \div 0.333) = 12 + 24 = 36$

58. $\left(\dfrac{9}{2/3}\right)^2 + \left(\dfrac{8}{3/4}\right)^2 = (9 \div 0.667)^2 + (8 \div 0.75)^2$

$$= 13.5^2 + 10.667^2 = 182.25 + 113.785 = 296.03$$

59. $\dfrac{10 - (6^2/9)}{8} = \dfrac{10 - (36/9)}{8} = \dfrac{10 - 4}{8} = \dfrac{6}{8} = 0.75$

60. $\dfrac{104 - (12^2/6)}{5} = \dfrac{104 - (144/6)}{5} = \dfrac{104 - 24}{5} = \dfrac{80}{5} = 16.00$

61. $\dfrac{x - 4}{2} = 2.58, \quad x - 4 = 5.16, \quad x = 9.16$

62. $\dfrac{x - 21}{6.1} = 1.04, \quad x - 21 = 6.344, \quad x = 27.34$

63. $x = \dfrac{14 - 11}{2.5} = \dfrac{3}{2.5} = 1.20$

64. $x = \dfrac{36 - 41}{8.2} = \dfrac{-5}{8.2} = -0.61$

Aron, A., & Aron, E. N. (1994). *Statistics for psychology.* Englewood Cliffs, NJ: Prentice Hall.

Astin, A. W. (1995). *American freshmen: National norms for fall, 1995.* Los Angeles: American Council of Education.

Bahrick, H. P., Bahrick, L. E., Bahrick, A. S., & Bahrick, P. E. (1993). Maintenance of foreign language vocabulary and the spacing effect. *Psychological Science, 4,* 316–321.

Bandura, A., Blanchard, E. B., & Ritter, B. (1969). The relative efficacy of desensitization and modeling approaches for inducing behavioral, affective, and attitudinal changes. *Journal of Personality and Social Psychology, 64,* 173–199.

Bangert-Drowns, R. L. (1986). Review of developments in meta-analytic method. *Psychological Bulletin, 99,* 388–399.

Barber, T. X. (1976). Suggested ("hypnotic") behavior: The trance paradigm versus an alternative paradigm. In T. X. Barber (Ed.), *Advances in altered states of consciousness & human potentialities* (Vol. 1, pp. 175–259). New York: Psychological Dimensions.

Barker, R. B., Dembo, T., & Lewin, K. (1941). Frustration and regression: An experiment with young children. *University of Iowa Studies in Child Welfare, 18,* No. 1.

Benjamin, L. T., Cavell, T. A., & Shallenberger, W. R. (1984). Staying with initial answers on objective tests: Is it a myth? *Teaching of Psychology, 11,* 133–141.

Berger, R. W., & Hart, T. H. (1986). *Statistical process control.* New York: Marcel Dekker.

Berry, D. S., & McArthur, L. Z. (1986). Perceiving character in faces: The impact of age-related craniofacial changes on social perception. *Psychological Bulletin, 100,* 3–18.

Biffen, R. H. (1905). Mendel's laws of inheritance and wheat breeding. *Journal of Agricultural Science, 1,* 4–48.

Birch, H. G., & Rabinowitz, H. S. (1951). The negative effect of previous experience on productive thinking. *Journal of Experimental Psychology, 41,* 121–125.

Blair, R. C., & Higgins, J. J. (1985). Comparison of the power of the paired samples *t* test to that of Wilcoxon's signed-ranks test under various population shapes. *Psychological Bulletin, 97,* 119–128.

Blair, R. C., Higgins, J. J., & Smitley, W. D. S. (1980). On the relative power of the *U* and *t* tests. *British Journal of Mathematical and Statistical Psychology, 33,* 114–120.

Bowen, R. W. (1992). *Graph it! How to make, read, and interpret graphs.* Englewood Cliffs, NJ: Prentice-Hall.

Box, J. F. (1981). Gosset, Fisher, and the *t* distribution. *American Statistician, 35,* 61–66.

Bradley, D. R., Bradley, T. D., McGrath, S. G., & Cutcomb, S. D. (1979). Type I error rate of the chi-square test of independence in R × C tables that have small expected frequencies. *Psychological Bulletin, 86,* 1290–1297.

Bradley, J. V. (1978). Robustness? *British Journal of Mathematical Statistical Psychology, 31,* 144–152.

Bransford, J. D., & Franks, J. J. (1971). The abstraction of linguistic ideas. *Cognitive Psychology, 2,* 331–350.

Brehm, J. W., & Cohen, A. R. (1962). *Explorations in cognitive dissonance.* New York: Wiley.

Bureau of Labor Statistics. (December 1944). Union wages and hours of motor truck drivers and helpers. July 1, 1944. *Monthly Labor Review.*

Calhoun, J. P., & Johnston, J. O. (1968). Manifest anxiety and visual acuity. *Perceptual and Motor Skills, 27,* 1177–1178.

Camilli, G., & Hopkins, K. D. (1978). Applicability of chi-square to 2 × 2 contingency tables with small expected cell frequencies. *Psychological Bulletin, 85,* 163–167.

Campbell, S. K. (1974). *Flaws and fallacies in statistical thinking.* Englewood Cliffs, NJ: Prentice-Hall.

Christensen, L. B., & Stoup, C. M. (1991). *Introduction to statistics for the social and behavioral sciences* (2d ed.). Pacific Grove, CA: Brooks/Cole.

Cohen, I. B. (March 1984). Florence Nightingale. *Scientific American, 250,* 128–137.

Cohen, J. (1969). *Statistical power analysis for the behavioral sciences.* New York: Academic.

Cohen, J. (1988). *Statistical power analysis for the behavioral sciences* (2d ed.). Hillsdale, NJ: Erlbaum.

Cohen, J. (1992). A power primer. *Psychological Bulletin, 112,* 155–159.

Cohen, S., & Williamson, G. M. (1991). Stress and infectious disease in humans. *Psychological Bulletin, 109,* 5–24.

Coren, S., & Halpern, D. F. (1991). Left-handedness: A marker for decreased survival fitness. *Psychological Bulletin, 109,* 90–106.

David, F. N. (1968). Francis Galton. In D. L. Sills (Ed.), *International encyclopedia of the social sciences* (Vol. 6, pp. 48–53). New York: Macmillan.

Doll, R. (1955). Etiology of lung cancer. *Advances in Cancer Research, 3,* 1–50.

Dudek, F. J. (1981). Data sets having integer means and standard deviations. *Teaching of Psychology, 8,* 51.

Dunkling, L. (1993). *The Guinness book of names.* Enfield, Middx, UK: Guinness.

Durkheim, E. (1951, reprint). *Suicide.* New York: Free Press.

Edwards, A. L. (1972). *Experimental design in psychological research* (4th ed.). New York: Holt, Rinehart & Winston.

Edwards, A. L. (1984). *An introduction to linear regression and correlation* (2d ed.). San Francisco: Freeman.

Ellis, D. (1938). *A source book of Gestalt psychology.* London: Routledge & Kegan Paul.

Everitt, B., & Hay, D. (1992). *Talking statistics: A psychologist's guide to design & analysis.* New York: Halstead.

Ferguson, G. A., & Takane, Y. (1989). *Statistical analysis in psychology and education* (6th ed.). New York: McGraw-Hill.

Fisher, R. A. (1925). *Statistical methods for research workers.* London: Oliver and Boyd.

Fisher, R. A., & Yates, F. (1963). *Statistical tables for biological, agricultural, and medical research* (6th ed.). Edinburgh: Oliver and Boyd.

Flynn, J. R. (1987). Massive IQ gains in 14 nations: What IQ tests really measure. *Psychological Bulletin, 101,* 171–191.

Forbs, R., & Meyer, A. B. (1955). *Forestry handbook.* New York: Ronald.

Gaito, J. (1980). Measurement scales and statistics: Resurgence of an old misconception. *Psychological Bulletin, 87,* 564–567.

Galton, F. (1869). *Hereditary genius.* London: Macmillan.

Galton, F. (1889). *Natural inheritance*. London: Macmillan.

Galton, F. (1901). Biometry. *Biometrika, 1*, 7–10.

Gardner, P. L. (1975). Scales and statistics. *Review of Educational Research, 45*, 43–57.

Greene, J. E. (Ed.-in-chief). (1966). *McGraw-Hill modern men of science*. New York: McGraw-Hill.

Hacking, I. (1984). Trial by number. *Science 84, 5*, 69–70.

Harris, R. J. (1985). *A primer of multivariate statistics* (2d ed.). New York: Academic.

Harter, H. L. (1960). Tables of range and studentized range. *Annals of Mathematical Statistics, 31*, 1122–1147.

Hasler, A. D. (1966). *Underwater guideposts*. Madison: University of Wisconsin Press.

Hasler, A. D., & Scholz, A. T. (1983). *Olfactory imprinting and homing in salmon*. New York: Springer-Verlag.

Hasler, A. D., Scholz, A. T., & Horrall, R. M. (1978). Olfactory imprinting and homing in salmon. *American Scientist, 66*, 347–355.

Hellekson, C. J., Kline, J. A., & Rosenthal, N. E. (1986). Phototherapy for seasonal affective disorder in Alaska. *American Journal of Psychiatry, 143*, 1035–1037.

Howell, D. C. (1992). *Statistical methods for psychology* (3d ed.). Boston: Duxbury.

Howell, D. C. (1995). *Fundamental statistics for the behavioral sciences* (3d ed.). Belmont, CA: Duxbury.

Huff, D. (1954). *How to lie with statistics*. New York: Norton.

Hurlburt, R. T. (1994). *Comprehending behavioral statistics*. Pacific Grove, CA: Brooks/Cole.

Irion, A. L. (1976). A survey of the introductory course in psychology. *Teaching of Psychology, 3*, 3–8.

Jenkins, J. G., & Dallenbach, K. M. (1924). Obliviscence during sleep and waking. *American Journal of Psychology, 35*, 605–612.

Johnson, R. C., McClearn, G. E., Yuen, S., Nagoshi, C. T., Ahern, F. M., & Cole, R. E. (1985). Galton's data a century later. *American Psychologist, 40*, 875–892.

Johnston, J. O., & Maertens, N. W. (1972). Effects of parental involvement in arithmetic homework. *School Science and Mathematics, 72*, 117–126.

Jones, S. S., Collins, K., & Hong, H. (1991). An audience effect on smile production in 10-month-old infants. *Psychological Science, 2*, 45–49.

Keppel, G. (1982). *Design and analysis: A researcher's handbook*. Englewood Cliffs, NJ: Prentice-Hall.

Kirk, R. E. (Ed.). (1972). *Statistical issues: A reader for the behavioral sciences*. Pacific Grove, CA: Brooks/Cole.

Kirk, R. E. (1995). *Experimental design: Procedures for the behavioral sciences* (3d ed.). Pacific Grove, CA: Brooks/Cole.

Kirk, R. E. (1984). *Elementary statistics* (2d ed.). Pacific Grove, CA: Brooks/Cole.

Kirk, R. E. (1990). *Statistics: An introduction* (3d ed.). Fort Worth, TX: Holt, Rinehart, and Winston.

Kramer, C. Y. (1956). Extension of multiple range tests to group means with unequal numbers of replications. *Biometrics, 12*, 307–310.

Landau, D., & Lazarfeld, P. F. (1968). Quételet, Adolphe. In D. L. Sills (Ed.), *International encyclopedia of the social sciences* (Vol. 13, pp. 247–257). New York: Macmillan.

Lewin, K. (1958). Group decision and social change. In E. E. Maccoby, T. M. Newcomb, & E. L. Hartley (Eds.), *Readings in social psychology* (3d ed.). New York: Holt, Rinehart, and Winston.

Loewi, O. (1963). The night prowler. In S. Rapport, & H. Wright (Eds.), *Science: Method and meaning*. New York: Washington Square Press.

Loftus, E. F. (1979). *Eyewitness testimony.* Cambridge, MA: Harvard University Press.

Loftus, G. R., & Loftus, E. F. (1988). *Essentials of statistics* (2d ed.). New York: Random House.

Mackowiak, P. A., Wasserman, S. S., & Levine, M. M. (1992). A critical appraisal of 98.6°F, the upper limit of the normal body temperature, and other legacies of Carl Reinhold August Wunderlich. *Journal of the American Medical Association, 268,* 1578–1580.

Mann, H. B., & Whitney, D. R. (1947). On a test of whether one or two random variables is stochastically larger than the other. *Annals of Mathematical Statistics, 18,* 50–60.

Mayo, E. (1946). *The human problems of an industrial civilization.* Boston: Harvard University Press.

McCrae, R. R., & Costa, P. T. (1990). *Personality in adulthood.* New York: Guilford.

McGaugh, J. L., & Petrinovich, L. F. (1965). Effects of drugs on learning and memory. *International review of neurobiology* (Vol. 8). New York: Academic.

McKeown, T., & Gibson, J. R. (1951). Observation on all births (23,970) in Birmingham, 1947: IV. "Premature birth." *British Medical Journal, 2,* 513–517.

McMullen, L., & Pearson, E. S. (1939). William Sealy Gosset, 1876–1937. *Biometrika, 30,* 205–253.

Micceri, T. (1989). The unicorn, the normal curve, and other improbable creatures. *Psychological Bulletin, 105,* 156–166.

Michell, J. (1986). Measurement scales and statistics: A clash of paradigms. *Psychological Bulletin, 100,* 398–407.

Milgram, S. (1969). Note on the drawing power of crowds of different size. *Journal of Personality and Social Psychology, 13,* 79–82.

Minium, E. W., King, B. M., & Bear, G. (1993). *Statistical reasoning in psychology and education* (3d ed.). New York: Wiley.

Mischel, H. N. (1974). Sex bias in the evaluation of professional achievements. *Journal of Educational Psychology, 66,* 157–166.

M&M/Mars. (1993). *A little illustrated encyclopedia of M&M/Mars.* Hackettstown, NJ: Author.

Moore, D. S., & McCabe, G. P. (1993). *Introduction to the practice of statistics* (2d ed.). New York: Freeman.

Mosteller, F., & Wallace, D. L. (1989). Deciding authorship. In J. M. Tanur, F. Mosteller, W. H. Kruskal, R. F. Link, R. S. Picters, G. R. Rising, & E. L. Lehmann (Eds.), *Statistics: A guide to the unknown* (3d ed.). Pacific Grove, CA: Wadsworth & Brooks/Cole.

Nelson, N., Rosenthal, R., & Rosnow, R. L. (1986). Interpretation of significance levels and effect sizes by psychological researchers. *American Psychologist, 41,* 1299–1301.

Nijsse, M. (1988). Testing the significance of Kendall's τ and Spearman's r_s. *Psychological Bulletin, 103,* 235–237.

Overall, J. E. (1980). Power of chi-square tests for 2 × 2 contingency tables with small expected frequencies. *Psychological Bulletin, 87,* 132–135.

Paulhus, D. L., & Van Selst, M. (1990). The spheres of control scale: 10 years of research. *Personality & Individual Differences, 11,* 1029–1036.

Pearson, E. S. (1949). W. S. Gosset. *Dictionary of national biography: 1931–1940.* London: Oxford University Press.

Pearson, K., & Lee, A. (1903). Inheritance of physical characters. *Biometrika, 2,* 357–462.

Porter, T. M. (1986). *The rise of statistical thinking.* Princeton, NJ: Princeton.

Powers, E., & Witmer, H. (1951). *An experiment in the prevention of delinquency: The Cambridge-Somerville Youth Study.* New York: Columbia University Press.

Robinson, L. A., Berman, J. S., & Neimeyer, R. A. (1990). Psychotherapy for the treatment of depression: A comprehensive review of controlled outcome research. *Psychological Bulletin, 108,* 30–49.

Rokeach, M., Homant, R., & Penner, L. (1970). A value analysis of the disputed Federalist papers. *Journal of Personality and Social Psychology, 16*, 245–250.

Rosen, L. A., Booth, S. R., Bender, M. E., McGrath, M. L., Sorrell, S., & Drabman, R. S. (1988). Effects of sugar (sucrose) on children's behavior. *Journal of Consulting and Clinical Psychology, 56*, 583–589.

Rosenthal, R., & Rosnow, R. L. (1991). *Essentials of behavioral research* (2d ed.). New York: McGraw-Hill.

Rosenthal, R., & Rosnow, R. L. (1991). *Essentials of behavioral research: Methods and data analysis* (2d ed.). New York: Wiley.

Rosnow, R. L., & Rosenthal, R. (1989). Definition and interpretation of interaction effects. *Psychological Bulletin, 105*, 143–146.

Runyon, R. P. (1981). *How numbers lie.* Lexington, MA: Lewis.

Schachter, S., & Gross, S. P. (1968). Manipulated time and eating behavior. *Journal of Personality and Social Psychology, 10*, 98–106.

Schumacher, E. F. (1979). *Good work.* New York: Harper & Row.

Shedler, J., & Block, J. (1990). Adolescent drug use and psychological health. *American Psychologist, 45*, 612–630.

Sherif, M. (1935). A study of some social factors in perception. *Archives of Psychology.* No. 187.

Sherman, M. (1927). The differentiation of emotional responses in infants: The ability of observers to judge the emotional characteristics of the crying infants, and of the voice of an adult. *Journal of Comparative Psychology, 7*, 335–351.

Smith, R. A. (1971). The effect of unequal group size on Tukey's HSD procedure. *Psychometrika, 36*, 31–34.

Snedecor, G. W., & Cochran, W. G. (1967). *Statistical methods* (6th ed.). Ames: Iowa State University Press.

Spatz, T. S. (1991). Improving breast self-examination training by using the 4MAT instructional model. *Journal of Cancer Education, 6*, 179–183.

Spearman, C. (1930). Autobiography. In C. Murchison (Ed.), *History of psychology in autobiography.* New York: Russell & Russell.

Sprent, P. (1989). *Applied nonparametric statistical methods.* London: Chapman and Hall.

Statistical abstract of the United States: 1994 (114th ed.). (1994). Washington, D.C.: U.S. Bureau of the Census.

Stevens, S. S. (1946). On the theory of scales of measurement. *Science, 103*, 677–680.

Stewart, I. (July 1977). Gauss. *Scientific American, 237*, 123–131.

Tanur, J. M., Mosteller, F., Kruskal, W. H., Lehmann, E. L., Link, R. F., Pieters, R. S., & Rising, G. R. (Eds.). (1989). *Statistics: A guide to the unknown.* (3rd ed.). Pacific Grove, CA: Wadsworth & Brooks/Cole.

Taylor, J. (1953). A personality scale of manifest anxiety. *Journal of Abnormal and Social Psychology, 48*, 285–290.

Tobias, S. (1980). *Overcoming math anxiety.* Boston: Houghton Mifflin.

Townsend, J. T., & Ashby, F. G. (1984). Measurement scales and statistics: The misconception misconceived. *Psychological Bulletin, 96*, 394–401.

Tufte, E. R. (1983). *The visual display of quantitative information.* Cheshire, CT: Graphics.

Tufte, E. R. (1990). *Envisioning information.* Cheshire, CT: Graphics.

Wainer, H. (1984). How to display data badly. *American Statistician, 38*, 137–147.

Walker, H. M. (1929). *Studies in the history of statistical method.* Baltimore: Williams & Wilkins.

Walker, H. M. (1940). Degrees of freedom. *Journal of Educational Psychology, 31*, 253–269.

Walker, H. M. (1968). Karl Pearson. In D. L. Sills (Ed.), *International encyclopedia of the social sciences.* New York: Macmillan and Free Press.

Wallace, R. K., & Benson, H. (1972). The physiology of meditation. *Scientific American, 226,* 84–90.

Wallace, W. L. (1972). College Board Scholastic Aptitude Test. In O. K. Buros (Ed.), *The seventh mental measurements yearbook.* Highland Park, NJ: Gryphon.

Watson, J. B. (1924). *Psychology from the standpoint of a behaviorist* (2d ed.). Philadelphia: Lippincott.

White, E. B. (1970). *The trumpet of the swan.* New York: Harper & Row.

Wilcox, R. R. (1987). New designs in analysis of variance. *Annual Review of Psychology, 38,* 29–60.

Wilcoxon, F. (1945). Individual comparisons by ranking methods. *Biometrics, 1,* 80–83.

Wilcoxon, F., & Wilcox, R. A. (1964). *Some rapid approximate statistical procedures* (Rev. ed.). Pearl River, NY: Lederle Laboratories.

Winer, B. J., Brown, D. R., & Michels, K. M. (1991). *Statistical principles in experimental design* (3d ed.). New York: McGraw-Hill.

Winter, L., Uleman, J. S., & Cunniff, C. (1985). How automatic are social judgments? *Journal of Personality and Social Psychology, 49,* 904–917.

Wittrock, M. C. (1986). Students' thought processes. In M. C. Wittrock (Ed.), *Handbook of research on teaching* (3d ed.). New York: Macmillan.

Wittrock, M. C. (1991). Generative teaching of comprehension. *The Elementary School Journal, 92,* 169–184.

Wood, T. B., & Stratton, F. J. M. (1910). The interpretation of experimental results. *Journal of Agricultural Science, 3,* 417–440.

Woodworth, R. S. (1926). Introduction. In H. E. Garrett (Ed.), *Statistics in psychology and education.* New York: Longmans, Green.

Work, M. S., & Rogers, H. (1972). Effects of estrogen level on food-seeking dominance among male rats. *Journal of Comparative and Physiological Psychology, 79,* 414–418.

Yates, F. (1981). Sir Ronald Aylmer Fisher. In E. T. Williams & C. S. Nichols (Eds.), *Dictionary of national biography 1961–1970.* Oxford: Oxford University.

Youden, W. J. (1962). *Experimentation and measurement.* Washington, D.C.: National Science Teachers Association.

Zwick, R. (1993). Pairwise comparison procedures for one-way analysis of variance designs. In G. Keren & C. Lewis (Eds.), *A handbook for data analysis in the behavioral sciences: Statistical issues* (pp. 43–71). Hillsdale, NJ: Erlbaum.

TO THE OWNER OF THIS BOOK:

I hope that you have found *Basic Statistics: Tales of Distributions*, Sixth Edition, useful. So that this book can be improved in a future edition, would you take the time to complete this sheet and return it? Thank you.

School and address: _____

Department: _____

Instructor's name: _____

1. What I like most about this book is: _____

_____

_____

2. What I like least about this book is: _____

_____

_____

3. My general reaction to this book is: _____

_____

4. The name of the course in which I used this book is: _____

_____

5. Were all of the chapters of the book assigned for you to read? _____

   If not, which ones weren't? _____

6. In the space below, or on a separate sheet of paper, please write specific suggestions for improving this book and anything else you'd care to share about your experience in using the book.

_____

_____

_____

_____

_____

Optional:

Your name: _____ Date: _____

May Brooks/Cole quote you, either in promotion for *Basic Statistics: Tales of Distributions,*
Sixth Edition, or in future publishing ventures?

Yes: _____ No: _____

Sincerely,

Chris Spatz

FOLD HERE

NO POSTAGE
NECESSARY
IF MAILED
IN THE
UNITED STATES

BUSINESS REPLY MAIL

FIRST CLASS   PERMIT NO. 358   PACIFIC GROVE, CA

POSTAGE WILL BE PAID BY ADDRESSEE

ATT: _*Chris Spatz*_____

**Brooks/Cole Publishing Company
511 Forest Lodge Road
Pacific Grove, California 93950-9968**

FOLD HERE

DECISION TREE FOR DESCRIPTIVE STATISTICS

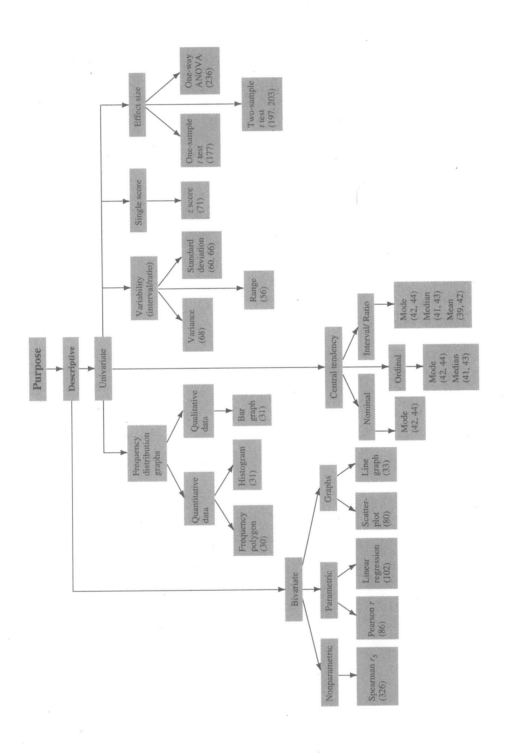